IMPROVING TEAMWORK
IN ORGANIZATIONS

Applications of Resource Management Training

IMPROVING TEAMWORK IN ORGANIZATIONS

Applications of Resource Management Training

Edited by

EDUARDO SALAS
CLINT A. BOWERS
University of Central Florida

ELEANA EDENS
Federal Aviation Administration

CRC Press
Taylor & Francis Group
Boca Raton London New York

CRC Press is an imprint of the
Taylor & Francis Group, an informa business

Reprinted 2010 by CRC Press

CRC Press
6000 Broken Sound Parkway, NW
Suite 300, Boca Raton, FL 33487

270 Madison Avenue
New York, NY 10016

2 Park Square, Milton Park
Abingdon, Oxon OX14 4RN, UK

Copyright © 2001 by Lawrence Erlbaum Associates, Inc.

Lawrence Erlbaum Associates, Inc., Publishers
10 Industrial Avenue
Mahwah, New Jersey 07430

Cover design by Kathryn Houghtaling Lacey

Library of Congress Cataloging-in-Publication Data

Improving teamwork in organizations : applications of resource management training /
Eduardo Salas, Clint A. Bowers, Eleana Edens.
  p. cm.
  Includes bibliographical references and indexes.
  ISBN 0-8058-2844-3 (cloth : alk. paper) – ISBN 0-8058-2845-1 (pbk. : alk. paper)
  1.   Teams in the workplace. 2. Employees—Training of. 3.
Airlines—Employees—Training of. I. Salas, Eduardo. II. Bowers, Clint A. III. Edens,
Eleana. IV. Title

HD66 .S255 2000

00-050400

Contents

PART V: FINAL OBSERVATIONS

Preface

Few of us can question the value of teamwork. Effective teamwork can increase organizational productivity and job efficiency. Teamwork can reduce human errors and promote job satisfaction. Teamwork can also help maintain safe conditions in complex and stressful environments. Indeed, teamwork can make organizations better and help accomplish their goals and missions. And so teamwork is a popular commodity these days. Many businesses, industries, agencies, and occupations demand it—from the military fields to the corporate rooms. But nowhere is teamwork as critical as in the cockpit of an airplane. Effective teamwork can make our skies safer. Since the 1980s the airlines (and the military as well) have invested considerable effort and resources to design and implement crew resource management (CRM) training in their organizations.

The airlines have been aided by a group of applied scientists and agencies (most of them represented in this book) whose aim has been to improve safety, reduce human error, and promote effective teamwork and decision-making in the cockpit. So after two decades of CRM training research and practice—progress has been made! We have made theoretical, methodological, and practical advances in the design, delivery, implementation, and evaluation of CRM training. We certainly hope that this book captures some of that progress.

We know more about how to promote teamwork and reduce errors in the cockpit than ever before and the preliminary results are encouraging. However, we strongly believe that CRM training benefits reach beyond the aviation community. CRM training (or team training) is also important (and needed) to the medical, emerging management, police, fire, and rescue domains (just to mention a few). We also hope that this book shows how CRM training can be applied and used in a variety of contexts, environments, and occupations.

While we are happy to document the progress, we also know that there are many complex issues still unresolved. And only more collaborative, interdisciplinary, and systematic research will provide practical solutions to organizations. We hope this book also engages, motivates, and energizes scientists and practitioners alike to continue working on understanding human performance in complex environments and providing training (and other interventions as well) solutions. More work is needed.

In closing, we would like to thank the thousands of aviators who have contributed to the development and institutionalization of CRM training and the scientists who pave the way most notably—Bob Helmreich, Clay Foushee, John Lauber, and Earl Weiner. We would also like to thank Anne Duffy for her continued support in this and other related efforts.

Eduardo Salas
Clint A. Bowers
Eleana Edens

1

An Overview of Resource Management in Organizations: Why Now?

Eduardo Salas
Clint A. Bowers
University of Central Florida

Eleana Edens
Federal Aviation Administration

Many people feel that the emergence of crew resource management (CRM) in the aviation industry has been one of the greatest successes of aviation and human factors psychology (e.g., Helmreich & Foushee, 1993). Many factors seemed to have contributed to this conclusion. First, it represents one of the best partnerships between those interested in science and those concerned with practice and applications. Scientists have done a credible job of providing both empirical and theoretical foundations for CRM development. By the same token, practitioners have used this information to develop state-of-the-art crew coordination training. This type of partnership, we believe, has been all too rare in our business.

A second success of CRM has been in the relationship that science has developed with industry. Despite the financial and organizational challenges, the aviation industry has been wonderfully accepting of the science and practice of CRM. In fact, this research and training have become a standard practice in the industry (see Wiener, Kanki, & Helmreich, 1993). Furthermore, the industry has continued to cooperate with researchers attempting to improve the training. Their support of our research is commendable and has allowed the field to progress much more quickly than one might have guessed.

Finally, CRM represents a scientific success because we have been able to establish its efficacy (e.g., Helmreich & Foushee, 1993; Leedom & Simon, 1995;

Salas, Fowlkes, Stout, Milanovich, & Prince, 1999; Stout, Salas, & Fowlkes, 1997). Too often in training research we are able to design and develop the training but not allowed to evaluate it. CRM has included evaluation throughout its evolution, so we have been able to develop the training interactively, allowing for the best possible training to result (Gregorich & Wilhelm, 1993).

Despite these many successes, however, one notable limitation has been the inability to transfer the lessons of CRM to other, nonaviation settings. In an era that has witnessed an increasing amount of research on and information about teamwork, team training, and group effectiveness (see Cannon-Bowers & Salas, 1998; Guzzo & Salas, 1995; Sundstrom, 1999; West, 1996) and its application to the design and delivery of (crew) resource management training, it is frustrating that one of our best developments has not reached mainstream corporate America—at least not yet. In fact, in the last decade we have seen an explosion of team training and performance research and applications, and most of the findings, conclusions, and principles from this research has, we believe, direct relevance to those interested in designing and delivery of resource management training to increase teamwork in different settings and solve organizational problems. Thus, we challenged the best CRM researchers in the world to think about how their research could apply to nonaviation businesses. The results of that challenge are contained within this book, which documents decades of thinking, researching, and application of CRM training. As the reader will see, many lessons were learned. (One is that it is difficult for us to leave the comfortable confines of aviation and CRM but, with some gentle prodding, the authors managed it.) We believe that these lessons, discussed in each of the chapters, can improve team performance in a variety of industries. We believe that many of the principles and assumptions behind (crew) resource management apply to many domains: medical, law enforcement, air traffic control, manufacturing, and industry. Although (crew) resource management was born in the aviation world, we now believe and know it applies to other settings; therefore, we wanted this book to illustrate this.

What about the label? Is it *crew resource management* or just *resource management*? In this book we and the authors have purposely avoided (as much as possible) using *CRM*. We wanted this book to have a broad appeal, and hence we dropped (for the purposes of the book) the *crew* label. As the reader will see, some of the authors could not avoid using *CRM*; it was inherent in what they were addressing. Other authors were more general in their approach and used *resource management* as an umbrella for their work.

Why now? Why a book on resource management training? Because we believe there is a need—the need to know where we are now after 20 years of researching and practicing (crew) resource management training. We (the scientists, practitioners, developers, users, and sponsors) need to know who is doing what and what is being applied. We need to know what works and what does not. We need to know what progress has been made scientifically and practically. Furthermore, there is a need to document the findings, conclusions, successes, failures, guidelines, and lessons learned from all this research and applications so that instructional developers in many diverse industries and organizations can benefit from this information. Therefore, our objectives in this book are twofold. First, we would like to provide those interested in designing and delivery of resource management training with useful and practical information containing the latest thinking and guidance available. Second, we seek to launch CRM training into a wide variety of industries and organizations as a viable intervention that can be used to enhance teamwork and organizational effectiveness as well as minimize human error.

ORGANIZATION OF THE BOOK

The book is organized into four main sections. The first section addresses the foundations of resource management training; that is, it focuses on uncovering, outlining, and managing the basic competencies (i.e., knowledge, skills, and attitudes [KSAs]) required for enhancing teamwork, communication, and coordination. The chapters cover methodologies and approaches for specifying the competencies to be trained and provide detailed discussions on a few specific skills (e.g., assertiveness, communication, stress management—which are at the core of resource management training) that have been subject to extensive research. In chapter 2 Seamster and Kaempf present a framework for uncovering resource management skills. They draw from their experience with airline pilots and offer a broader-based skill analysis process for resource management training. They extend the instructional system development process to include the job performance context and offer a number of useful suggestions for how to conduct a skill analysis process. Chapter 3, authored by Smith-Jentsch, Baker, Salas, and Cannon-Bowers, compares and contrasts the results of their analysis of air traffic control operation with findings from team performance and training research in other domains. In particular, they describe the team-related KSA requirements for air traffic control teams and highlight similarities between these and the requirements for crew coordination. They offered guidelines for linking specific KSAs to appropriate training methods. Next, Driskell, Salas, and Johnston analyze the importance of stress management in teams and for teamwork. On the basis of their research on stress exposure training they outline how stress can be managed in complex and dynamic environments. They offer guidelines for designing and developing stress training that can be adapted to various occupational environments where teams (and of course, individuals) are present. Jentsch and Smith-Jentsch, in chapter 5, focus on the role and importance of assertiveness in aviation teams. They draw their conclusions from many vivid historical examples. They dispel misconceptions regarding assertiveness and describe how this behavior affects the performance of many other types of teams. They offer several recommendations for measuring and improving team performance-related assertiveness. Finally, Kanki and Smith focus on communication skills. They identify the communications process skills that should be taught in an aviation setting. They also present a series of principles and guidelines that should be taken into account when designing and delivering communication.

Part II focuses on the tools needed for the design and delivery of resource management training. In chapter 7 Baker, Mulqueen, and Dismukes research literature on rater training and develop a series of guidelines for training raters to evaluate resource management skills. In addition, they reviewed four strategies that have been used to train supervisors who conduct performance appraisals, and they present evidence of each strategy's effectiveness. Finally, they review and summarize their findings with set guidelines for developing rater training in the future. Next, Prince and Jentsch discuss how low-fidelity training devices can be used in CRM skill training. These types of devices are not accepted in aviation environments across the world. The authors offer some insights on how to apply these devices to allow for skill practice in resource management training. Then, in chapter 9, Holt, Boehm-Davis, and Beaubien outline steps required to evaluate the effectiveness of a resource management training program and highlight various practical and theoretical issues that arise during this process.

The third section focuses on applications of resource management training in several industries and domain. In chapter 10 Boehm-Davis, Holt, and Seamster report on the experiences of their research team in developing, implementing, and evaluating resource management procedures at two airlines. The authors conclude this chapter with a summary of the lessons learned from these efforts and by offering a set of guidelines that can be applied to the development of resource management programs for teams that require a high level of coordination. Next, Flin and O'Connor offer an outline of how the basic principles of CRM training and assessment as used in the aviation industry have been used as a basis for the design of offshore CRM courses. They briefly describe the United Kingdom's offshore oil and gas industry and explain why CRM training is deemed appropriate for this work environment. Robertson describes in chapter 12 the development of maintenance resource management (MRM) training and the performance issues and problems in aviation maintenance that maintenance resource management addresses. She discusses current practices and describes some real-life elements to be used in developing and implementing maintenance resource management training programs for other domains. Next,

Davies describes the application of CRM in the medical field. A brief history of the evolution of CRM is given as well as the differences between current CRM programs in medicine and those in aviation. The results of a medical CRM program are discussed with respect to those who are likely to be affected: patients, personnel, the organization, and regulators. Finally, in chapter 14 Oser, Salas, Merket, and Bowers provide an overview of their team's research program and of the CRM training methodology developed and used in the U.S. Navy. A set of lessons learned for professionals involved in designing, delivering, and evaluating resource management training are presented, offering the possibility of benefit to other environments in which coordination among multiple crew members is necessary.

Section IV looks at the global issues of resource management training. Helmreich, Wilhelm, Klinect and Merritt discuss in chapter 15 the influences of three cultures that are relevant to the cockpit, such as the professional culture of pilots, the culture of organizations, and national cultures surrounding individuals and their organizations. A model of threat-and-error management in aviation is presented, along with findings from audits of crew performance. The transfer of training models from aviation to other domains, such as medicine and shipping, are briefly considered. In the final chapter we—Salas, Bowers, and Edens—provide a summary of the preceding chapters and synthesize the chapters into a set of general observations.

REFERENCES

Cannon-Bowers, J. A., & Salas, E. (1998). *Making decisions under stress: Implications for individual and team training.* Washington, DC: American Psychological Association.

Gregorich, S. E., & Wilhelm, J. A. (1993). Crew resource management training assessment. In E. L. Wiener, B. G. Kanki, & R. L. Helmreich (Eds.), *Cockpit resource management* (pp. 173–196). San Diego, CA: Academic Press.

Guzzo, R. A., & Salas, E. (1995). *Team effectiveness and decision making in organizations.* San Francisco: Jossey-Bass.

Helmreich, R. L., & Foushee, H. C. (1993). Why crew resource management? Empirical and theoretical basis of human factors training in aviation. In E. L. Wiener, B. G. Kanki, & R. L. Helmreich (Eds.), *Cockpit resource management* (pp. 3–45). San Diego, CA: Academic Press.

Leedom, D. K., & Simon, R. (1995). Improving team coordination: A case for behavioral-base training. *Military Psychology, 7,* 109–122.

Salas, E., Fowlkes, J. E., Stout, R. J., Milanovich, D. M., & Prince, C. (1999). Does CRM training improve teamwork skills in the cockpit?: Two evaluation studies. *Human Factors, 41,* 326–343.

Stout, R. J., Salas, E., & Fowlkes, J. E. (1997). Enhancing teamwork in complex environments through team training. *Group Dynamics, 1,* 169–182.

Sundstrom, E. (1999). (Ed.). *Supporting work team effectiveness: Best management practices for fostering high performance.* San Francisco: Jossey-Bass.

West, M. A. (1996). (Ed.). *Handbook of work group psychology.* West Sussex, England: Wiley.

Wiener, E. L., Kanki, B. G., & Helmreich, R. L. (1993). *Cockpit resource management.* San Diego, CA: Academic Press.

FOUNDATIONS OF RESOURCE MANAGEMENT TRAINING

2

Identifying Resource Management Skills for Airline Pilots

Thomas L. Seamster
Cognitive & Human Factors

George L. Kaempf
Sun Microsystems, Inc.

There is a growing awareness of the importance of resource management skills related to decision making, team coordination, and planning, especially in organizations where work teams perform complex, time-constrained, and critical tasks. Much of the resource management research and implementation has focused on the performance of individuals within the context of teams. Organizations need a more effective approach to the analysis and training of complex resource management skills at both the individual and team levels.

In this chapter we present a framework for managing the development of skill training and a method for identifying resource management skills. The framework examines resource management skills as layered phenomena: individual, team, and organization. The skill identification methods described in this chapter provide an illustration of how several analytic methods can be combined to form an effective strategy for analyzing resource management skills in a complex work environment. These methods were tested in a joint effort between airlines and researchers. They extend the traditional instruction system development (ISD) process beyond job tasks to include the job performance context. These methods were developed not as the final solution to the identification of resource management skills but as a step toward a broader-based skill-analysis process.

INTRODUCTION TO RESOURCE MANAGEMENT SKILLS

Organizations, such as airlines, have had difficulty identifying resource management skills that can provide individuals, teams, and the organization with the guidance and standards for systematic training and assessment. Part of this difficulty is due to traditional job and task analysis methods with their emphasis on the individual in the specification of job or task knowledge and skills. With the growing awareness of the importance of the team and the organization on skill training and development, the concept of individual skill is being expanded to the team and organizational level, and methods for skill identification need to address the needs of all levels of the organization.

Resource management skill identification and specification has been problematic, with many of the difficulties showing up when organizations try to integrate the identified skills with their training and performance assessment programs. Airlines have attempted to identify resource management skills over the last 10 years. Some widely used lists, such as the skill markers (see Law & Wilhelm, 1995) tend to be more general and were developed primarily for assessment and not as part of a detailed training curriculum. At a more detailed level, some airlines have started working with observable behaviors. The observable behaviors are much more specific than the general markers, but again, these elements are primarily for assessment purposes (Seamster, Hamman, & Edens, 1995). The status of resource management skills at many airlines is that they have been identified in the form of general markers, and some skills have been specified as observable behaviors to be assessed, but the actual skills have been neither identified nor are they being systematically trained. From a review of an existing comprehensive skill list from one airline (Lanzano, Seamster, & Edens, 1997), just 13 out of about 400 cognitive skills were specified as being related to resource management. In addition, most of the skills in that listing had been identified by just one or two subject matter experts (SMEs). The organization was aware of numerous gaps, especially in the listing of resource management skills. That review concluded that organizations need methods to identify a broader set of resource management skills.

Aviation resource management, referred to as *crew resource management* (CRM), has its roots in social psychology, the attitude component of crew coordination (see Gregorich, Helmreich, & Wilhelm, 1990), and early resource management familiarization courses were based on attitude and personality inventories and training. However, there have been a growing number of references to resource management skills (e.g., Brannick, Prince, Salas, & Stout, 1995; Cohen, Freeman, & Thompson, 1995; Goldsmith, Johnson, Seamster, & Edens, 1995; Jentsch, Bowers, & Holmes, 1995), indicating a shift from knowledge and attitude training to skill training. With this shift there is a growing need for organizations to provide focused resource management skill training rather than the more general knowledge and concept training currently provided in familiarization courses. Although the line oriented flight training (LOFT) development process (Hamman, Seamster, & Edens, 1995) allows for skill practice and assessment, airlines have yet to identify the key resource management skills and specify how those skills are best trained.

The need for more detailed understanding of resource management skills is highlighted by Proctor and Dutta's (1995) argument that the less we know about the skills, the greater the need for higher fidelity simulation. This is the current state of resource management training in aviation. Little is understood about the specific resource management skills, and consequently much of the training is done in the relatively expensive full flight simulators. Full flight simulators allow airlines to provide their crews with realistic simulations and the opportunity to practice ill-defined resource management skills, but they do not always provide the most efficient form of skill training.

From an operational perspective, organizations need a method that will allow them to identify resource management skills within the context of the individual, the teams composed of those individuals, and the organization itself. Once skill identification can be done with some precision, the organizations will be in a much stronger position to address how those skills should be trained. The approach presented here is for the identification of resource management skills, based on a broader context extending beyond the traditional ISD process to one that encompasses the entire organization. This identification method is designed to be part of a complete skill analysis that can be used by organizations committed to improving their resource management skill training and assessment program.

The methods presented in this chapter should be viewed as preliminary steps toward a broader understanding and use of resource management skills. These methods can help organizations clearly specify resource management skills so that they can then determine how to best train those skills. Because resource management skills have a substantial cognitive component, organizations will ultimately need some form of cognitive task analysis to provide better descriptions of the skills that are particularly difficult to train or that are closely linked with performance problems. The methods presented here take organizations one step closer to cognitive analyses while sidestepping some of the more controversial issues about the nature and content of cognitive skills. Some researchers model cognition as predominantly knowledge based and distinguishable from skills that are the "do" part of human performance (Salas & Cannon-Bowers, 1997). The approach advocated here is based on a cognitive skill acquisition framework (Van-Lehn, 1996) in which cognition has both a knowledge and a skill component and includes a substantial number of skills that are trainable.

LEVELS OF RESOURCE MANAGEMENT SKILLS

An important new direction in research on job performance and skill is the growing inclusion of job context (Arvey & Murphy, 1998). Resource management skills should be considered from multiple levels to ensure that the broader job context is reflected in the skill identification process. From a performance perspective (Guzzo & Dickson, 1996), those levels are (a) the individual operator or worker, (b) the team, and (c) the organization (see Fig. 2.1 for a representation of the three levels). The next three sections address resource management skills from the perspective of each of these levels to provide a comprehensive background for the skill identification process. Once that has been established, methodological considerations are presented, followed by the skill identification methods developed at the airlines.

Individual Level

Traditionally, skill identification has focused on the individual performer, because job and task analysis methods have concentrated on individual knowledge and skill requirements. The emphasis in resource management

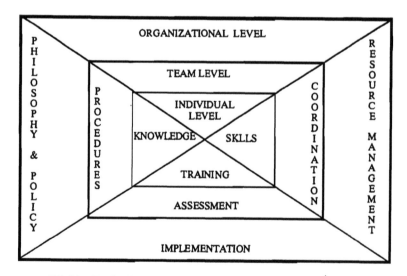

FIG. 2.1. Key levels in the identification of resource management skills.

has broadened over the last few years to include team and organizational considerations, sometimes at the expense of individual skills (see Hackman, 1993, for an example). The individual is still very important, for several reasons. First, most task analyses serving curriculum development evolved from analyses performed at the individual level. Any new resource management skills must have links back to those task listings and the individual level. Second, resource management skills have substantial cognitive elements, and cognitive analysis and modeling have strong roots in the individual performer. Finally, the individual skill level should remain a focus of training because skill development starts at the individual, not at the team or organization level. Although some resource management training takes place at the team level, the focus of the skill training will remain at the individual level until more effective team and organizational methods have been developed and successfully implemented.

A closer look at these reasons helps to emphasize the importance of the personal level both in the identification of the skills and the ultimate implementation of a resource management program. Historically, the individual level has been the origin of skill identification, and many training programs still emphasize that level. Any new methods must take that origin into account and must ensure that the resulting skills can fit back into the traditional ISD framework. For example, some methods for analyzing teams evolved from the ISD-based techniques by moving beyond the task analyst or SME to groups of SMEs representing different groups within the training department.

In addition, skill identification methods not only must link back to the traditional task analysis framework but they also must link forward to more precise cognitive analysis methods. Cognitive task analysis has moved from a supplemental tool for training systems (Tannenbaum & Yukl, 1992) to methods tailored to the analysis of complex tasks such as those found in aviation (Seamster, Redding, & Kaempf, 1997). With established cognitive analysis methods emphasizing the individual level of analysis, it is useful to work with skill identification methods that have some links to the individual level. If one were to work with methods that identified only team or organizational skills, it would be more difficult to perform a cognitive analysis of those types of skills. In short, the individual level still plays an important part both as input to and output from the skill identification process.

Team Level

The *team*, which has become a preferred term in organizational psychology (Guzzo & Dickson, 1996), refers to small groups that work together on a common set of tasks. Teams are a particular type of work group. A key distinguishing element of a team is that the team has both individual-participant and team accountability (Katzenback & Smith, 1993). In other types of work groups, accountability focuses on the individual, with each participant responsible for his or her own actions and performance. In a team environment, where the overall performance is highly coordinated, accountability lies with the team.

In the context of resource management, a team comprises a group of workers who have complementary skills and a common set of performance goals and standards and who act with mutual responsibility and accountability (Katzenback & Smith, 1993). The team is a level above the individual. At the team level, with an emphasis on team accountability, a primary concern is performance and skill assessment, especially in commercial aviation where resource management assessment shifts from the individual to team evaluation. Therefore, resource management skill identification methods should address the growing needs of the team level assessment. In aviation, the more traditional assessments look at individual performance. As long as a team member did his or her job, that individual achieved a satisfactory assessment independent of team performance. Under the newer form of resource management skill assessment, each member has responsibility for overall team performance, and ratings are assigned in large part on the basis of team effort.

As performance assessment continues to move to the team level, especially in cases where team coordination is critical to overall job or mission success, the team components of resource management skills need to be more fully addressed. Without a deeper understanding of resource management skills and how to assess and debrief them, coaches and trainers are limited in their ability to improve team performance. Most assessors can point out what is wrong with the performance of a team, but they cannot specify the optimal training intervention to improve that performance. For example, an evaluator may tell team members that they did not attend to a specific task in a timely manner under certain conditions. Such feedback tells the team that there was a specific problem, but it does not help the team members understand what they need to do to better perform under those or related conditions in the future. Why did the crew not start the task earlier? Could they have assigned tasks differently? What other conditions might cause the same problem? What should the team and individuals practice so that they can improve their performance under those conditions? These are diagnostic questions that can be answered only with a deep understanding of the resource management skills.

At present, most team feedback and debriefings are limited to what the team did incorrectly and fail to specify how the team can ameliorate its performance or to inform the team members about what they should practice in order to improve their resource management skills. For example, in aviation simulator training specific skill practice is readily given for flight performance problems such as when a pilot is having difficulty controlling the aircraft during an engine failure inserted at critical takeoff speed. In such a case, the individual is given specific instructions on what should be done to perfect the maneuvers. Then the pilot practices either certain components or the full maneuver until he or she reaches the desired level of proficiency. But that same level of feedback and practice is not available for team coordination or other resource management problems that occur in training sessions. When debriefing resource management performance, the instructor and crew may discuss what was done incorrectly, but the crew rarely has the opportunity to practice, develop, or refine their management skills. When identifying new resource management skills, organizations need to address this component, making sure that the resulting skills can be assessed and diagnosed at the team level.

Organizational Level

Organizations and researchers pay increasingly greater attention to the effects of the organization on job performance and skill development. Organizations, as well as industry groups, support the identification of skill standards and have started to play a more active role in the specification of skills. Furthermore, recent research addresses some of the organizational activities that affect job performance, and working within that broader context should lead to more accurate predictions of job performance (Arvey & Murphy, 1998). For example, the effects of the social and organizational environment also affect research on skill acquisition (Voss & Wiley, 1995). Researchers are starting to address the fact that skill acquisition takes place within an organizational and sociocultural environment that helps create a need for specific skill developments. This is particularly true in the area of resource management skills, where an organization's operating philosophy and policies can have a strong influence on training and motivating the development of individual and team skills.

The Goals 2000 legislation established a National Skills Standards Board to identify clusters of occupations that will define national skill standards (Sheets, 1995). The national skill standards intend to support worker training and certification as well as career counseling and worker transition assistance. The skill standards movement has generated new ideas about the nature and role of skills within organizations. To underscore the old and new perspective, Merritt (1996) compared the traditional skill-components model with the newer professional model. Under the skill-component model, one supported by ISD methods, workers have limited roles, and the worker's effectiveness is based on how well he or she performs on a list of predefined tasks. Under this model it is the managers who have control over the identification and development of skill standards, and traditionally management had an analyst or SME perform the skill analysis. The professional model assumes that the worker is in charge and able to make decisions and handle nonroutine situations. Under this model it is the worker who is in control of the work, and the organizational context must be considered in the identification of skills. Thus, organizations should be shifting to the professional model, and the skill identification methods they use should reflect this new outlook on the role of skill standards in the workplace.

Coincident with the greater emphasis on the organizational level, skill analysis methods are expanding from the traditional ISD approach involving one or two analysts to a broader approach involving a number of representatives throughout the organization. There are at least two approaches that can be taken in the formation of the resource management skill identification team. The first is to conduct a set of preliminary rating sessions with SME representatives from the key departments and select only those team members who most closely agree with each other about the definition and specification of resource management skills. The second approach is to conduct a preliminary session to build a consensus, working with all participants, and use reliability data to identify areas of agreement and areas where additional consensus must be established. The first approach may be more efficient, but it can also exclude some parts of the organization from the skill identification process. The second approach, in which everyone works as a team to develop a good level of agreement, ensures broad organizational representation and should lead to better acceptance of the team's results.

The organization level must be actively involved in the skill identification process, and the methods used must be sensitive to the organizational context. The organizational level plays a central part in ensuring that the resource management skills, once identified, will be properly trained and assessed. Furthermore, the organization needs to set the appropriate philosophy, policies, and reward systems to ensure that individuals and teams are properly motivated to develop and maintain their resource management skills.

Operational Implications

For the practitioner, there are several key points that should be used in formulating an overall approach to resource management training and assessment. The individual level is essential to skill and knowledge training because a substantial part of resource management skill development—the actual acquisition and refinement of skills—takes place within the individual operator, pilot, or team member. Keep track of the individual skill level when selecting skill identification methods to ensure that the resulting skills can be easily linked to task structures, structures generally specified at the individual worker level.

It is at the team level that establishing skill assessment is most important, especially in work environments where resource management assessment has shifted from the individual to team evaluation. In such environments each member has responsibility for overall team performance, and ratings are assigned in large part on the basis of the team effort. Skills must also be addressed at the team level so that teams are given adequate opportunity to practice, develop, and refine their management skills. When identifying new resource management skills, organizations need to address the assessment, diagnosis, feedback, and additional practice required to strengthen team-level skills.

Finally, the organizational level is essential to the proper implementation of resource management skills. The organization plays a central part in ensuring that the resource management skills, once identified, will be properly trained and assessed in a thorough and ongoing manner. Furthermore, the organization needs to establish and maintain an appropriate philosophy, policies, procedures, and system of rewards to ensure that individuals and teams are properly motivated to develop and maintain their resource management skills.

SKILL ANALYSIS TECHNIQUES

This section provides a brief review of methods for identifying resource management skills. The methods comprise both data collection and analysis techniques, with an emphasis on those steps that have been evaluated at several airlines. The following discussions provide a brief description of general data collection techniques followed by a discussion of the four analytic steps that should be considered when identifying resource management skills.

Data Collection Techniques

Existing skill data collection techniques provide a starting point in formulating skill identification methods that address the needs of individual, team, and organizational levels. The main data collection techniques that have been used in a skill analysis are listed in Table 2.1 (Levine, Thomas, & Sistrunk, 1988), which shows the range of data collection types that can be used individually or in combination to identify and analyze skills.

A brief review of these methods helps to identify the ones best suited for the identification of resource management skills. As the name implies, *direct observation* is the most immediate form of data collection; it can provide very detailed, firsthand information about a job. A major disadvantage of using direct observation in the analysis of resource management skills is that it can be very difficult to analyze the data because of the cognitive nature of many those skills. Questionnaires include a range of instruments, such as checklists, ratings, and surveys. Questionnaires can be very useful, especially if they are structured to collect data that can be analyzed for interrater reliability (IRR). Questionnaires of this type are particularly useful

TABLE 2.1
General Skill Data Collection Techniques

Technique	Description
Direct observation	An analyst observes and records continuous job performance. Because of the time and cost associated with direct observation, it usually is limited to a small sample size.
Questionnaires	Checklists, surveys, or other types of instruments are used to systematically collect job information. The use of questionnaires allows an organization to collect job data from a relatively large number of pilots of SMEs. This technique can be useful in group settings.
Work diaries/logs	Diaries and logs are filled out by incumbents to record job activities. This form of data collection is used early in analyses, when not much is known about the job or skills.
Individual interviews	Individual interviews take place in the form of a meeting between an incumbent and an analyst. For skill analyses, a structured interview that allows some flexibility is recommended.
Group interviews	This involves a meeting between SMEs and a facilitator. In a group interview, the focus should be on a single job. This technique, when used with multiple SMEs, can permit collection of a great amount of job data in a shorter period of time.
Analysis of training systems documents	Analysis of job and organizational documents can provide preliminary job information. Document analysis is often performed in situations where little task information is available.
SME conferences	SMEs collect data about tasks or task elements such as skills. SMEs work together to establish some level of agreement prior to providing the job data. If SMEs can be brought together at the same time and place, a conference can be an efficient way to collect job data.
Job performance by analyst	Requires as analyst to learn and perform the job of interest. It has limited application in complex task environments where a substantial amount of experience is required to perform the tasks.

Note. SME = subject matter expert.

when administered in a group setting where SMEs can then review their results and determine what the team must do to achieve greater consensus.

Work diaries and logs are most useful in the early phases of skill identification; this technique does not easily lend itself to group consensus. The individual and group interview may be unstructured or structured. For the purposes of a skill analysis, the structured form is more desirable, because the emphasis is on the more formal data collection methods. The advantages of the interview include familiarity, flexibility, and the interactive nature of this form of data collection (Kirwan & Ainsworth, 1992). The analysis of documents is another useful technique at the beginning of the skill identification process, but it has limited application in group settings other than providing data that can be used in questionnaires or as part of other group activities.

SME conferences clearly lend themselves to group data collection and, with their emphasis on broader organizational involvement, offer an efficient way to collect skill data and to reach team or group agreement. Finally, having the analyst perform the job, although useful in some basic job environments, has little application in complex environments, such as aviation or operating rooms, where a substantial amount of experience and skill is required to perform the job. The best techniques for resource management data collection are those that have flexibility, allow for group interaction, and can be used by a team to reach consensus about the skills. The questionnaires combined with group interviews and SME conferences offer the best data collection techniques as organizations move away from their dependence on ISD toward methods that allow them to account for the broader organizational context.

Resource Management Skill Analysis

Researchers working with teams from two airlines evaluated approaches to the analysis of resource management skills (Seamster, Prentiss, & Edens, in press). The four steps, which are presented next, are: (a) definition of resource management skills, (b) identification of resource management skill verbs, (c) identification of resource management skills, and (d) categorizing resource management skills.

These methods were based on the questionnaire and SME conference data collection techniques. Once the data were collected they were analyzed by a set of five IRR data analysis techniques and presented back to the team with the goal of attaining a pre-established level of consensus. Researchers initially worked with all five of the IRR measures (Williams, Holt, & Boehm-Davis, 1997). Those measures included an agreement index, congruency with the group, rater intercorrelations, sensitivity, and significant differences between individual raters and group means. These measures have been used to help evaluators improve their reliability in rating observable behaviors and other elements of team performance.

Working with all five measures of skill identification overwhelmed the team members; however, with experience they found the agreement index and congruency to be the most helpful in establishing team consensus. The agreement index identified specific items that did not meet the established benchmark for group agreement. Congruency, the comparison of individual rating distributions with the team distributions, informed team members where they were in relation to the team. This helped individuals determine if they were being either too strict or too lenient in their interpretation of the criteria or the rating scales.

The four steps were integrated into a skill analysis that served two purposes. First, the analysis generated a list of resource management skills. Second, the process involved personnel from the organization in the systematic identification of skills, personnel that would ultimately be responsible for the training and assessment of those skills. The methods used rating and agreement index data to help representatives from different departments form a consensus about the nature and identity of resource management skills. The airline used SMEs from three to five departments representing the full range of individuals responsible for developing, assessing, and implementing resource management training. In addition, 5 to 10 SMEs were involved in each of the methods to obtain a broad consensus and to distribute the workload across several individuals. This reduced some of the bottlenecks encountered in traditional skill analyses when one or two SMEs are asked to perform most of the analysis.

In the following sections we present the four skill analysis steps with a general description of the activity, a description of the number of SMEs (or participants), and the form of data analysis. The results of the study are presented in the next section.

Step I: Definition of Resource Management Skill.   Resource management skills can be understood through their relationship to knowledge. Even though knowledge can be defined so broadly as to include skills (see Cooke, 1994) and skills may also be defined in general terms to include most knowledge (see Royer, Cisero, & Carlo, 1993), this lack of distinction is detrimental, especially in resource management training and assessing. One of the difficulties in this area has been the lack of concise definitions (Jonassen & Hannum, 1995). Even though researchers do not clearly distinguish between knowledge and skills, it makes sense to create a clear, not arbitrary, distinction between these two important elements of the skill identification process (Seamster, Redding, & Kaempf, 1997).

Aviation training programs provide general definitions of knowledge and skills, making it difficult to distinguish between the two elements (Federal Aviation Administration, 1991), with *knowledge* defined as the information required to develop skills and attitudes and *skill* defined as the ability to perform an activity or action. These definitions should be more operationally relevant so that organizations could better train and assess these two different elements of human performance (Lanzano et al., 1997).

From an operational perspective, what is the difference between a knowledge and a skill? Knowledge is something that can be trained and assessed independently from full job performance. Thus, knowledge is primarily informational or conceptual in nature and not tightly coupled with performance. On the other hand, skills are tightly linked with a specific set of subtasks and are best trained and assessed as part of job performance. The skills are clearly performance based and trained through a practice-and-feedback cycle in which trainees perform in a part- or full-task job environment. Proctor and Dutta (1995) provided a good starting point for an operational definition of *skill*: "Skill is a goal-directed, well-organized behavior that is acquired through practice and performed with economy of effort" (p. 18). Because skills are acquired through practice, they can be trained generally by means of extended practice. In addition, skills are evaluated through performance characterized by an economy of effort. From an operational perspective, skills are actions or behaviors performed to a specified level of proficiency that occur in one or more subtasks of a specific job domain.

To be of use at the individual, team, and organizational levels, the definition of resource management skills requires operational criteria that directly address training and assessment issues. For example, Seamster, Prentiss, and Edens (1997) proposed that resource management skills are:

Trainable (through a practice-and-feedback cycle).

Related to one or more resource management components.

Assessed through observable behaviors.

Improve performance in a team setting.

Named at a consistent level.

The initial step in defining resource management skill was to review existing definitions for skills and knowledge, and a total of 19 criteria were identified that could be used to specify a resource management skill. Next, an SME conference with 5 to 11 participants was convened to gather group support for keeping or deleting the criteria. This group narrowed that list to the 5 most important criteria. Those 5 criteria were then evaluated by having SMEs check each criterion against a set of 20 potential resource management skill statements. Each criterion was refined until the group of SMEs, working individually, were able to reach a benchmark level of agreement about the number of criteria that applied to each skill statement. A description of the agreement index follows.

The SMEs first worked in a group with a facilitator to refine the list of criteria. They then worked individually to evaluate the criteria against specific skill statements. The agreement index was computed to determine whether the group had reached consensus on how to apply the criteria against skill statements. *Agreement* is the degree to which SMEs give the same rating for the same item based on a statistic, r_{wg}. By itself, the agreement index is not sufficient to establish reliability, but the mean agreement index across all items does provide an indicator of group consensus. Throughout the rating cycles, when the mean agreement index was .6 or greater (1.0 is perfect agreement), no additional IRR statistics were used, and the SMEs did not review the agreement index for individual items. In cases where the mean agreement fell below .6, the group of SMEs reviewed each item that had an agreement index below .6, and the congruency measure for each SME, to determine who was close to the group distribution and who needed to revise his or her rating strategy. For this first step, the SMEs needed several rating cycles with feedback before they were able to reach their benchmark of a mean agreement index of .6.

Step 2: Identification of Resource Management Skill Verbs. The second step in the skill analysis was to identify a set of verbs closely related to CRM skills. A listing of 400 resource management skills was reviewed, and 109 verbs were extracted from those skill statements. This step identified verbs that could be used to standardize resource management skill statements as well as to specify new skills. A rating instrument enabled SMEs to rate each verb on a 5-point scale where 1 meant *not related* and 5 signified *completely related* (to resource management skills). The results of this instrument were used to identify a set of resource management verbs that could be used to standardize skill specification.

This step was evaluated working with two different groups, one with 5 SMEs and one with 10 SMEs. The SMEs worked individually with the rating instrument to rate the set of verbs on the basis of how closely each verb was related to resource management skills. Before completing the rating exercise, the team established an agreement index benchmark and a rating cutoff. Verbs that met the benchmark and cutoff would be included in the team's verb listing. After the rating session, the rating and agreement index were computed for each verb, resulting in a list of 37 to 46 resource management verbs.

Step 3: Identification of Resource Management Skills. Once resource management skill criteria were defined and verbs identified, an extensive list of potential resource management skill statements was examined to determine which met the skill criteria. The initial list of 400 potential resource management skills was refined by deleting and changing the skill statements that were not based on the skill verbs identified. This resulted in 320 uniform skill statements. An instrument was developed to enable SMEs to specify which skill criteria were met by each of these skill statements. SMEs worked individually, placing a 1 in each cell where the criterion was met and leaving the cell blank where the criterion was not met by the skill statement. If a skill statement satisfied all the criteria, it received a score of 5.

This step was presented to two groups of SMEs, one with 6 members and a second with 12. The SMEs worked individually to determine how well each skill statement satisfied the CRM skill criteria. To reduce SME workload, the skill statement listing was broken up into segments such that each SME rated approximately 100 skills. All SMEs rated a list of 10 common skill statements to facilitate computation of IRR measures across the entire rating team. The mean rating and agreement index were computed for each skill statement. Statements with a mean greater than or equal to the pre-established cutoff and an agreement index greater than or equal to the benchmark were determined to be resource management skills.

Step 4: Categorizing Resource Management Skills. After a list of CRM skills was identified, the final step in the analysis phase consisted of grouping these skills into categories on the basis of existing CRM topics or categories. The categories were initially derived from the National Aeronautics and Space Administration/University of Texas model of CRM markers (see Helmreich, Wilhelm, Kello, Taggart, & Butler, 1990) and then refined by a specific airline over time into the following six categories:

- Briefing and Communication
- Leadership and Teamwork
- Situation Awareness
- Decision Making/Planning
- Crew Self-Evaluation
- Automation and Technology

SMEs, working with an instrument that presented individual skill statements, rated the skills by specifying their relationships to each of the resource management categories. Again, the SMEs received instructions and practice and then were asked to work individually when rating. SMEs worked with a 3-point scale on which 1 = fully related, 2 = partially related, and 3 = not related at all.

Seven SMEs worked on this step individually to rate a subset of the resource management skills. The mean or average rating was computed for each skill across all six categories. This resulted in six means for each statement, and statements with a mean of 1.5 or less (fully to partially related) under a specific category were included in that category. The rating results were reviewed through group discussion, and in cases where a skill belonged to more than one category, group majority was used to determine the primary resource management topic (see Seamster et al., in press). It should be noted that most of the skill statements were rated as related to more than one skill category, but the SMEs, through discussion, were able to agree on the primary skill category for each resource management skill.

RESOURCE MANAGEMENT SKILL RESULTS

The result of the first step was a set of criteria for crew resource management skills. As pointed out in the discussion of the first step, existing skill definitions lack the criteria to help organizations train and assess these skills. The following criteria were used by the airline in subsequent steps to guide SMEs through the process of identifying specific skills. The criteria statements identified at one airline (Seamster et al., in press) were:

- Resource management skills are assessed through observable behaviors.
- Resource management skill is a measurable level of proficiency to perform a task—it requires practice to meet a standard of performance.
- Resource management skill is directly related to a knowledge of one or more CRM components.
- Resource management skill improves individual performance in a crew setting.
- Resource management skill enhances mission awareness.

These criteria are representative of the type that can be used by organizations to more clearly define and identify resource management skills. Other organizations may identify more criteria, or they may emphasize the training aspects over the assessment elements. As with all four of these steps, the process used by the SMEs and the organizational involvement in that process is as important as the specific results.

Identification of Skill Verb Results

The two different teams identified 37 to 46 verbs that should be used to specify resource management skills. A sample of the type of verbs identified are shown in Table 2.2.

One airline had 10 SMEs rate the verbs and used an agreement index of .5 and a mean of 4.0 (on a 5-point scale) for its cutoff. That group had a low mean agreement index of .44, in part because of the size of the team and in

TABLE 2.2
Sample of the 47 Verbs Rated as Skill-Related by One Airline

acknowledge	analyze	brief
adhere	ask	communicate
advocate	assign	contribute

part because of the changes in team membership between the administration of the first and second method. Consequently, a number of participants were new to the process and were not familiar with the skill criteria and how these criteria had been identified. The second airline, with just 5 SMEs, had substantially more verbs meet their benchmark, and their mean agreement index was .7. The composition of this team remained constant across the first and second method, and the team members established a common understanding of the criteria and verbs used in the identification and specification of resource management skills.

Identification of Resource Management Skills

One group of SMEs identified 120 resource management skills, and a second group identified 180, on the basis of their ratings. Those two skill lists contained duplicate skills and also had statements that were specified at too high or too low a level of detail, but the lists served as a preliminary listing. The first group consisted of 12 SMEs, and their mean agreement index was relatively low at .49. This low index was due in part to the fact that the group had added several new members. As happens in many organizations, when word of an activity spreads, pressure can be exerted to add SMEs to the group. In this case, a low mean agreement index resulted. The second group consisted of 6 SMEs and had a mean agreement index of .81, substantially higher than their initial benchmark of .6, reflecting an incremental improvement as the team developed stronger consensus as its members worked through the three steps.

Categorization Results

The categorization activity, followed by a combining of skill statements, resulted in 49 resource management skill clusters grouped into the six categories. The skill cluster statements listed in Table 2.3 are limited to the behavior part of the skill specification and do not include the level of proficiency, which is normally provided in the context of the subtasks where it is

TABLE 2.3
Partial Listing of Resource Management Skill Clusters

Category	Resource Management Skill Behaviors
Situation awareness (SA)	• Acknowledge when something doesn't feel right. Direct attention to information or conditions that may have been missed. • Ask for help when becoming confused, overloaded, or when SA is in doubt. • Consider contingency requirements that may affect flight status or safety. • Determine, plan, and discuss aircraft configurations, airport specific procedures, and performance issues. • Establish goals on the basis of current conditions. Base backup plans on anticipated conditions. • Identify and verbalize any reduction in or loss of personal SA. Monitor SA of others. • Identify situations where additional resources are needed. • Inform crew when conditions change or unplanned situations emerge. • Monitor developments (e.g., fuel, time, weather, traffic, air traffic control, etc.). • Recognize and act on "bottom lines" established for safety. • Recognize and act to counter the effects of stress or fatigue, both personally and in fellow crew members. • Test assumptions; confirm understandings to maintain a shared level of situation awareness.
Crew self-evaluation	• Conduct operational review after any "eventful task"—openly evaluate successes and mistakes. • Discuss/debrief key events, asking "How could we have done better?" Stress what is right, not who is right. • Give and take constructive feedback. • Assess personal strengths and weaknesses—continually self-correct.

performed. The category of Situation Awareness had the largest number of skill clusters—12—whereas Crew Self-Evaluation had the smallest number, just 4. This skill categorization provided a usable set of skill clusters for curriculum development and crew proficiency assessment.

GUIDELINES

Guidelines for the Identification of Resource Management Skills

The process and results of this resource management skill analysis yielded a number of guidelines that can help organizations identify resource management skills. These guidelines are presented as prescriptive statements that provide general information about how to proceed (Salas, Cannon-Bowers, & Blickensderfer, 1997) with the planning and identification of resource management skills. Collectively, these guidelines address all three levels of the job performance context: the individual, the team, and the organization.

1: Work With a Team of SMEs.   When identifying resource management skills, work with a team of SMEs (5–10) to ensure stable ratings, to reduce individual team member workload, and to promote greater participation in the specification of those skills. Team size is important: Fewer than 5 SMEs do not provide an adequate range of experience, and more than 10 SMEs become more difficult to manage, schedule, and keep together as a team.

2: Use Team Membership and Formation to Extend Organizational Involvement.   Select team members from key organizational elements to ensure the range of needed expertise and to extend organizational involvement. Go beyond the few individuals connected with training to include expertise and representation of the operators, supervisors, and managers.

3: Establish Team Membership Early in the Process.   Form the team or group of SMEs as early as possible in the identification process. Once the team membership has been established, work with the group to establish a predetermined level of agreement and congruency and then limit or restrict new team members. Adding or replacing team members once the group has established good agreement generally reduces the group's efficiency and common understanding.

4: Use Some Form of IRR Measures.   Use a formal set of measures, such as those available to compute IRR, to calculate the level of agreement, correlations, and congruency. Use measures familiar to the organization of SMEs, such as those used in training raters or observers of individual or team behavior. Providing quantitative feedback to the group through these measures is particularly important at the beginning of the process when the team or group of SMEs is establishing agreement.

5: Make the Definition of Skill Operationally Relevant.   One of the first steps the team should undertake is to make operationally clear the concept of skill. Consider working with concepts of knowledge and skill, and have the SMEs distinguish between those two. Then, have the SMEs, through a structured process, determine the criteria for resource management skills in training, assessment, and improvement of the operations.

6: Use a Consistent Form of Skill Statement.   Have the SMEs specify and then consistently use a form of skill statement. Skill statements may include the action or behavior followed by a level of proficiency along with subtasks in which the skill is performed. Select the elements of a skills statement that are essential for skill identification, and then specify the characteristics of those elements. For example, it may be useful to have the SMEs identify a set of verbs and the structure of the action or behavior statement.

7: Develop a Flexible Approach to Skill Analysis. Use a set of flexible techniques for skill identification that are applicable throughout the training and assessment cycle. No longer is a fixed skill set identified at the front end of a training program and then used, unchanged, throughout the life of the program. Individuals, jobs, the technology, and the organization are constantly changing, and skill identification has become an ongoing process.

8: Include the Organizational Context. Make sure that organizational constraints and mission requirements are reflected in the skill identification process. Go beyond microconsiderations of the task and subtask to those reflected in organizational policies and philosophy when identifying the full range of skill requirements.

9: Use a Combination of Analysis Techniques. Use a combination of techniques or steps rather than just a single method to identify resource management skills (Levine et al., 1988). It is important to propose several techniques that can be used in combination. Use the results from the different techniques to establish supporting evidence for the required skills. Evidence from several different data sources tends to be more convincing than evidence from a single source.

10: Consider Modifying Techniques and Using Them in New Ways. Existing ISD analysis methods can be modified and used in new ways, resulting in substantially broader applications (Jonassen & Hannum, 1995). Several of the methods, initially not intended to be used by a group of SMEs, can be applied in a group context to gain essential information about resource management skills. For example, applying IRR measures to methods used by an analyst can turn some individual analyst techniques into a group skill identification process.

CONCLUSIONS

Three levels of analysis—individual, team, and organization—are critical for the identification of resource management skills. The individual level is important for its links to traditional ISD structures and its compatibility in form with the more established cognitive analysis methods. The team level has important links to the newer forms of team assessment and the important contributions that these skills can make in diagnosing team performance. Ultimately, the significance of the organizational level lies in its support for broadening the work performance context, and it ensures a successful implementation of the resource management skill training and assessment program.

Also critical to identification of resource management skills is the use of a set of analytic techniques such as those described in here. These steps demonstrate how to implement a set of analytic methods using reliable SME ratings within the context of large training organizations.

ACKNOWLEDGMENTS

This research was supported in part by the Office of the Chief Scientific and Technical Advisor for Human Factors (AAR-100) at the Federal Aviation Administration through Grant 94-G-034 to George Mason University. We thank Dr. Eleana Edens at AAR-100 for continuing advice and support of this work.

REFERENCES

Arvey, R. D., & Murphy, K. R. (1998). Performance evaluation in work settings. *Annual Review of Psychology, 49,* 141–168.

Brannick, M. T., Prince, C., Salas, E., & Stout, R. (1995). Assessing aircrew coordination skills in the TH-57 pilots. In R. S. Jensen (Ed.), *Proceedings of the Eighth International Symposium on Aviation Psychology* (pp. 1069–1072). Columbus: Ohio State University.

Cohen, M. S., Freeman, J. T., & Thompson, B. (1995). Training metacognitive skills for decision making. In R. S. Jensen (Ed.), *Proceedings of the Eighth International Symposium on Aviation Psychology* (pp. 789–794). Columbus: Ohio State University.

Cooke, N. J. (1994). Varieties of knowledge elicitation techniques. *International Journal of Human-Computer Studies, 41,* 801–849.

Federal Aviation Administration. (1991). *Advisory circular 120-54: Advanced qualification program.* Washington, DC: Author.

Goldsmith, T. E., Johnson, P. J., Seamster, T. L., & Edens, E. S. (1995). Toward a cognitive analysis of CRM skills. In R. S. Jensen (Ed.), *Proceedings of the Eighth International Symposium on Aviation Psychology* (pp. 528–533). Columbus: Ohio State University.

Gregorich, S. E., Helmreich, R. L., & Wilhelm, J. A. (1990). The structure of cockpit management attitudes. *Journal of Applied Psychology, 75,* 682–690.

Guzzo, R. A., & Dickson, M. W. (1996). Teams in organizations: Recent research on performance and effectiveness. *Annual Review of Psychology, 47,* 307–338.

Hackman, J. R. (1993). Teams, leaders, and organizations: New directions for crew-oriented flight training. In E. L. Wiener, B. G. Kanki, & R. L. Helmreich (Eds.), *Cockpit resource management* (pp. 47–69). San Diego: Academic Press.

Hamman, W. R., Seamster, T. L., & Edens, E. S. (1995). The development and results of the line operational evaluation in the advance qualification program. In R. S. Jensen (Ed.), *Proceedings of the Eighth International Symposium on Aviation Psychology* (pp. 658–662). Columbus: Ohio State University.

Helmreich, R., Wilhelm, J., Kello, J., Taggart, W., & Butler, R. (1991). *Reinforcing and evaluating crew resource management: Evaluator/LOS instructor reference manual* (Technical manual 90-2). Austin: National Aeronautics and Space Administration/University of Texas.

Jentsch, F., Bowers, C. A., & Holmes, B. (1995). The acquisition and decay of aircrew coordination skills. In R. S. Jensen (Ed.), *Proceedings of the Eighth International Symposium on Aviation Psychology* (pp. 1063-1068). Columbus: Ohio State University.

Jonassen, D. H., & Hannum, W. H. (1995). Analysis of task analysis procedures. In G. J. Anglin (Ed.), *Instructional technology past, present and future* (2nd ed., pp. 197-214). Englewood, CO: Libraries Limited.

Katzenback, J. R., & Smith D. K. (1993). The discipline of teams. *Harvard Business Review, 71*, 111-120.

Kirwan, B., & Ainsworth, I. K. (Eds.) (1992). *A guide to task analysis.* London: Taylor and Francis.

Lanzano, J., Seamster, T. L., & Edens, E. S. (1997). The importance of CRM skills in an AQP. In R. S. Jensen & L. Rakovan (Eds.), *Proceedings of the ninth international symposium on aviation psychology* (pp. 574-579). Columbus, OH: The Ohio State University.

Law, J. R., & Wilhelm, J. A. (1995). Ratings of CRM skill markers in domestic and international operations: A first look. In R. S. Jensen (Ed.), *Proceedings of the Eighth International Symposium on Aviation Psychology* (pp. 669-674). Columbus: Ohio State University.

Levine, E. L., Thomas, J. N., & Sistrunk, F. (1988). Selecting a job analysis approach. In S. Gael (Ed.), *The job analysis handbook for business, industry, and government* (Vol. 1, pp. 339-352). New York: Wiley.

Merritt, D. (1996). *A conceptual framework for industry-based skill standards.* Berkeley: University of California, National Center for Research in Vocational Education.

Proctor, R. W., & Dutta, A. (1995). *Skill acquisition and human performance.* Thousand Oaks, CA: Sage.

Royer, J. M., Cisero, C. A., & Carlo, M. S. (1993). Techniques and procedures for assessing cognitive skills. *Review of Educational Research, 63*(2), 201-243.

Salas, E., & Cannon-Bowers, J. A. (1997). Methods, tolls, and strategies for team training. In M. A. Quiñones & A. Dutta (Eds.), *Training for 21st century technology: Applications of psychological research* (pp. 249-279). Washington, DC: American Psychological Association.

Salas, E., Cannon-Bowers, J. A., & Blickensderfer, E. L. (1997). Enhancing reciprocity between training, theory and practice: Principles, guidelines, and specifications. In J. K. Ford, S. W. J. Kozlowski, K. Kraiger, E. Salas, & M. S. Teachout (Eds.), *Improving training effectiveness in work organizations* (pp. 291-322). Mahwah, NJ: Lawrence Erlbaum Associates, Inc.

Seamster, T. L., Hamman, W. R., & Edens, E. S. (1995). Specification of observable behaviors within LOE/LOFT event sets. In R. S. Jensen (Ed.), *Proceedings of the Eighth International Symposium on Aviation Psychology* (pp. 663-668). Columbus: Ohio State University.

Seamster, T. L., Prentiss, F. A., & Edens, E. S. (1997). Methods for the analysis of CRM skills. In *Proceedings of the Ninth International Symposium on Aviation Psychology* (pp. 500-504). Columbus: Ohio State University.

Seamster, T. L., Prentiss, F. A., & Edens, E. S. (1999). Implementing CRM skills within crew training programs. In *Proceedings of the Tenth International Symposium on Aviation Psychology* (pp. 500-504). Columbus: Ohio State University.

Seamster, T. L., Redding, R. E., & Kaempf, G. L. (1997). *Applied cognitive task analysis in Aviation.* Aldershot, England: Ashgate.

Sheets, R. G. (1995). *An industry-based occupational approach to defining occupational/skill clusters.* Washington, DC: U.S. Department of Labor.

Tannenbaum, S. I., & Yukl, G. (1992). Training and development in work organizations. *Annual Review of Psychology, 43*, 399-441.

VanLehn, K. (1996). Cognitive skill acquisition. *Annual Review of Psychology, 47*, 513-539.

Voss, J. F., & Wiley, J. (1995). Acquiring intellectual skills. *Annual Review of Psychology, 46*, 155-181.

Williams, D. M., Holt, R. W., & Boehm-Davis, D. A. (1997). Training for inter-rater reliability: Baseline and benchmarks. In *Proceedings of the Ninth International Symposium on Aviation Psychology* (pp. 514-520). Columbus: Ohio State University.

3

Uncovering Differences in Team Competency Requirements: The Case of Air Traffic Control Teams

Kimberly A. Smith-Jentsch
Naval Air Warfare Center Training Systems Division

David P. Baker
American Institute for Research

Eduardo Salas
Janis A. Cannon-Bowers
Naval Air Warfare Center Training Systems Division

Since the late 1980s, much research has been conducted to identify teamwork behaviors that contribute to effective team performance (e.g., Kanki & Foushee, 1989), safety (e.g., Hartel, Smith, & Prince, 1991; Kern, 1997; Nagle, 1988), and decision making (e.g., Smith-Jentsch, Johnston, & Payne, 1998). A large portion of this research has focused on commercial and military aircrews (e.g., Wiener, Kanki, & Helmreich, 1993) and Navy command and control teams (e.g., Cannon-Bowers & Salas, 1998). However, very little research has examined the validity of teamwork models developed in one domain for other types of teams that vary in structure or composition. Such analyses are necessary to provide guidelines to practitioners regarding which strategies can be generalized easily from one environment to the next, which strategies may need significant modifications, and which strategies must be specifically tailored to their particular application. In this chapter we compare and contrast the results from research on air traffic control (ATC) teamwork with findings from team research in other domains. In particular, we describe the team-related knowledge, skill, and attitude (KSA) requirements for ATC teams and highlight similarities and differences between these requirements and requirements for aircrew coordination or crew resource management (CRM).

First, we highlight key characteristics that differentiate ATC teams from aircrews. Second, a series of team-related KSAs that have been identified as being critical for ATC team performance are defined and compared against those currently trained and assessed for aircrews. Finally, we offer guidelines for enhancing ATC teamwork by linking specific KSAs to appropriate training methods within the three stages of CRM training: awareness, practice and feedback, and continual reinforcement (Federal Aviation Administration [FAA], 1989, sec. 5c; Gregorich & Wilhelm, 1993). These guidelines have practical applications for ATC teams as well as other types of teams that share similar team and task characteristics.

CHARACTERISTICS THAT DIFFERENTIATE
AIR TRAFFIC CONTROL TEAMS FROM AIRCREWS

Air traffic control involves a highly coordinated set of entities (i.e., ATC tower cab, terminal radar approach control [TRACON], and en route facilities) that guide aircraft from one airport to another. This coordination is primarily sequential in nature. Civilian ATC begins with the filing of a flight plan with the FAA. General aviation flight plans are filed through a flight service station, whereas commercial air carriers maintain a direct computer linkage with the FAA. For a commercial flight, the first contact between the aircrew and ATC occurs at the gate when the crew requests clearance to depart. The tower cab crew updates weather conditions, clears the flight to its destination via a planned route, issues appropriate taxi instructions, and clears the flight for takeoff. As the aircraft departs the runway, radar contact is made by the terminal area radar (or TRACON, depending on the class of air space), and the terminal area radar controller issues appropriate instructions to direct the flight out of the airport's airspace. As the aircraft departs the terminal airspace, the aircrew contacts the air route traffic control center (ARTCC, or *en route center*). En route centers provide radar control services for the flight between terminals. The flight may proceed along designated airways or, in many cases, such as high altitude commercial flights, fly a fuel-efficient direct route between terminals. In either case, the flight is handed off from en route center to en route center as it proceeds on its path until nearing its destination. The last en route controller hands off the aircraft to the terminal radar controller, who sequences the flight for landing at the airport. As the aircraft nears the runway (i.e., within 5 miles, approximately), control is passed from the radar controller to the tower cab team. The tower cab team gives final clearance to land, ensuring that there are no other planes or ground vehicles on the runway. As the aircraft turns off the runway after landing, the ground controller in the tower cab then issues taxi instructions to the gate.

Given the above description, it is clear that teamwork plays a critical role in ensuring safe and efficient ATC. But what is ATC teamwork? And how does it differ from aircrew coordination, or CRM? The answers to these questions will influence the degree to which training designed to foster teamwork for aircrews will fully satisfy ATC team training needs. We begin this chapter by discussing three characteristics that differentiate ATC teams from aircrews and that thus may affect the generalizability of training across these two team task domains. These variables include: (a) the responsibility for and consequences of decisions, (b) status variability, and (c) the stability of team membership.

Responsibility for and Consequences of Decisions

One important distinction between ATC teams and aircrews involves the nature of the decision-making tasks performed by each. Although both types of teams must coordinate with one another to obtain the information necessary to make good decisions, the responsibility for and consequences of those decisions are much more individualized for ATC teams. Coordination among controllers is highly sequential in nature. Responsibility for a particular aircraft is passed from one controller to another. Each controller in the link has the potential to make the next controller's job easier or more difficult depending on the accuracy and completeness of the information he or she passes, the spacing he or she allows between aircraft, and the orderliness of traffic flow. However, although interdependencies within and across ATC teams abound, for the most part, individual controllers make individual decisions about individual aircraft within their designated airspace. In contrast, the information exchanged among aircrew members is typically used to support team-level decisions made by the aircraft commander. Furthermore, the consequences of these decisions affect the entire crew in a very personal way: life or death.

Differences in the degree to which responsibility for and consequences of decisions are shared by team members may influence the generalizability of CRM training. In particular, training programs designed to increase individuals' willingness to monitor and shift workload may need to be repackaged. For example, many aircrew CRM programs use slogans such as "Both sides of the cockpit burn equally well" to motivate such behavior. There is no analogous argument that is likely to be quite as effective at motivating controllers to be supportive toward one another. Training designed to change nonsupportive attitudes and behavior may be more effective if it emphasized the potential ripple effects of one controller's error on other controllers in terms of both safety and efficiency.

Status Variability

A second characteristic of ATC teams that differentiates them from aircrews is the degree of status variability among team members. In aircrews, the captain is viewed as the ultimate authority figure. However, no equivalent type of leadership exists in ATC teams. In an ATC tower cab, a "controller in charge" (CIC) is always assigned to manage operations such as scheduling breaks, opening or closing positions, coordinating with other facilities, directing major changes in operations (e.g., changing runways in use), and ensuring timely performance of critical tasks (e.g., directing the creation of a new automatic terminal information service [ATIS] message). However, team members take turns filling this role. Thus, CIC is a temporary position rather than a rank (e.g., captain, first officer, etc.) attached to any one individual. A few facilities require that a supervisor is always present to serve as CIC. Even in such circumstances, however, the dynamics between controllers and supervisors are far different from those between captains and first officers. Whereas captains usually have more technical experience as a pilot than their subordinates, ATC supervisors may be viewed by their teams as being less technically proficient than the average controller because they spend less time per month "on position" actually controlling traffic. Moreover, their role is far less directive in nature. As a result, leadership–follower issues that tend to be heavily emphasized in aircrew CRM programs are not likely to directly generalize to ATC.

For example, most aircrew CRM programs include a module on assertiveness. The focus of such training is typically to increase junior crewmembers' ability and willingness to offer concerns and suggestions to higher status crewmembers (Jentsch & Smith-Jentsch, chap. 5, this volume). The premise is that lower status crewmembers are often intimidated by higher status crewmembers and thus tend to hold back their opinions or to state them in an indirect manner. Thus, training is designed to move individuals from a passive or indirect communication style to a more assertive style. While individuals within an ATC team vary in their level of skill and experience, certified professional controllers are generally considered peers. Furthermore, the nature of the job tends to attract individuals who are competitive and self-confident—even aggressive. Thus, the challenge of resource management training for controllers may be more a matter of refining overly aggressive styles rather than overly submissive ones.

Stability of Team Membership

A third characteristic that differentiates aircrews from ATC teams is the stability of team membership. Whereas membership in commercial and military aircrews tends to be quite variable (Hackman, 1993), most controllers work with the same team members for extended periods of time. A recent survey of FAA ATC teams across the country found that 95% of the facilities sampled had scheduling policies designed so that controllers worked with the same teammates for the majority of their working hours (Smith-Jentsch, Kraiger, Salas, & Cannon-Bowers, 1999). Furthermore, although controllers had the opportunity to change their schedules (and therefore their teams) every 6 months, the controllers in the survey reported having worked with their current teammates for an average of 2 years.

Extended interaction with the same set of teammates introduces an additional set of attitudes and knowledge that affect team coordination and performance. These "teammate-specific competencies" (Cannon-Bowers, Tannenbaum, Salas, & Volpe, 1995) can have a positive or a negative impact on ATC teamwork as teams either develop mutual trust or resentment based on long-term personality conflicts. For example, Smith-Jentsch, Kraiger, Cannon-Bowers, and Salas (2000) found that the more knowledge controllers had about their teammates' areas of expertise (teammate-specific knowledge) and the more cohesive they felt toward those teammates (teammate-specific attitude), the more willing they were to ask for and accept assistance when working with those teammates. This has important implications for the generalizability of aircrew CRM training to an ATC environment. Specifically, aircrew CRM training does not typically target teammate-specific competencies, because aircrews do not remain together long enough to justify the investment. However, given that such competencies have been shown to play a significant role in ATC teamwork, they should be emphasized in ATC team training.

In the preceding sections of this chapter we have outlined three characteristics that distinguish ATC teams from aircrews and, therefore, may influence the generalizability of aircrew CRM training to an ATC environment. In the following sections we discuss the implications of these differences for the content and methods used to train ATC teamwork. In particular, we emphasize the unique team training requirements that exist for ATC teams that are due to their relatively stable membership. Finally, we offer guidelines for linking team competency requirements to appropriate training strategies. These guidelines can be applied by practitioners charged with creating or adapting training strategies for use in a variety of team domains.

DEFINING ATC TEAM COMPETENCIES

In recent years, a number of task analyses have been performed on the ATC specialist (ATCS) occupation (Ammerman, Becker, Bergen, et al., 1987; Ammerman, Becker, Jones, & Tobey, 1987; Ammerman, Fairhurst, Hostetler, & Jones, 1989; Nickels, Bobko, Blair, Sands, & Tartak, 1995; Smith-Jentsch et al., 1999). Although each of these analyses has contributed to our understanding of the KSAs possessed by effective ATCs, only one was conducted specifically for the purposes of identifying the appropriate content of ATC team training. This analysis spanned 3 years and included a review of existing ATC task analyses and training materials, observation of team coordination in a variety of ATC facilities, focus groups, and structured interviews with 50 controllers across the country (Baker & Smith-Jentsch, 1997; Milanovich, Smith-Jentsch, & Harrison, 1999; Smith-Jentsch, Kraiger, Cannon-Bowers, & Salas, 1998; Smith-Jentsch, Zeisig, Cannon-Bowers, & Salas, 1997). Finally, 330 controllers from 51 facilities participated in a computer-based survey (Smith-Jentsch et al., 1999). As a result, 14 KSAs were identified as being important for ATC teamwork. Several of these are similar to the KSAs identified in previous analyses of ATC and aircrew requirements. However, Smith-Jentsch et al. (1999) also uncovered a number of KSAs that were not addressed in these previous efforts.

The differentiating characteristics among aircrews and ATC teams described earlier are likely to explain the disparity between the KSAs identified by Smith-Jentsch et al. (1999) and those discussed in the aircrew CRM literature (e.g., Helmreich, Foushee, Benson, & Russini, 1986; Prince & Salas, 1993). Differences between the Smith-Jentsch et al. (1999) analysis and previous ATC task analyses are likely due to other factors. For example, previous ATC analyses have tended to focus on individual technical proficiencies rather than team competencies. Second, they have been conducted primarily for the purpose of selection rather than training; therefore, they did not address knowledge and attitudes about specific teammates that may influence performance and cannot be used to select individuals for an ATC job. The most recent and comprehensive of these analyses was the *separation and control hiring analysis*, which identified lists of tasks and worker requirements of ATC specialists in the jobs of en route, TRACON, tower cab, and flight service (Nickels et al., 1995). Although this is a detailed and relatively recent analysis of ATC positions, its primary focus is to uncover technical competencies that can be used to select individuals for an ATC job, not for determining team competencies that should be targeted for training.

In the following sections we describe the KSAs identified by Smith-Jentsch et al. (1999) as being critical for ATC teamwork. As a point of comparison, Table 3.1 lists KSAs that appear to be comparable identified by Nickels et al. (1995) for ATC positions and two aircrew CRM analyses (Helmreich et al., 1986; Prince & Salas, 1993).

Skills

Smith-Jentsch et al. (1999) identified four teamwork skills as being important for ATC team performance: information exchange, supporting behavior, team feedback skill, and flexibility. *Information exchange* involves passing rel-

TABLE 3.1
Comparison of Teamwork Knowledges, Attitudes, and Skills

Domain	ATC Analyses		Aircrew Analyses	
	Smith-Jentsch, Kraiger, Salas, & Cannon-Bowers (1999)	Nickels, Bobko, Blair, Sands, & Tartak (1995)	Prince & Salas (1993)	Helmreich, Foushee, Benson, & Russini (1986)
Skills	Supporting behavior skill	Thinking ahead Working cooperatively Self-awareness	Adaptability & flexibility	Recognition of stressor effects
	Information exchange	Oral communication Active listening Interpreting information Summarizing information	Communication Situational awareness	Communication & coordination
	Team feedback skill	Interpersonal tolerance	Assertiveness Leadership	Command responsibility
Attitudes	Flexibility	Flexibility	Adaptability & flexibility	
	Belief in the importance of teamwork Collective orientation Team cohesion Collective efficacy Mutual trust			Command responsibility Communication & coordination
Knowledge	Interpositional knowledge Knowledge of performance-related signs of stress Knowledge of the components of teamwork Knowledge of teammate characteristics Shared task expectations	Composure		Recognition of stressor effects

evant data to team members who need it, before they need it, and ensuring that the messages sent are understood as intended. In terms of receiving information, this involves effective listening and asking for clarification when needed. Information provided to others should describe current and projected states of events as well as a controller's plans and intentions. Effective information exchange helps controllers to build and maintain their own situation awareness as well as contribute to the team's shared understanding of the "big picture." As shown in Table 3.1, these same behaviors, although clustered under different skill labels, have been identified previously as being important in both the ATC and aircrew CRM literature.

The second skill, *supporting behavior*, involves offering and requesting assistance in an effective manner both within and across teams in the ATC system (e.g., tower cab, TRACON). When controllers demonstrate effective supporting behavior, their teams can maintain a high level of performance in complex, high-workload situations. Supporting behavior has two primary components: (a) requesting and accepting assistance and (b) providing assistance. Providing assistance effectively begins with monitoring one's teammates for signs and symptoms of stress. Once such signs have been noted, a controller should offer assistance without having to be asked. This is critical, because many controllers find it uncomfortable to admit that they could use some help. Moreover, when a controller is overloaded he or she may not have time to stop and ask for help. When offering assistance, a controller should be careful not to do so in a manner that is condescending and may cause the other person to resist accepting help. It is important to be clear on what type of assistance is needed; otherwise, a well-meaning controller may actually make a situation more confusing for the person he or she is trying to help. Finally, in the event that a controller suspects that one of his or her teammates could use some assistance but the controller is unable to provide it him- or herself, he or she should notify another controller or supervisor who may have the time to help.

Requesting assistance effectively involves monitoring oneself for signs and symptoms of stress and requesting assistance before it is too late. Controllers should attempt to specify the type of assistance they desire whenever possible. Moreover, they should persist in seeking assistance from multiple sources when needed if the first person they ask is too busy to help. The final aspect of supporting behavior involves being willing to accept assistance graciously. This means avoiding defensive reactions and thanking team members who offer assistance after the situation has subsided. This is important for positively reinforcing cooperation within the team. Skill dimensions comparable to supporting behavior, as defined here, have been described previously as being important in both an ATC and an aircrew environment (see Table 3.1).

The third skill, *flexibility*, involves the ability and willingness to adapt performance strategies quickly and appropriately to changing task demands.

This includes monitoring for cues that indicate that a change in strategy is needed, identifying viable alternatives for action, objectively considering input from others, and compromising when needed. Flexibility also involves quickly selecting a performance strategy with the greatest probability for success when faced with a number of viable alternatives. Effective flexibility allows an ATC team to deal with the unexpected and provide consistently safe and efficient service. As shown in Table 3.1, flexibility has been defined previously as an important teamwork skill in both the ATC analysis conducted by Nickels et al. (1995) and the aircrew CRM literature.

Finally, *team feedback skills* allow team members to communicate their observations, concerns, suggestions, and requests in a clear and direct manner without becoming hostile or defensive. This involves listening objectively to another controller's point of view and expressing oneself using statements that are specific, behavior-based versus personal, and solution oriented. When controllers use effective team feedback skills they are better able to correct and prevent errors, resolve conflicts, and continuously enhance their performance. Behaviors associated with team feedback skill, as defined here, have been identified previously as being important for aircrew CRM (Prince, Chidester, Bowers, & Cannon-Bowers, 1992; Prince & Salas, 1993; Smith-Jentsch et al., 1996). These behaviors tend to be grouped under the terms *leadership, assertiveness,* or both. Nickels et al. (1995) also referred to some of these behaviors as interpersonal tolerance.

Knowledge

Team-related knowledge helps controllers know when and how to apply the four teamwork skills just defined. Although the aircrew CRM literature has long discussed teamwork skills that are important for aircrew performance, much less has been written about the team-related knowledge that enables pilots to use those skills effectively. Nickels et al. (1995) specified several knowledge areas that are important for the job of a controller; however, this knowledge was technical in nature (e.g., knowledge of street physics)—team-related knowledge was not specified. As shown in Table 3.1, Smith-Jentsch et al. (1999) defined five categories of knowledge that support ATC teamwork. Three of these knowledge areas are teammate-generic in that a controller can generalize them to any group of controllers with whom he or she must work. The remaining two knowledge areas involve information that controllers learn about their individual teammates. These are referred to as teammate-specific knowledge competencies (Cannon-Bowers et al., 1995).

Teammate-Generic Knowledge. Smith-Jentsch et al. (1999) identified three teammate-generic knowledge competencies: inter-positional knowledge, knowledge about teamwork, and knowledge about the performance-related signs of stress. *Inter-positional knowledge* involves understanding the tasks performed by the other teams and team members with whom a controller must coordinate. This includes understanding the impact of actions taken on the ability of others to meet their goals and requirements. Specific components of inter-positional knowledge may include, for example, knowledge about the physical layout of the tower, TRACON, or cockpits of different aircraft. Inter-positional knowledge allows controllers to anticipate the information needs of others, support one another during high-workload periods, and avoid frustration and interteam conflicts. Inter-positional knowledge, as defined here, is not specified as a knowledge competency in either the analysis conducted by Nickels et al. (1995) or in the CRM literature that we reviewed.

Knowledge about teamwork helps controllers diagnose and correct coordination breakdowns. A controller who is knowledgeable about the components of ATC teamwork understands the impact that both teammate-generic and -specific attitudes and knowledge can have on the use of team skills. This aids them in adjusting their performance strategies on the fly and critiquing themselves after a critical incident. Finally, knowledge about ATC teamwork improves controllers' ability to provide effective on-the-job training. Although most CRM programs train knowledge about the components of teamwork in the cockpit, such knowledge has not typically been identified as a competency that enables crews to critique their own performance. Additionally, Nickels et al. (1995) did not discuss knowledge about the components of teamwork.

Knowledge about the performance-related symptoms of stress is critical for members of teams, such as ATC teams, that operate in environments characterized by time pressure, rapidly unfolding events, high information-processing demands, and severe consequences of error. This knowledge is necessary in order to determine when to offer or request assistance. Some of the signs of stress that have been identified as being typical in an ATC environment are arm gesturing, standing or sitting closer to the window or radar scope, nonresponsiveness, narrowing of attention, and irritability.

Although Nickels et al. (1995) did not specify the knowledge component necessary in order for controllers to recognize dangerously stressful situations, they did describe the "composure" needed to remain calm when such situations occur. Additionally, Gregorich, Helmreich, and Wilhelm (1990) discussed the recognition of stressor effects as being important for aircrew CRM. Again, however, the knowledge component is not stressed. Instead, Gregorich et al. emphasized the attitudinal component of monitoring for stress symptoms and being willing to compensate when needed.

Teammate-Specific Knowledge. Smith-Jentsch et al. (1999) identified two categories of teammate-specific knowledge as being important for ATC teamwork: knowledge about teammate characteristics and knowledge about team task expectations. As shown in Table 3.1, we found no mention of team-mate-specific knowledge competencies in the Nickels et al. (1995), Helmreich et al. (1986), or Prince and Salas (1993) analyses.

Knowledge about teammate characteristics helps controllers to be aware of situations in which individual teammates may require assistance and to anticipate what type of assistance those teammates prefer. This knowledge may include information about a teammate's strengths and weaknesses as a controller, previous experience, or preferences. *Knowledge about team task expectations* includes information regarding a specific team's preferred strategies or procedures for handling different types of situations. When a team lacks shared task expectations, team members may unknowingly work against one another, causing interpersonal and task-related conflicts.

Attitudes

Team-related attitudes affect team members' willingness to use effective teamwork skills. Smith-Jentsch et al. (1999) identified five team-related attitudes that support ATC teamwork. Two of these were teammate generic—in other words, they are expected to affect a controller's performance regardless of the particular teammates with whom he or she must work. The remaining three attitudes identified were teammate specific. As is the case with teammate-specific knowledge, these attitudes have meaning only in the context of a specific team of individuals.

Teammate-Generic Attitudes.   As shown in Table 3.1, Smith-Jentsch et al. (1999) identified two teammate-generic attitudes: belief in the importance of ATC teamwork and collective orientation. Both of these attitudes are similar to attitudes described by Gregorich et al. (1990) in regard to aircrew CRM. *Belief in the importance of ATC teamwork* simply refers to the opinion that teamwork skills are necessary to achieve the most effective and efficient performance as a controller. *Collective orientation* refers to the tendency to view oneself as part of a larger system. For example, when asked which positions comprised their team, collectively oriented tower cab controllers were more likely to include pilots, supervisors, and departure and approach controllers. Collectively oriented controllers are expected to be better able to provide effective supporting behavior because they are more likely to consider the impact of their actions on the workload of other team members.

Teammate-Specific Attitudes.   Smith-Jentsch et al. (1999) identified three teammate-specific attitudes that are important for effective ATC teamwork: collective efficacy, mutual trust, and team cohesion. As is shown in

Table 3.1, we found no comparable attitudes discussed by Nickels et al. (1995), Helmreich et al. (1986), or Prince and Salas (1993).

Collective efficacy can be viewed as a controller's confidence in the technical abilities of his or her individual teammates as well as the team's ability to coordinate and adapt to rapidly changing situations. *Mutual trust* involves a belief that one's teammates can be counted on to be honest and to act with good intentions toward one another. Teammates who trust one another are more likely to feel comfortable asking for help from one another, to confide in one another, and settle conflicts amicably when they arise. Finally, *team cohesion* refers to the desire to become or remain a member of a specific team of individuals. Controllers from cohesive teams often consider their teammates personal friends and may make special efforts to be scheduled together.

LINKING TEAM TRAINING CONTENT
TO APPROPRIATE METHODS

In the previous sections of this chapter we have described KSAs that have been identified as being critical for ATC teamwork. We recommend that these KSAs be targeted for ATC team training. Moreover, we have compared and contrasted these against KSAs identified as being important for aircrew CRM. This analysis highlighted some limits to the generalizability of aircrew CRM training to an ATC context. In the remaining sections of this chapter we offer guidelines for selecting training methods that are most appropriate for developing the various team-related KSAs defined above. CRM training has been defined as having three stages (FAA, 1989; Gregorich & Wilhelm, 1993): (a) awareness, (b) practice and feedback, and (c) continual reinforcement. The following sections are organized around these three phases.

Awareness

The first stage of CRM, *awareness*, involves communicating teamwork principles and concepts that are fundamental to a particular task domain and developing attitudes and beliefs that will motivate trainees to be receptive to those ideas (FAA, 1989; Gregorich & Wilhelm, 1993). Thus, in terms of the competencies described earlier in this chapter, the awareness stage of CRM should target teammate-generic knowledge and attitudes. In the following sections we describe methods that we consider most appropriate for training teammate-generic attitudes and knowledge. Previous research and practice in which these methods have been used to train controllers, aircrews, and other types of teams is reviewed. Finally, examples of specific strategies are described.

Team Training Workshops and Seminars. Team training workshops are generally 1- to 3-day courses held off-site in which an instructor provides information about teamwork concepts, offers demonstrations of effective and ineffective teamwork, and facilitates group discussions and role play exercises that focus on teamwork topics. Such workshops have been used extensively to support aircrew CRM awareness in both commercial and military aviation (Baker, Bauman, & Zalesny, 1991; Prince et al., 1992; Prince & Salas, 1999; Swezey, Llaneras, Prince, & Salas, 1991). Additionally, one nationally conducted workshop (Air Traffic Control Teamwork Enhancement [ATTE]) and several locally developed workshops (e.g., Controller Awareness Resource Training) have been developed to support ATC teamwork awareness.

Guideline 1: Team Training Workshops Should Be Used to Foster Awareness Through Teammate-Generic Attitudes and Knowledge

The group discussions stimulated by team training workshops are considered highly effective at challenging dysfunctional teamwork attitudes and increasing sensitivity toward teamwork concepts. This is particularly true when they focus on case studies of teamwork breakdowns that were causal factors in real-life accidents. On the basis of their review of the ATTE workshop, Milanovich et al. (1999) noted that it addressed both of the teammate-generic attitudes listed in Table 3.1 (i.e., collective orientation, belief in the importance of teamwork). However, only one of the three teammate-generic knowledge categories was emphasized (i.e., knowledge about teamwork). In fact, data from their survey of 330 controllers across the country indicated that those who had attended the ATTE workshop scored higher than others on a measure of knowledge about teamwork. However, these controllers were not found to have greater inter-positional knowledge or knowledge about the performance-related signs of stress. Therefore, it was recommended that the next-generation ATC team training workshop incorporate information directly tied to these two knowledge categories. This type of knowledge is needed to provide controllers with context-specific cues they can look for to determine how and when to offer assistance and pass information to others in the ATC system.

The ATTE workshop currently discusses stress symptoms (e.g., sleeplessness, weight gain or loss) and stress reducers (e.g., exercise, cultivating interests outside of work) that are general and personal in nature. This type of information may help controllers deal with the long-term effects of stress on their lives; however, it does not specify symptoms (e.g., hand gesturing, talking faster) they can look for to determine when a teammate may be at risk of making an error because of high workload levels or strategies for shifting that workload. Previous research has demonstrated that providing individuals with this type of information improved their performance in a series of stressful scenarios (Inzana, Driskell, Salas, & Johnston, 1996). Team training workshops offer an ideal forum for increasing controllers' knowledge about the performance-related signs of stress through lectures, demonstrations, and group discussions in which trainees share their own experiences and helpful hints.

Although team training workshops are recommended for training teammate-generic knowledge and attitudes, they are not considered to be the most appropriate method for developing teammate-specific knowledge or attitudes. This is because trainees do not necessarily attend these workshops together with the individuals with whom they regularly work. Moreover, when teammates do attend team training workshops together they are usually joined by individuals from many other teams as well. Thus, the emphasis is not on the particular issues that a given team is having but on concepts that are relevant to all teams. Finally, because of class size, team training workshops generally do not allow for the individualized practice and feedback necessary for building teamwork skills (Smith-Jentsch et al., 1996). Although the role plays and group exercises used in this forum are considered useful for demonstrating teamwork concepts, they must be followed up by other strategies for skill building (described in the *Practice and Feedback* section) to achieve lasting behavior change.

Computer-Based Instruction. Computer-based instruction [CBI] enables individual trainees to be presented with information about teamwork concepts and demonstrations of those concepts in a standardized format. Thus, one of the benefits of CBI is that the quality of a training presentation is not dependent on the skill or mood of an instructor. CBI is also generally more cost effective than a team training workshop, because an instructor or facilitator is not needed and each trainee can complete his or her training at a time that is most convenient for him or her. This is particularly useful in an ATC environment, where it is often difficult to take multiple controllers off the schedule at the same time.

A second benefit of CBI is that trainees can progress through lessons at their own pace. Thus, quick learners and more experienced trainees are not forced to listen to detailed descriptions that may be necessary to convey principles to less experienced trainees. Conversely, less experienced trainees are not rushed through material in order to keep up with others in the class. These unique properties of CBI make it an effective method for imparting teammate-generic knowledge that may be too detailed to be conveyed effectively in a team training workshop. In particular, CBI is expected to be more efficient than a workshop for training large numbers of controllers who vary in their baseline level of teammate-generic knowledge. Thus we proceed to our second guideline.

Guideline 2: CBI Should Be Used to Develop Teammate-Generic Knowledge in Situations Where the Targeted Trainees Vary Substantially in Their Baseline Knowledge Levels

CBI might be particularly useful, for example, for developing controllers' inter-positional knowledge. Smith-Jentsch et al. (1999) reported that tower controllers' knowledge about the task demands and limitations faced by TRA-CON controllers, pilots in the cockpit, or both, varied substantially. Thus, although most controllers could benefit from some level of training on inter-positional knowledge, it may be difficult to present a single workshop on the topic that will have an appropriate level of detail for all controllers at a particular facility. However, CBI could be developed that is flexible enough to allow trainees with varying types of experience to spend as much or as little time as necessary to fill in the gaps of their own inter-positional knowledge.

Such a tool has been developed to enhance inter-positional knowledge for teams onboard naval warships (Duncan et al., 1996). This CBI presented information about the requirements of each team position as well as graphical illustrations of the physical layout of spaces where each team member is located. In addition, trainees could view displays and listen to communications that are available to these teammates during a series of simulated performance scenarios. Results from a validation study demonstrated that teams whose members received this training displayed more effective communication patterns than did a group of no-treatment controls. These results suggest that CBI for inter-positional knowledge may be an effective strategy for improving ATC teamwork. Such training could be used by itself, as a follow-up to a team training workshop, or to prepare individual controllers for their participation in such a workshop.

Milanovich et al. (1999) reported that 62% of the facilities they surveyed used CBI for technical training such as procedures for handling difficult weather situations. However, none of the facilities reported having used CBI to train individuals about teamwork concepts. Furthermore, there has been relatively little use of CBI to train CRM concepts within the aviation community. Although we recommend CBI as a method for training teammate-generic knowledge, it is not expected to be an effective method for improving teammate-specific knowledge or any of the attitudes listed in Table 3.1. The reason for this is that CBI does not provide the interaction among teammates and between trainees, and an instructor that is considered crucial for training such competencies.

Practice and Feedback

The second stage of CRM training, *practice and feedback*, involves developing the skills necessary to apply the concepts, introduced in the awareness stage, to on-the-job situations. This stage is critical because "individuals may accept, in principle, abstract ideas of [CRM concepts] but may find it difficult to translate them into behavior" on-the-job (Helmreich & Foushee, 1993, p. 26). We recommend that simulation and on-the-job training (OJT) are effective methods for developing the four teamwork skills described earlier as being critical for ATC performance.

Simulation. Simulation training allows individuals to practice using team skills to accomplish their tasks in an environment that is artificial but mirrors the actual work environment. Simulation exercises can vary substantially in physical fidelity, or the degree to which cues, tools, and tasks are identical to those used in the actual job environment. For example, cockpit CRM programs have incorporated simulation exercises that range from role-played situations in a classroom setting, to personal computer (PC)-based simulation, to full-motion simulators (Baker, Prince, Shrestha, Oser, & Salas, 1993; Bowers, Salas, Prince, & Brannick, 1992; Hays, Jacobs, Prince, & Salas, 1992; Prince, Oser, Salas, & Woodruff, 1993). When used to train teamwork skills, the physical fidelity of a simulation is less important than the psychological fidelity of pre-scripted scenario events (Baker et al., 1993; Bowers et al., 1992; Prince et al., 1993). Realistic scenario events that elicit CRM skills have been derived from accident and incident reports, observation of teams in a naturalistic environment, or structured interviews (Baker, Brannick, & Chidester, 1997; Baker & Smith-Jentsch, 1997; Prince et al., 1993).

Simulation has been used extensively to train developmental air traffic controllers in both technical and teamwork skills. For example, the FAA Air Traffic Control Academy uses simulation that ranges in physical fidelity from a 360-degree high-fidelity tower simulator to PC-based simulation. Additionally, most TRACON facilities have the ability to run simulation exercises on operational radar scopes. This provides new controllers with the opportunity to practice their skills in the actual work environment alongside certified professional controllers without endangering aircraft. However, Milanovich et al. (1999) found that at the time of their survey none of the facilities reported using simulation for recurrent training of ATC teamwork skills.

Smith-Jentsch et al. (1999) found that experience as an air traffic controller was not significantly correlated with a controller's use of effective teamwork skills. This suggests that even experienced controllers can benefit from opportunities to practice and receive feedback on their teamwork skills in the context of low-fidelity simulation exercises. Such simulation can be used as a stand-alone tool or to supplement existing team training workshops. In fact, team training workshops are unlikely to produce significant behavior change on-the-job if they do not provide controllers with the opportunity to practice applying CRM concepts and to receive feedback on their performance in some sort of simulation exercise. Much has been learned over the last 10 years about how best to design simulation exercises that foster effective teamwork (Beard, Salas, & Prince, 1995; Fowlkes, Dwyer, Oser, & Salas, 1998; Prince et al., 1993; Salas, Bowers, & Rhodenizer, 1998). Findings from this body of research have direct application to simulation-based training for ATC. In particular, it has been argued that in simulation-based training the scenario is the curriculum and as such should be directly tied to learning objectives. Thus, our third guideline:

Guideline 3: Role Play and Simulation-Based Exercises Used to Develop Teamwork Skills Should Incorporate Realistic Events That Are Specifically Designed to Elicit Particular Teamwork Skills

Critical incidents requiring ATC teamwork can be used to develop role-play, PC-based, or high-fidelity simulations. These events should be scripted so as to target the four ATC teamwork skills listed in Table 3.1. For example, Baker and Smith-Jentsch (1997) described the development of a library comprising modular event sets that can be strung together to form simulation exercises. Interviewers asked 50 certified professional controllers to describe actual incidents where a controller's teamwork skills either saved the day or led to a coordination breakdown. These critical incidents were then coded so as to indicate which of the ATC teamwork skills they required. The resulting event set library specifies the cues necessary to work these events into a realistic simulation exercise. For example, a classroom-based role-play exercise may be used to simulate a break-room situation requiring the use of effective feedback skills to resolve a conflict that occurred previously in the tower. On the other hand, PC-based simulation may be used to practice and receive feedback on flexibility and supporting behavior skills. These same events could be introduced into higher fidelity simulation training provided to new controllers at the FAA Air Traffic Control Academy.

Because many controllers work with the same individuals on a regular basis, simulation-based training that allows teammates to practice coordinating together and to provide one another feedback is likely to be most effective. Smith-Jentsch et al. (1997) found that controllers reported being more motivated to learn from training involving interaction with their regular teammates. Additionally, controllers who participate in simulation-based training with their own teammates have the opportunity to learn about one another's preferred task strategies for handling critical situations that may rarely occur on the job but require a rapid response when they do. Finally, collective efficacy should increase after controllers are given the opportunity to demonstrate, in a simulated environment, that they as a team can overcome such situations. Thus, our fourth guideline states:

Guideline 4: For Teams With Stable Membership, Simulation-Based Training Will Be Most Effective if Teammates Have the Opportunity to Practice and Receive Feedback Together

OJT. Much of the initial training that controllers receive is OJT. ATC OJT involves pairing a controller who is new to a facility with two or more controllers who are experienced at that facility. These experienced controllers are referred to as *on-the-job training instructors* (OJTIs). OJTIs are responsible for monitoring and coaching developmental controllers each time they work a position. This continues until it is deemed by the OJTIs that the developmental controller is qualified to work the position, or is "checked out."

OJT provides an excellent opportunity for controllers to practice and receive feedback on ATC teamwork skills. However, currently there is little guidance as to how OJTIs should accomplish this. Checklists provided to structure postperformance feedback sessions simply list coordination as a category to be discussed; specific behaviors are not defined. This is relatively common when it comes to OJT. Most OJT provided in organizational settings is informal and unstructured. As a result, the quality of instruction is highly dependent on the skill of the OJTI. This is also the case within an ATC environment. It has been noted that experts from a variety of domains have difficulty communicating with novices regarding their expertise on tasks that have become second nature to them (Zsambok, Crandall, & Militello, 1994). Consistent with this notion, Smith-Jentsch et al. (1999) reported that many of the experienced controllers whom they interviewed had difficulty specifying behaviors that promoted effective teamwork in the tower cab. However, researchers have begun to develop tools that aid OJTIs in "unpacking" such expertise and communicating it effectively to trainees following an OJT session (Smith-Jentsch, Zeisig, Acton, & McPherson, 1998; Zsambok et al., 1994).

One such tool was developed by Smith-Jentsch, Zeisig, et al. (1998), who described a debriefing guide designed to aid Navy instructors in training teamwork skills. This debriefing guide offers probes designed to solicit concrete examples of specific teamwork behaviors (e.g., passing information along before having to be asked) from trainees after a simulation exercise or OJT session. These probes are organized around four teamwork dimensions that serve as higher order learning objectives (e.g., information exchange). Instructors record examples of each component behavior within the debriefing guide while observing a trainee's performance. This helps to make the instructor's feedback more specific and solution oriented. However, the ultimate goal of the debriefing guide is to help trainees to identify and correct their own mistakes. This process is expected to lead to a deeper understanding of teamwork principles as well as increased task confidence and personal commitment to learning. A similar type of guide could be developed to facilitate OJT in an ATC environment. Thus, our fifth guideline states:

> *Guideline 5: Structured Debriefing Guides That Help Instructors Diagnose Trainees' Strengths and Weaknesses on Specific Teamwork Skills Should Be Used to Enhance the Quality of OJT*

Continual Reinforcement

The final stage of CRM training, *continual reinforcement*, involves repeated exposure to team concepts as well as on-the-job feedback and reinforcement of CRM concepts from multiple sources. One way in which continual reinforcement can be achieved is by explicitly addressing ATC team-related KSAs within supervisor and OJTI courses. Specifically, supervisors and OJTIs should be trained to diagnose teamwork-related problems and to facilitate a climate that supports effective teamwork within their teams.

Team Climate. In an environment such as ATC where team membership is relatively stable, continual reinforcement involves the development of teammate-specific attitudes and knowledge that enable individuals to apply generic team skills effectively when working with a particular set of teammates. One way this can be accomplished is by establishing and maintaining a team climate that facilitates the sharing of expertise, clarification of role expectations, and effective resolution of on-the-job conflicts. Consistent with this notion, tower cab team climate has been linked to team cohesion among controllers (Kraiger, Smith-Jentsch, & Harrison, 1999). Team cohesion, in turn, has been linked to controllers' willingness to ask for and accept support from a particular set of teammates (Smith-Jentsch et al., 2000).

There is also some evidence that the climate within a team or crew can have a significant impact on learning and transfer from a team training course. For example, Kraiger et al. (1999) found that controllers from teams with a supportive climate reported that their motivation to learn from a course would be increased if they were told that they would attend training with their teammates. Furthermore, Smith-Jentsch, Salas, and Brannick (in press) demonstrated that first officers' perceptions of the climate within their crews affected whether they transferred assertive communication skills from a CRM program. Thus, on the basis of previous research on team climate, team training, and the development of team competencies, our sixth guideline states that:

Guideline 6: Supervisors and Team Leaders Should Be Trained to Create and Maintain a Climate That Reinforces the Development and Use of Effective Team Competencies On The Job

Team Briefings.   Milanovich et al. (1999) reported that 83% of the facilities they surveyed provided opportunities for controllers to participate in regularly scheduled team briefings. The frequency with which these briefings were held varied across facilities from once a week to once a month, and the length of time allotted for each briefing varied from 30 minutes to 90 minutes. Facilities reported that team briefings were often used to convey new regulations or to train controllers on new procedures or equipment. However, the briefings also offered ATC teams an opportunity to discuss coordination problems and to share expectations and expertise.

Smith-Jentsch et al. (1999) demonstrated the positive effects of ATC team briefings on teammate-specific knowledge. They found that controllers from facilities that held more frequent team briefings had greater knowledge about their teammates' strengths and weaknesses (30-minute briefings held weekly were most effective). This knowledge, in turn, predicted which controllers would be more likely to ask for and accept assistance or feedback from members of their team. On the basis of these results, our seventh guideline states:

Guideline 7: Team Briefings Should Be Used to Develop Teammate-Specific Knowledge and Attitudes and to Reinforce the Use of Effective Teamwork Skills On The Job

Additional studies outside the ATC environment have demonstrated that the benefits of team briefings can be increased in a number of ways. First, Blickensderfer, Cannon-Bowers, and Salas (1997) found that teams developed greater shared task expectations after a team briefing if they were provided with preparatory training on how to state their feedback in a way that is clear and direct without being hostile. Second, Tannenbaum, Smith-Jentsch, and Behson (1998) demonstrated the importance of using a facilitator to solicit and reinforce constructive comments. They reported that teams whose briefings were facilitated by leaders who were trained in effective facilitation skills were more likely to admit mistakes, offer suggestions, and request clarification from one another. As a result, these teams outperformed those in a control group on a series of simulated missions. Finally, Smith-Jentsch, Zeisig, et al. (1998) demonstrated that individuals who used a debriefing guide to structure their team briefings around four teamwork skills developed greater shared knowledge about teamwork. We recommend that these learned lessons be used to further enhance the usefulness of ATC team briefings. Thus, our eighth guideline states that:

> *Guideline 8: Team Briefings Will Be Most Useful When Team Members Are Provided Training on How to State Their Comments Constructively, and When a Facilitator Structures the Team Discussion Using a Debriefing Guide and Encourages and Reinforces Participation*

CONCLUSION

ATC teams and aircrews have a number of similarities. Both work in environments that present high information-processing demands, where there are interdependencies among team members, and where the consequences of error can be a matter of life and death. However, there are also a number of key differences that may affect the generalizability of team training from one environment to the other. We have highlighted the impact of one of these: the stability of team membership. Much of the research on aircrew CRM and CRM training has clear applicability to an ATC environment. This research tends to focus on teammate-generic attitudes and skills. Additionally, research in other team environments (e.g., Navy CIC teams) has begun to explore methods for training teammate-generic knowledge. Findings from these efforts should be used to develop and test similar training strategies for both aircrews and ATC teams. However, because controllers frequently work with the same set of teammates, teammate-specific attitudes and knowledge have a significant impact on their performance as well. Furthermore, teammate-specific competencies are likely to affect the degree to which controllers transfer what they learn in team training back on the job. Thus, it is critical that practitioners involved in designing training programs for ATC teams or other teams with stable membership include strategies that focus on developing teammate-specific attitudes and knowledge. Such strategies (e.g., team briefings and training for supervisors & OJTIs) should facilitate the exchange of information among teammates about expectations, preferences, and areas of expertise as well as continual reinforcement and self-critique on important teamwork processes.

ACKNOWLEDGMENT

The views expressed here are those of the authors and do not reflect an official position of the organizations with which they are affiliated.

REFERENCES

Ammerman, H. L., Becker, E. S., Bergen, L. J., Clausen, C. A., Davies, D. K., Inman, E. E., & Jones, G. W. (1987). *FAA air traffic control operation concepts: Volume 5. ATCT/TCCC tower controllers* (Rep. No. DTF-A01-85-Y-01034). Washington, DC: U.S. Department of Transportation, Federal Aviation Administration.

Ammerman, H. L., Becker, E. S., Jones, G. W., & Tobey, W. K. (1987). *FAA air traffic control operations concepts: Volume 1. ATC background and analysis methodology* (Rep. No. DOT/FAA/AP-87-01). Washington, DC: U.S. Department of Transportation, Federal Aviation Administration.

Ammerman, H. L., Fairhurst, W. S., Hostetler, C. M., & Jones, G. W. (1989). *FAA air traffic control task knowledge requirements: Volume 1. ATCT tower controllers* (Rep. FAA/ATC-TKR). Washington, DC: U.S. Department of Transportation, Federal Aviation Administration.

Baker, D. P., Bauman, M., & Zalesny, M. (1991). Development of aircrew coordination exercises to facilitate training transfer. In R. S. Jensen (Ed.), *Proceedings of the Sixth International Symposium on Aviation Psychology* (Vol. 1, pp. 314–319). Columbus, OH: The Ohio State University.

Baker, D. P., Brannick, M. T., & Chidester, T. R. (1997). Developing scenario event sets for LOE: A data-driven approach. In R. S. Jensen & L. Rakovan (Eds.), *Proceedings of the Ninth International Symposium on Aviation Psychology* (Vol. 2, pp. 1172–1177). Columbus, OH: The Ohio State University.

Baker, D. P., Prince, C., Shrestha, L., Oser, R., & Salas, E. (1993). Aviation computer games for CRM skills training. *International Journal of Aviation Psychology, 3,* 143–156.

Baker, D. P., & Smith-Jentsch, K. A. (1997). A methodology for simulation training and assessment of tower cab teamwork skills. In R. S. Jensen & L. Rakovan (Eds.), *Proceedings of the Ninth International Symposium on Aviation Psychology* (Vol. 1, pp. 181–184). Columbus, OH: The Ohio State University.

Beard, R. L., Salas, E., & Prince, C. (1995). Enhancing transfer of training: Using role-play to foster teamwork in the cockpit. *International Journal of Aviation Psychology, 5,* 131–143.

Blickensderfer, E., Cannon-Bowers, J. A., & Salas, E . (1997, April). *Does overlap in team member knowledge predict team performance?* Paper presented at the 12th annual meeting of the Society for Industrial and Organizational Psychology, St. Louis, MO.

Bowers, C. A., Salas, E., Prince, C., & Brannick, M. T. (1992). Games teams play: A methodology for investigating team coordination and performance. *Behavior Research Methods, Instruments, and Computers, 24,* 503–506.

Cannon-Bowers, J. A., & Salas, E. (1998). Individual and team decision making under stress: Theoretical underpinnings. In J. A. Cannon-Bowers & E. Salas (Eds.), *Making decisions under stress: Implications for individual and team training* (pp. 17–38). Washington, DC: American Psychological Association.

Cannon-Bowers, J. A., Tannenbaum, S. I., Salas, E., & Volpe, C. E. (1995). Defining competencies and establishing team training requirements. In R. Guzzo & E. Salas (Eds.), *Team effectiveness and decision making in organizations* (pp. 333–380). San Francisco: Jossey-Bass.

Duncan, P. C., Rouse, W. N., Johnston, J. H., Cannon-Bowers, J. A., Salas, E., & Burns, J. J. (1996). Training teams working in complex systems: A mental model-based approach. In W. B. Rouse (Ed.), *Human/technology interaction in complex systems* (Vol. 8, pp. 173–231). Greenwich, CT: JAI Press.

Federal Aviation Administration. (1989). *Cockpit resource management training* (Advisory Circular No. 120-51). Washington, DC: Author.

Fowlkes, J. E., Dwyer, D., Oser, R. L., & Salas, E. (1998). Event based approach to training (EBAT). *International Journal of Aviation Psychology, 8,* 209–221.

Gregorich, S. E., Helmreich, R. L., & Wilhelm, J. A. (1990). The structure of cockpit management attitudes. *Journal of Applied Psychology, 75,* 682–690.

Gregorich, S. E., & Wilhelm, J. A. (1993). Crew resource management training assessment. In E. K. Wiener, B. G. Kanki, & R. L. Helmreich (Eds.), *Cockpit resource management* (pp. 173–198). San Diego, CA: Academic Press.

Hackman, J. R. (1993). Teams, leaders, and organizations: New directions for crew-oriented flight training. In E. L. Wiener, B. G. Kanki, & R. L. Helmreich (Eds.), *Cockpit resource management* (pp. 47–70). San Diego, CA: Academic Press.

Hartel, C., Smith, K., & Prince, C. (1991, April). *Defining aircrew coordination: Searching mishaps for meaning.* Poster presented at the Sixth International Symposium on Aviation Psychology, Columbus, OH.

Hays, R. T., Jacobs, J. W., Prince, C., & Salas, E. (1992). Requirements for future research in flight simulation training: Guidance based on a meta-analytic review. *International Journal of Aviation Psychology, 2,* 143–158.

Helmreich, R. L., & Foushee, H. C. (1993). Why crew resource management? Empirical and theoretical bases of human factors training in aviation. In E. L. Wiener, B. G. Kanki, & R. L. Helmreich (Eds.), *Cockpit resource management* (pp. 3–45). San Diego, CA: Academic Press.

Helmreich, R. L., Foushee, H. C., Benson, R., & Russini, W. (1986). Cockpit management attitudes: Exploring the attitude–behavior linkage. *Aviation, Space, and Environmental Medicine, 57,* 1198–1200.

Inzana, C. M., Driskell, J. E., Salas, E., & Johnston, J. H. (1996). The effects of preparatory information on enhancing performance under stress. *Journal of Applied Psychology, 81,* 429–435.

Kanki, B. G., & Foushee, H. C. (1989). Communication as group process mediator of aircrew performance. *Aviation, Space, & Environmental Medicine, 60,* 402–410.

Kern, T. (1997). *Redefining airmanship.* New York: McGraw-Hill.

Kraiger, K., Smith-Jentsch, K. A., & Harrison, R. J. (1999, April). Validation of a technique for defining collective climates: A tool to support team training needs analysis. In K. A. Smith-Jentsch (Chair), *Climate analysis as a tool for diagnosis and employee development.* Symposium conducted at the 14th annual conference of the Society for Industrial and Organizational Psychology, Atlanta, GA.

Milanovich, D. M., Smith-Jentsch, K. A., & Harrison, J. R. (1999, May). *Training air traffic control teamwork: Current practices & future directions.* Paper presented at the Tenth International Symposium on Aviation Psychology, Columbus, OH.

Nagle, D. C. (1988). Human error in aviation operations. In E. L. Wiener & D. C. Nagle (Eds.), *Human factors in aviation* (pp. 263–303). New York: Academic Press.

Nickels, B. J., Bobko, P., Blair, M. D., Sands, W. A., & Tartak, E. L. (1995). *Separation and control hiring and assessment (SACHA) final job analysis report* (Contract No. DTFA01-91-00032). Washington, DC: Federal Aviation Administration.

Prince, C., Chidester, T. R., Bowers, C. A., & Cannon-Bowers, J. A. (1992). Aircrew coordination: Achieving teamwork in the cockpit. In R. W. Swezey & E. Salas (Eds.), *Teams: Their training and performance* (pp. 329–353). Norwood, NJ: Ablex.

Prince, C., Oser, R., Salas, E., & Woodruff, W. (1993). Increasing hits and reducing misses in CRM/LOS scenarios: Guidelines for simulator scenario development. *International Journal of Aviation Psychology, 3,* 69–82.

Prince, C., & Salas, E. (1993). Training research for teamwork in the military aircrew. In E. L. Wiener, B. G. Kanki, & R. L. Helmreich (Eds.), *Cockpit resource management* (pp. 337–366). San Diego, CA: Academic Press.

Prince, C., & Salas, E. (1999). Team processes and their training in aviation. In D. Garland, J. Wise, & D. Hopkin (Eds.), *Aviation human factors* (pp. 193–213). Mahwah, NJ: Lawrence Erlbaum Associates.

Salas, E., Bowers, C. A., & Rhodenizer, L. (1998). It is not how much you have but how you use it: Toward a rational use of simulation to support aviation training. *International Journal of Aviation Psychology, 8,* 197–208.

Smith-Jentsch, K. A., Johnston, J. H., & Payne, S. (1998). Measuring team-related expertise in complex environments. In J. A. Cannon-Bowers & E. Salas (Eds.), *Making decisions under stress: Implications for individual and team training* (pp. 61–87). Washington, DC: American Psychological Association.

Smith-Jentsch, K. A., Kraiger, K., Cannon-Bowers, J. A., & Salas, E. (1998, April). A data driven model of precursors to teamwork. In K. Kraiger (Chair), *Team effectiveness as a product of individual, team, and situational factors.* Symposium conducted at the 13th annual conference of the Society of Industrial and Organizational Psychology, Dallas, TX.

Smith-Jentsch, K. A., Kraiger, K., Cannon-Bowers, J. A., & Salas, E. (2000). *Familiarity breeds teamwork: A case for training teammate-specific competencies.* Manuscript submitted for publication.

Smith-Jentsch, K. A., Kraiger, K., Salas, E., & Cannon-Bowers, J. A. (1999, April). *Teamwork in the control tower: How does it differ from cockpit resource management?* Paper presented at the Tenth International Symposium on Aviation Psychology, Columbus, OH.

Smith-Jentsch, K. A., Salas, E., & Baker, D. P. (1996). Training team performance-related assertiveness. *Personnel Psychology, 49,* 909–936.

Smith-Jentsch, K. A., Salas, E., & Brannick, M. (in press). To transfer or not to transfer: The combined effects of trainee characteristics and team transfer environments. *Journal of Applied Psychology.*

Smith-Jentsch, K. A., Zeisig, R. L., Acton, B., & McPherson, J. A. (1998). Team dimensional training. In J. A. Cannon-Bowers & E. Salas (Eds.), *Making decisions under stress: Implications for individual and team training* (pp. 271–297). Washington, DC: American Psychological Association.

Smith-Jentsch, K. A., Zeisig, R. L., Cannon-Bowers, J. A., & Salas, E. (1997). Defining and training tower cab teamwork. In R. S. Jensen & L. Rakovan (Eds.), *Proceedings of the Ninth International Symposium on Aviation Psychology* (pp. 201–206). Columbus, OH: The Ohio State University.

Swezey, R., Llaneras, R., Prince, C., & Salas, E. (1991). Instructional strategy for aircrew coordination training. In R. S. Jensen (Ed.), *Proceedings of the Sixth International Symposium on Aviation Psychology* (Vol. 1, pp. 302–307). Columbus, OH: The Ohio State University.

Tannenbaum, S. I., Smith-Jentsch, K. A., & Behson, S. J. (1998). Training team leaders to facilitate team learning and performance. In J. A. Cannon-Bowers & E. Salas (Eds.), *Decision making under stress: Implications for individual and team training* (pp. 247–270). Washington, DC: American Psychological Association.

Wiener, E. L., Kanki, B. G., & Helmreich, R. L. (Eds.). (1993). *Cockpit resource management.* San Diego, CA: Academic Press.

Zsambok, C. E., Crandall, B., & Militello, L. (1994). *OJT: Models, programs, and related issues.* (Contact Number: MDA903-93-C-0092 for the U.S. Army Research Institute for the Behavioral and Social Sciences, Alexandria, VA). Fairborn, OH: Klein Associates, Inc.

4

Stress Management: Individual and Team Training

James E. Driskell
Florida Maxima Corporation

Eduardo Salas
University of Central Florida

Joan Johnston
Naval Air Warfare Center Training Systems Division

Managers and other professionals in industry often have considerable difficulty with the term *stress*. They are likely to be no more comfortable with the phrase *stress management*. Given that managers, trainers, and other professionals in industry are the target audience for this chapter, we attempted to discover why this problem existed. This turned out to be a relatively easy task. First, we searched the PsycLIT database using the search term *stress management* and uncovered more than 1,900 articles related to stress management published since 1990. Second, we scanned the abstracts of these articles to identify what the authors meant by the term *stress management*. We found that stress management included interventions such as relaxation, aerobic conditioning, biofeedback, yoga, music therapy, hypnosis, play, humor, diet management, and transcendental meditation.

It is no surprise that those people who are concerned with worker performance and productivity, who may be responsible for designing training to enhance performance, and who may have to justify these training expenditures to others may be a bit concerned with this mixed bag of psychological treatments and interventions. It seems that the concept of stress management has been used so broadly as to mean almost anything.

Therefore, we attempt to simplify matters by summarizing our perspective on stress management in the following three points:

- Many task environments involve, on occasion, high-stress or high-demand conditions. Personnel may be faced with multiple tasks that must be performed under extreme time pressure and under complex and often ambiguous conditions. Furthermore, these types of critical or emergency conditions, when events "heat up," are when effective performance is most needed.
- High-stress or high-demand conditions exact a price on performance. Stress can result in increased errors, reduced speed of response, and narrowed attention, all of which may lead to poor performance.
- Normal training procedures do not provide pre-exposure to the stress environment or the special skills training required to maintain effective performance under stress. The purpose of stress training is to prepare individuals and teams to maintain task performance under demanding operational conditions.

Simply stated, *stress training* is a type of training (or a modification of existing training) in which the training designer attempts to anticipate the events that the trainee is likely to face in the operational environment. Because the term *stress* evokes such varied responses, we could as easily call this type of intervention *emergency training* or *training for high-demand conditions*. In this chapter we discuss guidelines for designing and developing stress training that can be adapted for various occupational environments.

WHAT IS STRESS, AND WHY DO WE WANT TO MANAGE IT?

To illustrate what we mean by the term *stress*, we offer the following example:

> On April 28, 1988, an Aloha Airlines Boeing 737 was on a scheduled flight from Hilo to Honolulu, Hawaii. Suddenly, the pilots heard a loud "whooshing" sound, followed by wind noise. The captain looked up and "there was blue sky where the first-class ceiling had been." The fuselage had separated, and 18 feet of the cabin's exterior had peeled off of the airplane. The pilots donned oxygen masks because of the rapid decompression, and were forced to use hand signals to communicate because of the ambient noise. As they maneuvered for an emergency landing, the No. 1 engine failed. The pilots were able to land the airplane at Maui's Kahului Airport, making a normal touchdown and landing. (National Transportation Safety Board, 1989)

Most observers would define this as a high-stress event. Because we assume at least some agreement on this point, it may be useful to consider informally some of the characteristics of this event that lead us to label it as high stress. First, there were sudden and unexpected demands that disrupted normal procedures. Second, the pilots had to perform multiple tasks under distractions such as wind, noise, and time pressure. Third, the consequences of poor performance were severe. Finally, a successful outcome required considerable skill—quite likely, different skills than those required under normal operating conditions.

Now that we have an intuitive feel for what we mean by high stress, we can define stress more precisely. Salas, Driskell, and Hughes (1996) defined *stress* as a process by which certain environmental demands evoke an appraisal process in which perceived demand exceeds resources and that results in undesirable physiological, psychological, behavioral, or social outcomes.

Although considerable research has examined individual reactions to stress, comparatively little work has examined the effects of stress on team or crew performance (see Driskell & Salas, 1991). Some research has examined the effects of stress on social behavior. For example, Mathews and Canon (1975) found that individuals were less likely to help or assist others when exposed to loud ambient noise. Rotton, Olszewski, Charleton, and Soler (1978) found that loud noise reduced subjects' ability to discriminate among people occupying different roles. Wegner and Giuliano (1980) found that increased arousal led to greater self-focused attention. Some studies indicate that time pressure is likely to inhibit joint problem solving (Walton & McKersie, 1965). Yukl, Malone, Hayslip, and Pamin (1976) found that under high time pressure, team members reached agreements sooner, but they made fewer offers and reached poorer joint outcomes.

S. Cohen (1980) attempted to explain these results by arguing that the effect of stress on interpersonal behavior is a consequence of the narrowing of attention that occurs under stress. The classic arousal perspective argues that stress results in heightened arousal, and these increased demands lead to a narrowing of attention (Easterbrook, 1959). As attention narrows, peripheral (less relevant) task cues are first ignored, followed by restriction of more central or task-relevant cues. Team tasks require attention to both direct task-related activities and interpersonal or teamwork activities such a coordination and communication. Thus, the narrowing of attentional focus under stress may have both cognitive and social effects. As important social or interpersonal cues (such as attention to others' requests or actions) are neglected, team performance suffers. Thus, one effect of stress is a narrowing of attentional capacity, and this narrowing of attention may lead to a neglect of social or interpersonal cues and impaired social behavior. In fact, Driskell, Salas, and Johnston (1999) found that team members were less likely to maintain a broad team perspective under stress and were more likely to shift to a more individualistic self-focus, resulting in poorer overall team performance.

In summary, what do we mean by *stress*? Stress is a high-demand, high-threat situation that results in degraded performance. It is time limited: Stress conditions occur suddenly, and often unexpectedly; quick and effective task performance is critical; and consequences of poor performance are immediate and often catastrophic.

STRESS TRAINING

Evidence indicates that the effects of stress are costly in terms of individual performance and organizational productivity, and considerable effort has been devoted to developing stress training interventions to overcome these effects. It is important that we distinguish between *training* and *stress training*. The primary goal of training is skill acquisition and retention. Therefore, most training takes place under conditions designed to maximize learning, such as a quiet classroom and the practice of task procedures under predictable and uniform conditions. In this manner, the traditional "classroom" training format typically does a good job of promoting initial skill acquisition.

However, some tasks must be performed in conditions quite unlike those encountered in the training classroom. For example, high-stress environments include specific task conditions (such as time pressure, increased task load, distractions) and require specific responses (such as the flexibility to adapt to novel and often-changing environmental contingencies) that differ from those found in a normal or more benign environment. The primary purpose of *stress training* is to prepare the individual to maintain effective performance in a high-stress environment.

There have been numerous attempts to implement different types of stress training techniques, and these approaches have met with various degrees of success (see Lipsey & Wilson, 1993, for an overview of the effectiveness of various psychological interventions). However, most studies have examined the effectiveness of isolated training techniques, such as training attentional-focusing skills (Singer, Cauraugh, Murphey, Chen, & Lidor, 1991) or overlearning (Driskell, Willis, & Copper, 1992). What has been missing is an integrated approach to developing stress training. This approach provides a structure for developing stress training programs rather than implementing individual techniques. An integrated model of stress training must incorporate two critical components of stress training.

- It must provide a means to provide pre-exposure to the high-stress or emergency conditions that may be faced by the trainee. This pre-exposure can both be informational (e.g., providing information regarding the stress environment) and behavioral (e.g., providing the trainee the opportunity to practice under simulated stress conditions).

- Training must incorporate specialized skills training (such as training decision-making skills) to impart those skills required to maintain effective performance in high-stress environments.

Driskell and Johnston (1998) developed an integrated stress training approach, termed *stress exposure training* (SET), that provides a structure for designing, developing, and implementing stress training. The SET approach is defined by a three-stage training intervention:

1. *Information Provision*: An initial stage in which information is provided regarding stress and stress effects. The purpose of this phase of training is to provide trainees with basic information on stress, stress symptoms, and the likely stress effects in the performance setting.
2. *Skills Acquisition*: A skills training phase in which specific cognitive and behavioral skills are taught and practiced. These are called *high performance* skills, because they represent the skills required to maintain effective performance in the stress environment.
3. *Application and Practice*: This stage involves the application and practice of these skills in a graduated manner in a simulated stress environment. Allowing a trainee to practice skills in a graduated manner across increasing levels of stress (from moderate-stress scenarios or exercises to higher stress exercises) ensures successful task performance.

Note that SET is a model for stress training rather than a specific training technique. The SET model describes three stages of training, each with a specific overall objective. However, the specific content of each stage will vary according to the specific training requirements. For example, in the Information Provision stage the trainer will provide information on the specific stressors that are likely to be faced in a particular environment. Likewise, any number of stress training techniques—such as attentional training, overlearning, or physiological control—can be implemented in the Skills Acquisition phase of the SET approach. In other words, SET does not prescribe one type of training that must be applied in all settings but provides a model to guide the design of stress training for any given task.

There are special considerations regarding stress training for team tasks. For team tasks, stress training should take place in the team setting. Team members must develop not only confidence in their own capability to perform under stress conditions but also confidence in the team. Furthermore, many stressors, such as time pressure, may affect team processes such as coordination and communication. Therefore, stress training for team tasks should be conducted with intact teams and should address those team processes that may be degraded under stress (see Driskell & Salas, 1991; Kanki, 1996).

DESIGNING STRESS TRAINING

In the following sections we present specific guidelines for designing and implementing stress training. Note that this is not intended as a general overview of training design. Basic principles of training design and implementation are provided in more general texts (see I. L. Goldstein, 1986; Quinones & Ehrenstein, 1997; Wexley & Latham, 1991).

Needs Analysis. The primary goal of stress training is transfer to the real-world operational environment. Therefore, training is context specific and is designed to provide pre-exposure to the anticipated stress conditions that are likely to be encountered in the operational environment. Accordingly, Johnston and Cannon-Bowers (1996) noted the importance of designing stress training on the basis of an analysis of the task environment. This analysis becomes the basis for the development of training content, including the specific tasks to be trained and the types of stress to which trainees are exposed in training. A needs analysis is an important initial step in designing any type of training program. The end result of a needs analysis is the development of specific training objectives. General guidelines for conducting a needs analysis are readily available (I. L. Goldstein, 1986).

A second type of needs analysis is often called a *person analysis* and addresses the question of who is to be trained. Generally speaking, any individual or team that may be called on to perform under operational conditions (such as time pressure, increased threat, and high workload) that differ from those normally encountered in training should receive specialized stress training.

Training Development. One question that should be addressed in the early stages of training design is: Who will develop the training? One effective means to develop stress training is through the team approach. Because stress training is an instructional procedure designed to prepare individuals to perform in a specific real world environment, the development team should include not only experts in instructional development but also subject-matter experts who are familiar with the operational setting.

Sequencing of Training. When should stress training be implemented? The first issue to be considered is where stress training should be placed in the training schedule. Ideally, stress training should be integrated into the overall training curriculum. In the integrated approach, trainees would first receive initial skills training in the classroom and then practice these skills in a simulated operational setting incorporating the stressors that are likely to be found in that environment. In this case, stress training would be implemented as a component of normal technical training.

A second, related question is the timing or manner in which stress exposure is to be introduced during training. Exposure to stressors too early in training may interfere with initial skill acquisition. Thus, the high demand, ambiguity, and complexity of the stress environment may not be conducive to the early stages of learning. There is some evidence to support the effectiveness of *phased training*, in which skills training and exposure to stressors occur in two separate phases of training, with the stress training introduced after initial skill acquisition (Friedland & Keinan, 1986). Therefore, if stress training is presented as an integrated component of technical training, it should be introduced following initial skills acquisition. The introduction of stress training too early in the training curriculum may interfere with initial skill acquisition.

Fidelity. One dilemma facing the training designer is how to introduce realistic stressors in training. High-fidelity stressors are those that are just like the ones encountered in the operational environment. Some researchers argue for high-fidelity simulation of stress in training (Terris & Rahhal, 1969), whereas others suggest that a low-fidelity approach is more effective, arguing that intense stress during training may intensify fears or interfere with skill acquisition (Lazarus, 1966).

Given the complexity and high stresses inherent in real-world environments, one can never reproduce these stressors in the training setting. However, real-world stressors can be simulated in a training environment at a lower or moderate level of fidelity. Although fidelity is important in stress training, attempting to achieve too high a level of fidelity may be counterproductive; both for safety reasons and because if stressors are presented at too high a level of fidelity, little actual skills training or practice may take place. Furthermore, research suggests that stressors such as time pressure and task load can be effectively simulated in a moderate-fidelity training setting (Johnston, Driskell, & Salas, 1997). The training designer must balance the desire for high fidelity with requirements for safety and training effectiveness.

Feedback. Feedback or knowledge of results is critical for both learning and motivation (Wexley & Latham, 1991). Positive performance feedback is especially critical for stress training. When trainees practice a task during training in a simulated stress environment, by the end of the training session they must receive feedback that they are performing the task effectively. Because the stress environment is an extremely high-demand performance environment, individuals will develop either positive or negative expectations regarding their capacity to perform in that environment. It is important that training supports the development of positive performance expectations, for one critical reason: Individuals who develop positive expectations will have more confidence in their ability to perform their duties and will be more resistant to negative stress effects. On the other hand, individuals who are exposed to the stress environment during training and receive feedback that leads them to conclude that they are likely to fail in the operational setting are perhaps worse off than those who are not trained at all. This again suggests a graduated approach to stress training, which promotes the development of positive trainee expectations by proceeding from simple exercises to more complex and realistic training scenarios.

Evaluation and Follow-Up. A well-designed training evaluation is required to assess trainees' reactions to training, to assess whether training is achieving the desired objectives, and to gather other data that are critical to training success. Moreover, in designing stress training it is essential that continuing or recurrent training be considered. Stress training, by definition, is training that incorporates features of the stress environment. In most cases this environment is one that is rarely encountered in everyday situations. Therefore, unless continuing stress training is provided, trainees will not have the opportunity to practice and apply the skills learned, and these skills will decay over time. Follow-up training should be implemented at appropriate intervals following initial training.

IMPLEMENTING STRESS TRAINING

The specific events that take place in any given stress training intervention will be specific to the requirements and characteristics of that operational setting. For example, the types of stress that are relevant, and whether the focus of skills training should be on decision making, team coordination, or other skills are questions that are unique to the specific task environment. Therefore, what is most important is that we are able to establish a structure for stress training, a framework that describes how stress training should be implemented.

In the following sections we adopt the three stages of the SET approach (Driskell & Johnston, 1988) as a means to structure stress training. Our intent is to outline the three-stage stress training approach that has been shown to be overall an effective stress training intervention (Saunders, Driskell, Johnston, & Salas, 1996) and to describe the types of content that may compose each stage.

Phase 1: Information Provision

The first component of Phase 1 is trainee indoctrination. Trainees need to know why they are there, the objectives of training, and why stress training is important. This type of indoctrination is standard procedure but is particularly relevant for stress training because stress training is training "above and beyond" basic technical training, and thus its value must be clearly established for the trainee. Indoctrination may be provided by discussing operational incidents or case histories in which stressors such as extreme time pressure and task load had a significant impact on performance.

The second and primary component of Phase 1 is the provision of preparatory information. Performing under high-demand, high-stress conditions results in several negative consequences. One effect of stress is physiological. Physiological reactions include increased heart rate, sweating, shallow breathing, muscle tension, and other reactions. A second category of stress effects are performance effects, including distraction, narrowing of attention, tunnel vision, decreased search activity, and so on. These effects are well documented and typically result in degraded overall performance. However, the typical task performer is relatively naive and unknowledgeable regarding both performance and physiological effects of stress.

Research suggests that providing trainees with preparatory information about the stress environment may have several beneficial consequences. Preparatory information may increase a sense of controllability and increase a person's confidence in his or her ability to perform. Preparatory information enables the individual to form accurate expectations regarding stress reactions and events that are likely to occur in the stress performance environment. Finally, preparatory information decreases the distraction involved in attending to novel sensations and activities in real time in the stress environment, thus increasing attention devoted to task-relevant stimuli.

A recent study conducted by Inzana, Driskell, Salas, and Johnston (1996) demonstrated the value of providing preparatory information as a part of stress training. In this study, trainees were given preparatory information before engaging in a stressful military task simulation. The information included knowledge of the stressors inherent in the task environment (i.e., increased task load, auditory distraction, and time pressure), information on how these stressors might make the participants feel (e.g., physical sensations, such as a pounding heart and sweating palms) and, finally, information on how these stressors may affect task performance. The results indicated that task performers who were given preparatory information prior to task performance made fewer errors under stress, were less likely to report feeling stressed, and were more confident in their ability to perform the task.

Phase 2: Skills Acquisition

The effects of stress on the task performer are well documented. Stress may result in physiological changes, such as increased heartbeat, labored breathing, and trembling (Rachman, 1983); emotional reactions, such as fear, anxiety, frustration (Driskell & Salas, 1991), and motivational losses (Innes &

Allnutt, 1967); cognitive effects, such as narrowed attention (Easterbrook, 1959), decreased search behavior (Streufert & Streufert, 1981), longer reaction time to peripheral cues and decreased vigilance (Wachtel, 1968), and performance rigidity (Staw, Sandelands, & Dutton, 1981); and changes in social behavior, such as a loss of team perspective (Driskell et al., 1999) and a decrease in prosocial behaviors, such as helping (Mathews & Canon, 1975).

The second phase of stress training focuses on the acquisition of skills required to counter these negative stress effects. Driskell and Johnston (1998) described two types of stress training techniques. First, one can attempt to make the task performer less reactive to stress. For example, if the trainee overlearns the task so that responses are automated, then performance is less likely to be disrupted by increased demands. A second approach is to train the individual to compensate for or overcome the expected decrements imposed by stress. For example, decision-making training approaches attempt to train effective decision-making strategies that are appropriate to high-demand environments.

Therefore, the goal of training at this stage is to build high performance skills that are resistant to degradation under stress. Although the training content (the specific training techniques) implemented in this phase of training will depend on the requirements of the task, we describe in the following several candidate training techniques.

Cognitive Control Techniques. *Cognitive restructuring* is an elaborate term for a relatively simple but useful training technique. In a stress environment, performance suffers as attention is divided between task-relevant and task-irrelevant cognitions. The focus of this training approach is to train the individual to regulate emotions (e.g., worry and frustration), regulate distracting thoughts (self-oriented cognitions), and maintain task orientation (Wine, 1980). Although applied research is relatively sparse, cognitive control techniques may be effective in enhancing task performance under stress. However, note that other cognitive techniques, such as self-talk or imagery, in which trainees are taught to invoke a positive thought, phrase, or image in response to stress, may be distracting and detrimental to task performance in a high-demand, time-limited task setting.

Physiological Control Techniques. Some training techniques attempt to provide the trainee with control over negative physiological reactions to stress. Relaxation training has proven to be a successful stress reduction technique, although it may be difficult to implement in many applied settings because of the connotations associated with the term (no one wants personnel "relaxed" in a critical task situation). Nevertheless, the value of this training is that it attempts to train the responses characteristic of effective or high performers: being calm, relaxed, and under control. Training that enhances physiological control (awareness and control of muscle tension, breathing, etc.) may promote effective task performance under stress.

Modeling.   A number of studies have also examined observational practice, or behavioral modeling techniques (see A. P. Goldstein & Sorcher, 1974), and proponents argue the value of this approach for stress training. The opportunity to observe or model a team responding effectively in a realistic stress simulation (live or on videotape) may increase trainee familiarity with the performance setting and allow trainees to observe key behaviors that characterize effective performance in that setting.

Overlearning.   *Overlearning* refers to deliberate overtraining of a performance beyond the level of initial proficiency (Driskell et al., 1992). Overlearning can provide trainees with a set of well-learned responses that are less vulnerable to stress decrement. However, the training designer must ensure that the task that is overlearned is the task called for in the actual performance setting; thus, overlearning of skills that will be performed in a high-stress criterion setting should take place in a simulated stress environment.

Attentional Training.   Singer et al. (1991) examined whether attention-focusing skills could be directly trained and found that attentional training resulted in improved task performance when participants worked under conditions of noise stress. This approach included awareness training to describe when, why, and how attention may be distracted during task performance. This was followed by practice in performing the task under high-demand conditions, focusing attention, and refocusing attention after distraction. Training that concentrates directly on enhancing attentional focus may overcome the distraction and perceptual narrowing that occur in stress environments.

Training Time-Sharing Skills.   In an emergency situation a person may have to perform a primary task, deal with a second unexpected task, delegate a third task, monitor a fourth, and so on. Research suggests that multiple tasks can be performed effectively if they are practiced concurrently (Hirst, Spelke, Reaves, Caharack, & Neisser, 1980). Therefore, time sharing is a task-specific skill, and tasks that are likely to be performed concurrently in the operational environment must be practiced concurrently in training.

A second concern is training prioritization skills in multiple-task environments. An example of a commercial aviation emergency illustrates the problem: A commercial jet was on an approach for landing, a period of very high workload, when a landing gear light failed to illuminate. Over the next 4 minutes of flight, the crew was so preoccupied with this malfunction that they failed to monitor other critical flight activities and literally flew the plane into the ground (National Transportation Safety Board, 1973). In high-workload conditions, individuals, often by necessity, focus on some tasks to the exclusion of others, and often attention is devoted to low-priority or irrelevant tasks. Training that addresses the prioritization of tasks in high-workload environments may ensure that the most critical tasks are not neglected.

Decision-Making Training.   Formal, analytic decision-making approaches require the decision maker to carry out an elaborate and exhaustive procedure characterized by a systematic, organized information search, thorough consideration of all available alternatives, evaluation of each alternative, and re-examination and review of data before making a decision. Although this procedure is often taught as the decision-making ideal, some researchers have argued that under high task demands, decision makers do not have the luxury of adopting a time-consuming analytic strategy. Moreover, encouraging the decision maker to adopt an analytic model could undermine behavior that may more adequately fit the requirements of the task situation (see Cannon-Bowers & Salas, 1998). Johnston et al. (1997) found that on a time-pressured, naturalistic task, individuals who had been trained to use a less analytic strategy performed more effectively than those who used a formal, analytic decision strategy. These results emphasize the importance of adaptability and flexibility in decision making and suggest that one goal for training is to enhance flexibility in adapting decision-making strategies to task demands. Specialized training on decision-making skills in a stressful environment should ensure that strategies appropriate to high-demand task conditions are trained.

Enhancing Flexibility.   Research indicates that stress leads to greater problem-solving rigidity (Cohen, 1952). *Rigidity* refers to the tendency to approach a problem with a restricted attentional focus on a given set of cues and an expectancy that there is a single solution that does not vary. Flexible behavior includes attention to many task cues and the expectation that the correct problem solution may differ from situation to situation. Flexibility leads to more efficient performance under complex conditions in which more than one solution is possible or in novel task conditions in which solutions must be made under varied task contingencies.

Certain training procedures can enhance flexible behavior. Practice of a narrow range of examples in training can lead to response rigidity and poor transfer to more complex, real world environments. Gick and Holyoak (1987) found that positive transfer was more likely when a variety of different examples were provided during training. Schmidt and Bjork (1992) referred to this as *practice variability*, noting that intentional variation during skills practice can enhance the transfer of training. Thus, presenting training material or training activities in various contexts, from different perspectives, and with diverse examples can result in more flexible use of a skill under novel task conditions.

Phase 3: Application and Practice

Effective performance requires not only that skills are learned in training but also that they are transferred to the operational setting. The novelty of performing even a well-learned task in a high-stress environment can cause severe degradation in performance. Research has shown that, for some tasks, normal training procedures (training conducted under normal, non-stress conditions) often do not improve task performance when that task has to be performed under stress conditions (Zakay & Wooler, 1984). Therefore, the final phase of stress training requires the application and practice of skills learned under conditions that approximate the operational environment. Allowing skills practice to proceed in a graduated manner across increasing levels of stress (from moderate stress scenarios or exercises to higher stress exercises) allows pre-exposure to the conditions that are likely to be faced in a high-stress or emergency situation.

Graduated exposure to stress events in training serves several purposes. First, it serves as an adjunct to the preparatory information provided in Phase 1 of training, allowing the trainee to experience likely real world operational conditions. This reduces uncertainty and anxiety regarding these events and increases confidence in the ability to perform in this setting. Second, allowing trainees to perform tasks in a simulated stress environment increases familiarity with the type of performance problems inherent in this setting. Trainees can then be allowed the opportunity to bring performance back to baseline levels using skills learned in Phase 2. Finally, events that have been experienced during training will be less distracting when faced in the operational environment.

Pre-exposure to the stress environment can be accomplished in Phase 3 in a number of ways. Normal training exercises can be adapted by incorporating stressors such as increased time pressure, noise, or other distractors; trainees can role play emergency conditions; or more advanced simulations, incorporating realistic scenarios, can be used. In one training exercise, Johnston et al. (1997) incorporated stress into decision-making training by playing multitract recordings of task-related chatter over trainees' headphones, increasing the pace of training events, and having exhortations and interruptions occur at regular intervals (all of these events were relevant to the operational task environment).

One concern that inevitably arises among training professionals is the question of intensity or fidelity of stress training. *Fidelity* refers to the degree to which characteristics of the training environment are similar to those of the criterion setting. Some argue that if a training setting is not real-istic—it does not look and feel just like the real setting—then it is not useful. The tendency for those who share this view is to "turn up all the knobs" in training, to create an environment that is just as complex and stressful as the real thing. However, the danger in this approach is that the increased complexity will overload the trainee and prove detrimental to training. Regian, Shebilske, and Monk (1992) claimed that it is not necessarily true that higher fidelity always leads to better training and that many training strategies reduce fidelity early in training to reduce complexity. Keinan and Friedland (1996) noted that allowing skills practice in a graduated manner across increasing levels of stress satisfies three important requirements: It allows the individual to become more familiar with relevant stressors without being overwhelmed, it enhances a sense of individual control and builds confidence, and gradual exposure to stress is less likely to interfere with the acquisition and practice of task skills than would exposure to intense stress.

SUMMARY: GUIDELINES FOR STRESS TRAINING

Those interested in the topic of stress and performance are faced with a field of study that includes both fact and fads, interests that range from crew per-formance to surgery, and a bewildering array of stress interventions (see Driskell & Salas, 1996). No doubt stress is a difficult topic, but for those con-cerned with how people perform in demanding, real-world situations, it is a topic that is of utmost importance. Based on our own research and the work of others, we are able to derive some general principles for stress training. Although modest, the following guidelines are intended to assist those developing stress management training for individuals and teams.

Guideline I: Make Sure Everyone Is On the Same Page

When you mention "stress training" to 10 different people, you are likely to get 10 different interpretations of what you mean. It is critical that those in upper management positions and the trainees themselves clearly under-stand the goals of stress training. We offer the following four points:

- Stress is a high-demand, high-threat situation that results in degraded performance. It is time limited, stress conditions occur suddenly and often unexpectedly, quick and effective task performance is critical, and consequences of poor performance are immediate and severe.

- Stress conditions can impose negative physiological, cognitive, emotional, and social effects, all of which may contribute to impaired task performance.
- Technical skill is a necessary but not sufficient condition to support effective performance in the stress environment. Normal training procedures do not provide pre-exposure to the stress environment or the special skills training required to maintain effective performance under stress.
- The key goals of stress training include providing pre-exposure to the high-demand conditions that may be faced by the trainee in the operational environment and providing the specialized skills training required to maintain effective performance under stress conditions.

Guideline 2: Stress Training Must Be Based On a Comprehensive Needs Analysis

A careful needs analysis is required to develop training content, defining the specific tasks to be trained and the types of stress to which trainees are exposed in training. The training development team should include experts in instructional development and subject matter experts who are familiar with the operational setting.

Guideline 3: Stress Training Should Address Both Individual and Team Tasks

Because stress may degrade team processes such as coordination and communication, stress training for team tasks should be conducted with intact teams.

Guideline 4: Stress Training Should Be Implemented After Initial Skills Training

Stress training should be introduced into the training curriculum after initial skills are developed. The introduction of stress training too early may interfere with initial skill acquisition.

Guideline 5: Although Stress Training Must Approximate the Operational Setting, Absolute Fidelity Is Not Necessary

Given the complexities inherent in the real-world environment, absolute fidelity in training is neither possible nor necessarily desirable. The training designer may provide exposure to the stress environment in a graduated manner, from moderate to higher stress exercises.

Guideline 6: Stress Training Must Be Provided On a Continuing Basis

Emergency or high-demand conditions occur relatively rarely. Unless continuing training is provided, trainees may not have the opportunity to practice and apply the skills learned in stress training. Follow-up training should be planned at appropriate intervals.

Guideline 7: The Three-Stage SET Approach Has Been Shown to Be An Effective Training Procedure, Incorporating Information Provision, Skills Acquisition, and Application and Practice

The primary goal of the information-provision phase of training is to provide trainees with basic information on stress, stress symptoms, and likely stress effects in the performance setting. The second phase of training, skills acquisition, focuses on building the skills that are required to counter the negative effects of stress on performance. The third stage of training, application and practice, provides the trainee the opportunity to practice skills learned under simulated stress conditions. Stress should be introduced in a graduated manner, from moderate to higher stress training exercises.

REFERENCES

Cannon-Bowers, J. A., & Salas, E. (Eds.). (1998). *Making decisions under stress: Implications for individual and team training.* Washington, DC: American Psychological Association.

Cohen, E. L. (1952). The influence of varying degrees of psychological stress on problem-solving rigidity. *Journal of Abnormal and Social Psychology, 47,* 512–519.

Cohen, S. (1980). Aftereffects of stress on human performance and social behavior: A review of research and theory. *Psychological Bulletin, 88,* 82–108.

Driskell, J. E., & Johnston, J. H. (1998). Stress exposure training. In J. A. Cannon-Bowers & E. Salas (Eds.), *Making decisions under stress: Implications for individual and team training* (pp. 191–217). Washington, DC: American Psychological Association.

Driskell, J. E., & Salas, E. (1991). Group decision making under stress. *Journal of Applied Psychology, 76,* 473–478.

Driskell, J. E., & Salas, E. (Eds.). (1996). *Stress and human performance.* Hillsdale, NJ: Lawrence Erlbaum Associates.

Driskell, J. E., Salas, E., & Johnston, J. H. (1999). Does stress lead to a loss of team perspective? *Group Dynamics, 3,* 1–12.

Driskell, J. E., Willis, R., & Copper, C. (1992). Effect of overlearning on retention. *Journal of Applied Psychology, 77,* 615–622.

Easterbrook, J. A. (1959). The effect of emotion on cue utilization and the organization of behavior. *Psychological Review, 66,* 183–201.

Friedland, N., & Keinan, G. (1986). Stressors and tasks: How and when should stressors be introduced during training for task performance in stressful situations? *Journal of Human Stress, 12,* 71–76.

Gick, M. L., & Holyoak, K. J. (1987). The cognitive basis of knowledge transfer. In S. M. Cormier & J. D. Hagman (Eds.), *Transfer of training: Contemporary research and applications* (pp. 9–46). New York: Academic Press.

Goldstein, A. P., & Sorcher, M. (1974). *Changing supervisory behavior.* New York: Pergamon.

Goldstein, I. L. (1986). *Training in organizations: Needs assessment, development, and evaluation.* Monterey, CA: Brooks/Cole.

Hirst, W., Spelke, E. S., Reaves, C. C., Caharack, G., & Neisser, U. (1980). Dividing attention without alternation or automaticity. *Journal of Experimental Psychology: General, 109,* 98–117.

Innes, L. G., & Allnutt, M. F. (1967). *Performance measurement in unusual environments* (Institute of Aviation Medicine Technical Memorandum No. 298). Farnborough, England: Royal Air Force Institute of Aviation Medicine.

Inzana, C. M., Driskell, J. E., Salas, E., & Johnston, J. (1996). Effects of preparatory information on enhancing performance under stress. *Journal of Applied Psychology, 81,* 429–435.

Johnston, J. H., & Cannon-Bowers, J. A. (1996). Training for stress exposure. In J. E. Driskell & E. Salas (Eds.), *Stress and human performance* (pp. 223–256). Mahwah, NJ: Lawrence Erlbaum Associates.

Johnston, J., Driskell, J. E., & Salas, E. (1997). Vigilant and hypervigilant decision making. *Journal of Applied Psychology, 82,* 614–622.

Kanki, B. G. (1996). Stress and aircrew performance: A team-level perspective. In J. E. Driskell & E. Salas (Eds.), *Stress and human performance* (pp. 127–162). Hillsdale, NJ: Lawrence Erlbaum Associates.

Keinan, G., & Friedland, N. (1996). Training effective performance under stress: Queries, dilemmas and possible solutions. In J. E. Driskell & E. Salas (Eds.), *Stress and human performance* (pp. 257–277). Mahwah, NJ: Lawrence Erlbaum Associates.

Lazarus, R. S. (1966). *Psychological stress and the coping process.* New York: McGraw-Hill.

Lipsey, M. W., & Wilson, D. B. (1993). The efficacy of psychological, educational, and behavioral treatment: Confirmation from meta-analysis. *American Psychologist, 48,* 1181–1209.

Mathews, K. E., & Canon, L. K. (1975). Environmental noise level as a determinant of helping behavior. *Journal of Personality and Social Psychology, 32,* 571–577.

National Transportation Safety Board. (1973). *Aircraft accident report: Eastern Airlines, Inc. L-1011, N310EA Miami Florida* (NTSB/AAR-73/14). Washington, DC: Author.

National Transportation Safety Board. (1989). *Aircraft Accident Report: Aloha Airlines, Flight 243, Boeing 737-200, N73711, Near Maui, Hawaii, April 28, 1988* (NTSB/AAR-89-03). Washington, DC: Author.

Quinones, M. A., & Ehrenstein, A. (Eds.). (1997). *Training for a rapidly changing workplace.* Washington, DC: American Psychological Association.

Regian, J. W., Shebilske, W. L., & Monk, J. M. (1992). Virtual reality: An instructional medium for visual–spatial tasks. *Journal of Communication, 42,* 136–149.

Rotton, J., Olszewski, D., Charleton, M., & Soler, E. (1978). Loud speech, conglomerate noise, and behavioral aftereffects. *Journal of Applied Psychology, 63,* 360–365.

Salas, E., Driskell, J. E., & Hughes, S. (1996). Introduction: The study of stress and human performance. In J. E. Driskell & E. Salas (Eds.), *Stress and human performance* (pp. 1–45). Hillsdale, NJ: Lawrence Erlbaum Associates.

Saunders, T., Driskell, J. E., Johnston, J., & Salas, E. (1996). The effect of stress inoculation training on anxiety and performance. *Journal of Occupational Health Psychology, 1,* 170–186.

Schmidt, R. A., & Bjork, R. A. (1992). New conceptualizations of practice: Common principles in three paradigms suggest new concepts for training. *Psychological Science, 3,* 207–217.

Singer, R. N., Cauraugh, J. H., Murphey, M., Chen, D., & Lidor, R. (1991). Attentional control, distractors, and motor performance. *Human Performance, 4,* 55–69.

Staw, R. M., Sandelands, L. E., & Dutton, J. E. (1981). Threat-rigidity effects in organizational behavior: A multi-level analysis. *Administrative Science Quarterly, 26,* 501–524.

Streufert, S., & Streufert, S. C. (1981). *Stress and information search in complex decision marking: Effects of load and time urgency* (Tech. Rep. No. 4). Arlington, VA: Office of Naval Research.

Terris, W., & Rahhal, D. K. (1969). Generalized resistance to the effects of psychological stressors. *Journal of Personality and Social Psychology, 13*, 93–97.

Wachtel, P. L. (1968). Anxiety, attention, and coping with threat. *Journal of Abnormal Psychology, 73*, 137–143.

Walton, R. E., & McKersie, R. B. (1965). *A behavioral theory of labor negotiation: An analysis of a social interaction system.* New York: McGraw-Hill.

Wegner, D. M., & Giuliano, T. (1980). Arousal-induced attention to self. *Journal of Personality and Social Psychology, 38*, 719–726.

Wexley, K. N., & Latham, G. P. (1991). *Developing and training human resources in organizations* (2nd ed.). New York: HarperCollins.

Wine, J. D. (1980). Cognitive–attentional theory of test anxiety. In I. G. Sarason (Ed.), *Test anxiety: Theory, reseaech, and application* (pp. 349–385). Hillsdale, NJ: Lawrence Erlbaum Associates, Inc.

Zakay, D., & Wooler, S. (1984). Time pressure, training and decision effectiveness. *Ergonomics, 27*, 273–284.

5

Assertiveness and Team Performance: More Than "Just Say No"

Florian Jentsch
University of Central Florida

Kimberly A. Smith-Jentsch
Naval Air Warfare Center Training Systems Division

On the morning of February 19, 1996, a passenger aircraft operated by a large U.S. air carrier landed wheels up on Runway 27 at Houston International Airport. The airplane slid almost 7,000 feet before coming to rest adjacent to the runway. The cabin filled with smoke, and an emergency evacuation was conducted. Fortunately, no one was killed or seriously hurt. Twelve of the 82 passengers, however, sustained minor injuries, and the aircraft was seriously damaged (National Transportation Safety Board [NTSB], 1997).

What had happened? In its report on the investigation of the accident, the NTSB (1997) determined that the crew had failed to properly complete the in-range checklist. This resulted in a lack of hydraulic pressure needed to lower the gear and flaps. Next, the crew had failed to properly conduct the before-landing checklist. This would have included a check of whether the gear was down. Finally, the captain had failed to initiate a go-around when the aircraft's ground-proximity warning system was activated shortly before landing.

In addition to these causal and contributing factors, however, the NTSB noted in its analysis that throughout the final phases of the approach, the first officer (who was at the controls) had doubts about the configuration of the airplane. In fact, 30 seconds before touchdown, the first officer asked the captain whether he wanted "to take it around?" When the captain answered "no, that's all right," the first officer did not challenge the captain's statement further or ask for clarification. Explaining his behavior to the NTSB after the accident, the first officer noted that he had experienced an occurrence 2 years earlier in which a captain had complained about him. This occurrence had resulted in a suspension from duty for 60 days and evaluation of the first officer by a psychiatrist. Afterward, the first officer was afraid that "his career would be in jeopardy if another captain complained to management about him" (NTSB, 1997, p. 46). Consequently, he had "adopted a cautious and deferential mode of interaction with captains to prevent a recurrence" (NTSB, 1997, p. 46). In analyzing the behavior of the first officer, the NTSB stated its concern "that a pilot was disinclined to assertively challenge another pilot's decision, [. . .] because he feared reprisal" (NTSB, 1997, p. 47).

Two years prior, the NTSB had reviewed 37 major air carrier accidents (NTSB, 1994) and concluded that in over 80% of these accidents the first officer had failed to adequately monitor and challenge actions by the captain. The NTSB authors noted that the first officer was the last person who had the opportunity to prevent most of these accidents from occurring. In fact, nonassertiveness had been identified as a problem as early as the 1920s and 1930s. For example, Sir Gordon Taylor, Australian pioneer aviator, vividly illustrated the struggle between trust and fear and his resulting behavior the first time he flew as second pilot for Australian National Airlines in 1930 (Taylor, 1991, pp. 32–34):

> From the corner of my eye I watched [the captain]. Was this just mad bravado, or did he really know how to do it? [. . .] At 2000 feet, I had more or less made up my mind that a crash was inevitable, but I hoped that when the trees appeared out of the cloud in front, there might just be time to open the throttles and climb away; but actually I knew there wouldn't be, unless a miracle happened.

Although the examples we have provided so far vividly illustrate the dire consequences of nonassertive behavior in an aviation environment, there are many other types of teams for which assertive team members can prevent life-threatening mistakes (e.g., medical teams, air traffic control teams, police and firefighting teams). Moreover, assertive behavior affects the performance of many other types of teams (e.g., task forces, quality circles, autonomous work groups) that pool their ideas, observations, and concerns to reach decisions that have far-reaching financial implications for their organizations. As a result, a wide variety of organizations today have adopted assertiveness-training programs with the hope that they will enhance the productivity and creativity of their workforce (Ruben & Ruben, 1989). For example, nearly every cockpit (later *crew*) resource management (CRM) program in the last 20 years has included a module on assertiveness. Two recent studies have demonstrated that this type of training indeed increases the ability and willingness of junior crewmembers to challenge errors made by their captains during simulated flights (Jentsch, 1997; Smith-Jentsch, Jentsch, Payne, & Salas, 1996). Furthermore, pilots who participate in such training often cite anecdotal evidence that they were able to avert a potential mishap by applying their newfound assertive skills in the cockpit (Salas, Fowlkes, Stout, Milanovich, & Prince, 1999). However, some captains complain that first officers are "too assertive" after training, that they usurp authority, and that they often do not know the right answer and consequently should not assert themselves (cf. Buck, 1994).

Such criticisms may stem in part from misconceptions about the goals of assertiveness training. Thus, we begin this chapter by first presenting a definition of team performance-related assertiveness and distinguishing it from alternative approaches for conflict resolution (i.e., passivity, aggressiveness). Second, we make the case that assertiveness is both an attitude and a skill. Third, we summarize research that has investigated the multidimensionality of assertiveness. Fourth, we highlight a number of contextual factors that have been used to demonstrate the situation specificity with which individuals apply assertiveness.

The second major section of the chapter provides guidelines for measuring team performance-related assertiveness. This section is followed by guidelines for improving assertiveness in teams through training, organizational policies and procedures, and team climate.

DEFINING TEAM PERFORMANCE-RELATED ASSERTIVENESS

The following sections summarize lessons learned from research that has sought to better understand assertiveness. In particular, we seek to dispel some common misconceptions about the construct. These misconceptions may lead organizations to implement training strategies that yield limited results on-the-job or to make bad selection decisions on the basis of inappropriate inferences drawn from assertiveness measures.

Assertiveness Is Not Aggressiveness

In defining assertiveness, it is useful to contrast it with two alternative methods for resolving conflict: passivity and aggressiveness. These three communication styles can be arrayed on a continuum according to the degree to which they reflect a concern for one's own well-being at the possible expense of others, or vice versa (Fig. 5.1). On the one end of the continuum are passive responses. Passive responses are often worded in the form of a question, conveying a level of uncertainty that does not reflect the communicator's true feelings. For example, a team member who knows for a fact that his or her leader has just made a mistake may attempt to point this out by stating "Did we recommend something different in our progress report last week?" rather than

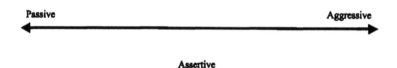

FIG. 5.1.   Continuum of communication styles.

stating "To me, this recommendation seems to conflict with statements we made in the last progress report." In other cases, individuals may make vague statements, "beating around the bush" or "sugarcoating" their critique of a particular strategy. Oftentimes a passive statement will indicate a reluctance to take personal responsibility for a belief or feeling; for example: "Other people might think that we should have done that first." A common element of passive statements is that they are an indirect and often watered-down expression of what an individual really feels or believes. As such, a very valid point stated in a passive manner often does not catch the attention of teammates or does not carry the weight that it should in the decision-making process.

On the opposite end of the continuum, aggressive statements are usually quite direct and unambiguous. However, they also include some form of hostility, defensiveness, or intrusion or imposition on another person's rights or needs. These statements are often manipulative, accusatory, disrespectful, or rude, and they reflect a clear disregard for the feelings, needs, or goals of others. As a result, even aggressive statements that, at their core, contain a grain of truth generally do not have the desired effect on team members. A good example of this involves the communication of negative feedback among team members. In the most effective teams, team members continuously learn from one another. This can happen only if they are open to feedback on their performance and to suggestions for improvement. When a team member makes an aggressive statement such as, "No, you're wrong . . . , and that's basic pilot knowledge" to another teammate, more often than not he or she will elicit a defensive reaction rather than careful consideration of the validity of his observation.

Assertiveness, on the other hand, involves communicating one's feelings, concerns, ideas, and needs to others in a clear and direct manner, but without demeaning or infringing on the rights of others. These statements suggest a sense of personal responsibility for one's own thoughts and feelings, as well as a sense of honesty and fairness. As such, many assertive statements begin with "I." For example, "I don't feel comfortable taking off under these weather conditions," or "I notice that we appear to be deviating from our designated altitude." A critical component of assertiveness that sets it apart from aggressiveness is an absence of the hostility and defensiveness that often prevents team members from considering and ultimately benefiting from aggressive comments.

It is the failure to make this critical distinction between aggressive and assertive communication styles that represents the first common misconception we would like to dispel about team performance-related assertiveness. Although most assertiveness training programs focus on moving individuals from a passive to a more assertive communication style, in certain team environments (e.g., air traffic control) it may be that training is also needed to refine overly aggressive communication styles. Furthermore, when previous authors have written about the potential dangers of assertiveness training gone awry (e.g., Buck, 1994), they appear to be concerned about team members adopting what we would describe as an aggressive communication style (e.g., usurping the authority of their leaders). Thus, it is important to distinguish among passive, assertive, and aggressive communication. It is assertiveness, and not aggressiveness, that we advocate as an effective approach for fully utilizing the human resources available within a team.

Assertiveness Is Both an Attitude and a Skill

Assertiveness has been treated by some as primarily an attitude (e.g., Helmreich, Foushee, Benson, & Russini, 1986), by others as a skill (e.g., Prince, & Salas, 1993), and by others still as a multidimensional skill with an attitudinal component (e.g., Lorr & More, 1980). We subscribe to the latter interpretation of assertiveness. This has important implications for both the measurement and training of assertiveness for team members. As implied by the previous discussion, individuals clearly differ in the degree to which they are willing to advance their own agenda and the forcefulness with which they are prone to do so. However, we argue that generating an assertive statement also requires a significant degree of skill. Team members must walk a fine line between being clear and direct and not putting others on the defensive, if they want their ideas and observations to be carefully considered. In addition to the words chosen to communicate an issue, posture, tone of voice, and eye contact also play a role in the effective use of assertiveness. Thus, training programs that attempt to change attitudes without offering opportunities to build assertive communication skills are likely to produce limited behavior change on-the-job. In the following section we delineate a number of behavioral response classes, or dimensions, within the overall construct of assertiveness.

Assertiveness Is a Multidimensional Skill

A large body of research to date has demonstrated that assertiveness is a multidimensional construct (Chan, 1993; Gambrill & Richey, 1975; Henderson & Furnham, 1983; Lazarus, 1973; Lorr & More, 1980; Nevid & Rathus, 1979; Northrop & Edelstein, 1998; Wills, Baker, & Botvin, 1983). Specifically, factor analyses of assertiveness inventories tend to reveal multiple subscales that are only moderately correlated with one another ($r = .30–.40$, on average). Although the number of assertiveness dimensions and labels used to describe them have varied somewhat from study to study, a review of the major works in this area reveals a substantial amount of overlap (see Table 5.1.)

Much of the research on the multidimensionality of assertiveness was conducted in the 1970s and 1980s. The majority of the inventories developed as part of this research have been validated using populations of clinically nonassertive subjects (e.g., battered wives, schizophrenics, clinically depressed patients; cf. Ruben & Ruben, 1989; children, adolescents, and young people who are isolated or rejected; cf. Argyle, 1995). However, a scale developed by Lorr and More (1980) consisting of four relatively independent factors has recently been used to predict team performance-related assertiveness in four studies (see Smith-Jentsch, Salas, & Baker, 1996; Smith-Jentsch, Salas, & Brannick, in press). For this reason, we discuss this scale and the dimensions within it in more detail.

Lorr and More (1980) defined four dimensions of assertiveness: (a) independence, (b) directiveness, (c) social assertiveness, and (d) defense of interests or rights. Simply put, independence can be thought of as the ability and willingness to resist conformity. This dimension is assessed with items such as "My opinions are not easily changed by those around me" and "I follow my own ideas even when pressured by a group to change them." Directiveness, on the other hand, involves being able and willing to initiate action and to accept or take on responsibility. Sample items include "In an emergency I get people organized and take charge" and "I work best in a group when I'm the person in charge." Social assertiveness can be described as an ability and willingness to initiate personal relationships. Items loading on this factor include "When I meet new people I usually have little to say" (reverse coded) and "It is easy for me to make small talk with people I've just met." Finally, defense of interests essentially involves the ability and willingness to refuse unreasonable requests and to let others know when they have violated your rights. Items contributing to scores on this dimension include, for example, "When someone repeatedly kicks the back of my chair in a theater I don't say anything" (reverse coded) and "If a friend betrays a confidence I express my annoyance to him/her."

Lorr and More (1980) argued that the moderate correlations among these four dimensions should be taken to mean that transfer from one dimension to another is unlikely. For example, it cannot be assumed that a person who is willing and able to "just say no" to an unreasonable demand will necessarily demonstrate directiveness in an emergency situation. Furthermore, it has been argued that although social assertiveness can enhance team performance through workplace friendships and that defending one's own interests may at times be advantageous to a team as well (e.g., refusing to take off in bad

TABLE 5.1
Dimensions of Assertiveness

	Lorr & More (1980)	Lazarus (1973)	Gambrill & Richey (1975)	Nevid & Rathus (1979)	Henderson & Furnham (1983)	Wills, Baker, & Botvin (1989)	Chan (1993)	Northrup & Edelstein (1998)
Directiveness			Handling a bothersome situation; Handling service situations	Assertive business dealings	Request change in another's irritating behavior			
Social assertiveness	Social assertiveness	Ability to initiate, continue, and terminate general conversation	Initiating interactions; Engaging in "happy talk"; Complimenting others	Avoid public confrontation; Verbal fluency	Initiating and maintaining interaction with nonintimate others	Social assertiveness	Readiness to express, initiate, and maintain interaction with nonintimate others	Initiating, maintaining, and terminating conversation
Defense of rights and interests	Ability to ask for favors or to make requests; Ability to say "no"		Confronting others; Responding to criticism; Turning down requests	Complaining to rectify injustice; General argumentativeness; Arguing over prices	Standing up for personal rights in a public situation; Ability to refuse requests and ask for personal information	Rights assertiveness	Confrontational or nonempathic assertive responses; Interpersonal situations in which one avoids hurting others' feelings yet requests changes	Standing up for one's rights; Offering and receiving favors and compliments; Initiating and refusing requests

(continued)

TABLE 5.1
(Continued)

Lorr & More (1980)	Lazarus (1973)	Gambrill & Richey (1975)	Nevid & Rathus (1979)	Henderson & Furnham (1983)	Wills, Baker, & Botvin (1989)	Chan (1993)	Northrup & Edelstein (1998)
Independence	Ability to express positive or negative feelings	Resisting pressure to alter one's consciousness Giving negative feedback	Spontaneity Insensitive self-expressiveness				Expressing positive or negative emotions or feelings Expressing personal opinions, including disagreements Offering and receiving criticism
		Admitting personal deficiencies		Ability to deal with criticism and pressure Positive assertion Unassertive acceptance General expressiveness Display of feelings	Substance assertiveness General assertiveness		Admitting a personal inability, ignorance, or mistake

weather), the assertive behaviors most closely tied to effective team performance (e.g., pointing out discrepancies, initiating solutions, stating and maintaining opinions) appear to fall within Lorr and More's dimensions of independence and directiveness (Smith-Jentsch, Salas, & Baker, 1996).

One implication of this notion is that self-report measures of independence and directiveness should be better predictors of team performance-related assertiveness than defense of interests or social assertiveness measures. In support of this proposition, Smith-Jentsch, Salas, and Baker (1996) found that only scores on Lorr and More's (1980) independence and directiveness dimensions correlated significantly with ratings of team member assertiveness demonstrated during a flight simulation. The implications for training are clear: If the primary objectives of assertiveness training are to improve team performance, then training time and resources should be allocated so as to maximize opportunities to practice behaviors consistent with independence and directiveness in role play exercises. Furthermore, the developmental feedback given to trainees should focus specifically on behaviors within these two response classes.

Assertiveness Is a Situation-Specific Attitude

We argued earlier that assertiveness has an attitudinal component. Thus, a training program can teach trainees to be skillful at generating assertive statements, but if the trainees are not convinced that it is appropriate to use such statements outside the training environment they are unlikely to transfer their newfound ability to the workplace. This notion has been supported by a number of studies that found individuals demonstrate more assertiveness in role play situations than in identical situations that appear to be "natural" or uncontrived (e.g., Frisch & Higgins, 1986; Gorecki, Dickson, Anderson, & Jones, 1981; Higgins, Alonso, & Pendleton, 1979; Smith & Marion-Landais, 1993). These findings suggest that individuals turn assertiveness off and on from one situation to the next. In other words, assertiveness appears to be highly situation specific. In the following sections of this chapter we briefly summarize research that has sought to identify those contextual factors that determine when an individual will be assertive.

Type of Interpersonal Relationship. There is some evidence that an individual's use of the same assertive response (e.g., stating an unpopular opinion) may vary as a function of the type of relationship that exists between him- or herself and the receiver. In this regard, Kolotkin (1980) delineated three types of interpersonal relationships: (a) stranger (e.g., sales transactions), (b) personal (e.g., spouse, friend, sibling), and (c) work related (e.g., team coordination). Adopting this framework, Smith-Jentsch, Salas, and Baker (1996) suggested that a person may have a relatively stable tendency to state and maintain opinions when interacting with strangers; however, he or she may find it extremely difficult to do so with a friend or family member. Thus, the predictive validity of self-report measures of assertiveness should be highest when the context in which items are framed is directly parallel to the context in which assertive behavior is desired.

To test this notion, Smith-Jentsch, Salas, and Baker (1996) asked business students, who were required to work on team projects over the course of a college semester, to complete three self-report scales and to rate their teammates' use of team performance-related assertiveness. Two of the self-report scales (i.e., Lorr & More, 1980; Rathus, 1973) had been used in many previous studies. However, items within these scales were either framed in the context of stranger or personal interaction or did not specify interpersonal context at all. The third self-report measure (Smith, Marion-Landais, & Blume, 1993) was developed specifically to assess team performance-related assertiveness, and thus each item was framed in a work-team context. The results indicated that although each of the three scales measured the same classes of assertive behavior on which peer ratings were based (i.e., independence and directiveness), only the work-team-focused scale was a significant predictor of behavior. These findings support the contention that an individual's tendency toward team performance-related assertiveness may be relatively independent from his or her tendency to respond assertively in situations involving strangers or people with whom a personal relationship exists.

Teammate Gender. There is some evidence that the gender configuration of a team is another contextual factor that can affect team performance-related assertiveness. However, the exact nature of these gender effects remains unclear. For example, some studies have found that individuals are more willing to be assertive toward another individual of the same gender and more likely to respond favorably when someone of the same gender behaves assertively toward them (e.g., Stebbins, Kelly, Tolor, & Power, 1977; Wilson & Gallois, 1985). In contrast, Smith-Jentsch, Salas, and Baker (1996) found that both men and women demonstrated greater assertiveness toward women in a team task environment. In sum, it appears clear that individuals consider the gender of the receiver when determining whether (or to what degree) they are willing to assert themselves. However, additional work in this area is needed to better specify the nature of gender-related effects on assertiveness.

Status Differences. Another contextual factor that has been hypothesized to dictate when an individual will choose to be assertive involves status differences among team members. A number of researchers have reported evidence to support this notion in an aviation environment. For example, in a review of 216 military aviation mishap reports, Härtel, Smith, and Prince (1991) found that as the military rank of the captain increased so did the likelihood that nonassertiveness was listed as a causal factor. In another study, Komich (1985) found that one third of the pilots he interviewed on the subject of nonassertiveness in the cockpit were willing to accept a Federal Aviation Administration violation before asserting themselves to their captain.

Status theory (Milanovich, Driskell, Stout, & Salas, 1998; Torrance, 1955; Webster & Hysom, 1998) offers an explanation as to the cognitive processes responsible for status-typed differences in the willingness to assert oneself. According to this theory, salient status characteristics such as occupational title (e.g., captain vs. first officer, surgeon vs. nurse) lead team members to form assumptions or expectations regarding the competence of their teammates. This, in turn, leads them to adopt either a subordinate or superordinate behavior pattern or script.

As a partial test of this theory, Milanovich et al. (1998) asked a group of Navy aviators to review written descriptions of two hypothetical pilots and assign ratings of both task-relevant and general characteristics for both. With the exception of the pilots' position in the crew (i.e., captain or first officer) the background descriptions were virtually identical (e.g., time in military, college degree, marital status). The results indicated that participants rated the captain as significantly higher relative to the first officer on his ability to handle flight situations, general intelligence, verbal ability, and leadership. Furthermore, respondents appeared to have generalized their expectations for the hypothetical pilots to nonaviation-related situations, as evidenced by higher average ratings assigned to the captain on his ability to teach in an elementary school.

It has been suggested that status-typed expectations such as those illustrated in Milanovich et al.'s (1998) study may lead individuals to feel less responsible for their own actions and view themselves instead as an "agent" of their leader, simply executing his or her wishes (Milgram, 1974). Milgram (1974) argued that once an individual views him- or herself in this light, profound alterations in behavior occur. Specifically, an individual was said to be in "a state of mental organization which enhances the likelihood of obedience" (p. 134) and therefore reduces the likelihood that he or she will choose to assert him- or herself.

Team Climate. *Team climate* has been defined as "the norms and expectations that govern, both formally and informally, the performance of, and interaction between individual team members in the commission of a team task" (Smith, 1994, p. 10). As such, team climate is expected to have the potential to either facilitate or inhibit the use of assertiveness among team members. Thompson and Luthans (1990) applied social learning theory to the understanding of how norms and expectations are transmitted. These authors suggested that we develop "socially constructed realities" by observing antecedent–behavior–consequence transactions that convey accepted norms of behavior. Smith-Jentsch, Milanovich, and Salas (1999) examined the causal effects of team climate on crew member assertiveness in a simulated flight by staging such a transaction during a preflight brief. Participants were assigned the role of copilot and were not aware that the purpose of the simulation was to assess their assertiveness or that their captain and flight engineer were scripted to introduce errors during the flight. Half of the participants were randomly assigned to observe their captain positively reinforce the flight engineer after he asserted himself during the preflight brief, and the other half of the participants observed the captain respond negatively to the flight engineer's assertion. Results indicated that the 5-minute preflight brief had a significant impact on perceptions of team climate. Furthermore, perceptions of team climate were significant predictors of the copilots' use of assertiveness during the subsequent flight simulation.

These findings underscore the situation-specific nature of assertive behavior in teams. Specifically, it appears that subordinate crewmembers may demonstrate very different levels of assertiveness from one day to the next depending on the perceived climate that exists within each crew. It has been estimated that approximately 60% of pilots for major U.S. airlines regularly fly with new crewmembers (Sharp, 1993). Analogous situations are likely to exist within other types of teams as well (e.g., emergency medical teams, police units). In such situations, an individual's use of assertiveness may be quite variable as a function of the perceived climate within each team.

Thus far, we have summarized research that has sought to define assertiveness and the factors that influence its use, particularly in a work team environment. On the basis of this research, in the following sections we offer practical guidelines for measuring and improving team performance-related assertiveness.

MEASURING TEAM PERFORMANCE-RELATED ASSERTIVENESS

Lessons learned from the research described thus far have a number of implications for developing measures to be used for selection and the evaluation of training effectiveness. In the following sections we offer guidelines based on this research. We focus on two types of measures: self-report and behavioral.

Self-Report Measures

Self-report measures are often used to assess individuals' willingness to demonstrate assertiveness in a team setting. Some of these measures contain items that ask respondents to reflect on past behavior (e.g., "I almost always argue my opinion when I feel I am correct"). Others assess individuals' beliefs about the appropriateness of assertiveness in some hypothetical situation (e.g., "If a team is making a wrong choice, team members should make their opinions heard"). We offer the following recommendations for increasing the validity of such measures.

Guideline 1: Inventories Used to Predict Team Performance-Related Assertiveness Should Include Items That Reflect the Multidimensional Nature of the Construct

As we noted earlier, assertiveness has been repeatedly shown to be a multidimensional skill. Although the number of dimensions and labels used to describe it have varied, it is clear that assertiveness is more than "just saying no." Furthermore, there is some evidence to suggest that self-report measures of certain dimensions of assertiveness (e.g., social assertiveness) are not at all predictive of behavior consistent with other dimensions of assertiveness (e.g., independence, directiveness) that are important for effective team performance (Smith-Jentsch, Salas, & Baker, 1996; Smith-Jentsch et al., in press). Thus, inferring something about an individual's willingness to state opinions that differ from those of his or her teammates, for example, based on their tendency to refuse unreasonable requests may be inappropriate. Furthermore, collapsing scores from multiple subscales into an overall indicator of assertiveness is likely to reduce one's ability to predict behavior within a specific response class or dimension. Thus, it is recommended that those interested in developing, adapting, or using existing self-report scales to predict team performance-related assertiveness carefully specify the behaviors they are interested in predicting and ensure that scale items reflect those particular response classes.

Guideline 2: Items Within Inventories Used to Predict Team Performance-Related Assertiveness Should Specify a Work-Team Context

As described earlier, there is some evidence that the interpersonal context (e.g., stranger, personal, work-related interaction) within which self-report items are framed may have a significant impact on the predictive validity of those items (Smith-Jentsch, Salas, & Baker, 1996). Unfortunately, most of the published assertiveness inventories consist of items set in the context of personal or stranger interaction. Therefore, those interested in using a self-report measure of assertiveness either to select individuals for team positions or to validate the effects of a training intervention may find it necessary to adapt items from existing scales to specify a particular work-team context (e.g., air traffic control, emergency room).

Behavioral Assessments

Behavioral measures of assertiveness may be obtained through peer or supervisory ratings of on-the-job performance. They can also be obtained in the context of situational exercises that range in physical fidelity from a classroom role play to a high-fidelity full-motion flight simulator. Findings from the research reviewed in this chapter suggest that the reliability and predictive validity of assertiveness ratings assigned in a team-based situational exercise can be improved in a number of ways.

Guideline 3: Situational Exercises (e.g., Role Plays, Simulations) Used to Assess Team Performance–Related Assertiveness Should Include Multiple, Preplanned Opportunities to Perform the Skill

An individual's opportunity to demonstrate assertiveness in a team-based situational exercise is largely dependent on the actions of the teammates with whom they perform the exercise. For this reason, studies that have used confederate team members to introduce carefully scripted errors (e.g., Jentsch, 1997; Smith-Jentsch, Jentsch, et al., 1996) have tended to capture more consistent ratings of assertive behavior (e.g., willingness to point out another teammate's error) across situations than those that have not (e.g., Brannick, Prince, Prince, & Salas, 1995; Smith-Jentsch, Johnston, & Payne, 1998).

Guideline 4: The Degree of Skill Transparency Associated With a Situational Exercise Should Be Considered When Interpreting Behavioral Assessments

Kleinmann (1993) introduced the term *skill transparency* to describe participants' awareness of the targeted skill and associated behaviors being evaluated in an assessment exercise. Kleinmann argued that in an assessment situation, skill transparency introduces demand characteristics that remove variability that is due to disposition. In other words, even participants who would not normally use assertiveness to handle a particular type of situation may do well in an exercise where they know that they are being judged on their ability to be assertive. The resulting measure, then, may be an excellent indicator of pure skill at using assertiveness, or what one can do, while being a rather poor indicator of typical performance, or what one will do, in similar situations.

To test this notion, Smith-Jentsch (1996) examined correlations among behavioral ratings of team performance-related assertiveness obtained in a simulation exercise and parallel self-report measures of typical behavior prior to and after the simulation. In two studies, participants were randomly assigned to perform the identical simulation under one of two conditions. In each study, one group was simply instructed to coordinate effectively as a team the best they knew how (low skill transparency). The second group was instructed to use assertiveness to the best of their ability in order to handle conflict situations that would be imposed during the simulation by their confederate teammates (high skill transparency). Results indicated that ratings of assertiveness in the simulation were significantly related to participants' self-reported typical use of assertiveness prior to (Study 1) and after (Study 2) the simulation only for those in the low-transparency conditions.

The findings from this study suggest that people who develop situational exercises for the assessment of team performance-related assertiveness should consider their purpose when deciding what level of skill transparency to allow. It appears that behavioral measures based on exercises in which it is transparent to participants that their assertiveness is being evaluated can be taken only as an indication of skill or ability. Such measures may be most appropriate if one is interested in examining the effectiveness of a particular training method in building assertive skills. In contrast, measures taken in low-transparency situations are more likely to be predictive of typical on-the-job performance. A comprehensive transfer-of-training evaluation might use both types of measures (cf. Smith, 1994). In this way an organization can determine whether a particular program was successful at building skills, changing attitudes, or both, as well as the degree to which nonsupportive transfer environments are responsible for disappointing organizational results. This type of assessment would then allow for appropriate adjustments to be made (e.g., increase opportunities for practice and feedback during training, attempt to change transfer climate).

IMPROVING TEAM PERFORMANCE-RELATED ASSERTIVENESS

Lessons learned from previous research also have a number of implications for improving team performance-related assertiveness. The following sections offer guidelines derived from this research. These guidelines specify recommendations for improving assertiveness in teams through training, organizational policies and procedures, and team climate.

Training

Workplace assertiveness training has become widespread in many industries, ranging from health care to automobile production (Ruben & Ruben, 1989). However, the effectiveness of these assertiveness training programs varies as widely as the methods used. In the following sections we offer guidelines for maximizing the impact of assertiveness training designed to enhance team performance.

Guideline 5: Address Attitudes Regarding the Appropriateness of Assertiveness in a Variety of Workplace Settings; Highlight Factors (e.g., Status, Team Climate) That May Inhibit the Transfer of Assertiveness Training

Given results from previous research suggesting that individuals apply assertiveness in a highly situation-specific manner, strategies for improving team performance-related assertiveness are likely to be most effective if they address the contextual factors that threaten to inhibit transfer head on. In particular, trainees should be encouraged to discuss issues involving teammate gender, status, norms, and expectations.

Guideline 6: Provide Each Trainee With an Opportunity to Practice Using the Desired Assertive Behaviors in Role Played or Simulation-Based Workplace Situations and to Receive Feedback on Their Skill at Doing So

It is generally recognized that providing opportunities for active practice in experiential exercises is optimal when attempting to build interpersonal skills (Baldwin & Ford, 1988). Doing so, however, is usually more costly and time consuming than an attitudinally focused "awareness-type" seminar. Furthermore, it is generally more challenging for instructors to facilitate an effective role-play exercise and to provide developmental feedback to trainees than it is to simply lecture or facilitate a group discussion. Therefore, organizations often opt for the latter approach to training team performance-related assertiveness.

Unfortunately, results from a recent study suggest that active practice and feedback may be not only optimal but *essential* for improving team performance-related assertiveness. Smith-Jentsch, Salas, and Baker (1996) examined the additive benefits of the three instructional components involved in a strategy commonly used to train assertiveness: behavior role modeling. Behavior role modeling incorporates information, demonstration, practice, and feedback and has been used successfully to train a variety of interpersonal skills, such as leadership, communication, and negotiation (Baldwin & Ford, 1988).

Smith-Jentsch, Salas, and Baker (1996) used three experimental conditions. In each condition, participants received 1 hour of instruction. The first group spent the entire hour listening to information about (e.g., elements of an assertive statement), and persuasive arguments for, the use of team performance-related assertiveness. The second group received a condensed version of this information. In addition, however, they viewed models acting out passive, aggressive, and assertive responses to workplace situations on videotape and discussed the pros and cons of each approach. The third group received information and viewed the same behavioral models; however, instead of merely critiquing the models, these trainees also practiced responding to role-played situations and received feedback on their performance.

Results indicated that all three groups of trainees reported more positive attitudes toward team performance-related assertiveness after training than did a group of no-treatment controls. However, only the group who received active practice and feedback demonstrated more assertive behavior during a subsequent flight simulation. These findings imply that programs designed to enhance team performance-related assertiveness may not reap significant results in the workplace unless they incorporate elements of practice and feedback. Furthermore, assertiveness training programs that incorporate one or two role plays as demonstrations in front of the class may have a notable impact only on the few brave volunteers who probably needed the practice less than those not inclined to volunteer.

Guideline 7: Incorporate Metacognitive Strategies for Determining When to Be Assertive

Researchers and training practitioners alike have begun to focus their attention on strategies for enabling individuals to judge when to apply various behavioral strategies. This trend has developed in recognition of the fast pace of change in the workplace and the accompanying necessity for team members who can flexibly adapt to new environments and types of problems. Specifically, it has been argued that traditional strategies for imparting interpersonal skills should be supplemented with metacognitive techniques that allow individuals to thoughtfully apply those skills (Ford & Weissbein, 1997).

Jentsch, Bowers, Martin, Barnett, and Prince (1998) found that in nearly one third of the cases where junior aircrew members failed to assert themselves and challenge an in-flight discrepancy, they did so as a result of inaccurately judging the time criticality of the situation. In response to this finding, Jentsch (1997) developed a training program designed to help pilots identify assumptions, monitor their own decision making, and mentally simulate anticipated future states if action were not initiated. The training involved information and demonstration as well as practice judging the time criticality involved in several written vignettes. Jentsch (1997) compared the effects of this type of training to the effects of a more traditional skill-based assertiveness training course. Results indicated that both of the two training approaches increased monitoring and challenging behavior in a flight simulation. However, the specific effects of the two programs were quite different. Whereas the traditional skill-based assertiveness course improved behavior in scenario events for which the need for assertiveness was clear, only the metacognitively focused course led to improved behavior on events that required trainees to consider the timing of their response. On the basis of these results, Jentsch argued that the effectiveness of traditional skill-based approaches to assertiveness training could be enhanced by incorporating exercises that build metacognitive strategies for determining when to assert oneself.

Policies and Procedures

Guideline 8: Organizational Policies Should Support
Appropriate Assertion In the Workplace

Management policies (formal and informal) for dealing with situations in which subordinate team members assert themselves to their superiors have the potential to facilitate or inhibit team performance-related assertiveness. For example, management programs designed to increase the number of on-time flights and completed flights have been implicated as contributing to an unwillingness to assert crewmember concerns (e.g., Braniff International's "fast buck" campaign and subsequent 1968 crash [Nance, 1984; NTSB, 1969], or the Downeast Airlines accident in Rockland, Maine, in 1979 [NTSB, 1980]).
On the other hand, organizational policies can also increase the likelihood that team members will assert themselves when needed. One way of doing so may be to draft standard operating procedures designed to remove some of the ambiguity surrounding situations where a team member is expected to assert themselves. A good example of this is the Federal Aviation Administration's Advanced CRM (ACRM) program (Seamster, Boehm-Davis, Holt, & Schultz, 1998). Under ACRM, procedures are introduced into normal operations so that captains must specify the conditions under which a landing or go-around will be executed. This is designed to make it easier for junior crewmembers to assert themselves when thresholds have been reached. Ultimately, however, organizational rewards and punishments actually bestowed on those who do assert themselves are likely to make or break adherence to such policies and procedures.

Guideline 9: Team Members Should Be Assessed On
Their Effective Use of Team Performance-Related
Assertiveness

Another way that organizations can convey expectations regarding the use of assertiveness is to include it among the criteria on which team members are evaluated. Many airlines, for example, assess crewmember assertiveness as part of recurrent CRM simulator hops. Other organizations may collect supervisor, peer, and self-ratings of team performance-related assertiveness as part of their performance appraisal system.

Team Climate

As we argued earlier, team members often determine whether to assert themselves on the basis of their perceptions of team climate. Furthermore, the role of team leaders in creating perceptions of team climate has been described in numerous studies (e.g., Roullier & Goldstein, 1993; Smith-Jentsch et al., in press; Tannenbaum, Smith-Jentsch, & Behson, 1998; Tracey, Tannenbaum, & Kavanaugh, 1995).

Guideline 10: Train Team Leaders to Establish and Maintain a Climate That Supports Team Performance-Related Assertiveness

Assertiveness training for junior team members is unlikely to transfer to settings in which team leaders display a preference for deferential behavior from their subordinates. Smith (1994) demonstrated the chilling effect of a nonsupportive team leader on transfer climate and, ultimately, on behavior. Results indicated that trainees who observed their (confederate) captain respond negatively toward a third crewmember who asserted him- or herself during a preflight brief were no more assertive during a subsequent flight simulation than were participants who had not received assertiveness training.

It appears clear that if organizations are serious about fostering team performance-related assertiveness, they need to devote as much attention to training team leaders as they do to training team members. A comprehensive effort to increase team performance-related assertiveness should include training for leaders that: (a) makes them aware of how easily their behavior can lead to perceptions that assertiveness is desired or frowned on, (b) teaches them to identify team member attempts to be assertive, (c) provides them with practice and feedback encouraging and reinforcing such attempts, and (d) instructs them on how to deal with ineffective attempts at assertiveness (in other words, aggressiveness) without extinguishing all future attempts.

SUMMARY AND CONCLUSIONS

In this chapter we have discussed various issues related to assertiveness in teams. We began by highlighting the historical concern for the role of assertiveness in flight safety since the beginnings of aviation. We then tried to dispel a number of commonly held misconceptions regarding assertiveness and its application in the workplace. First, we attempted to discriminate assertiveness from aggressiveness. Second, we argued that assertiveness is both an attitude and a skill. Third, we noted that assertiveness is multidimensional. Fourth, we cited evidence that individuals appear to apply assertiveness in a highly situation-specific manner. Finally, we offered recommendations for measuring and improving team performance-related assertiveness. We noted that assertiveness training, despite its promises, is not a panacea. Continual reinforcement through policies and procedures as well as team-level climate is necessary to ensure transfer of training to the workplace.

ACKNOWLEDGMENT

The opinions expressed herein are those of the authors and do not necessarily represent the opinions of the organizations with which they are affiliated.

REFERENCES

Argyle, M. (1995). Social skills. In N. J. Mackintosh & A. M. Colman (Eds.), *Learning and skills* (pp. 76–103). London: Longman.

Baldwin, T. T., & Ford, J. K. (1988). Transfer of training: A review and directions for future research. *Personnel Psychology, 41,* 63–105.

Brannick, M. T., Prince, A., Prince, C., & Salas, E. (1995). The measurement of team process. *Human Factors 37,* 641–651.

Buck, R. N. (1994). *The pilot's burden—Flight safety and the roots of pilot error.* Ames: Iowa State University Press.

Chan, D. W. (1993). Components of assertiveness: Their relationships with assertive rights and depressed mood among Chinese college students in Hong Kong. *Behavior Research Therapy, 31,* 529–538.

Decker, P. J., & Nathan, B. R. (1985). *Behavior modeling training.* New York: Praeger.

Ford, J. K., & Weissbein, D. A. (1997). Transfer of training: An updated review and analysis. *Performance Improvement Quarterly, 10,* 22–41.

Frisch, M. B., & Higgins, R. L. (1986). Instructional demand effects and the correspondence among role-play, self-report, and naturalistic measures of social skill. *Behavioral Assessment, 8,* 221–236.

Gambrill, E. D., & Richey, C. A. (1975). As assertion inventory for use in assessment and research. *Behavior Therapy, 6,* 550–561.

Gorecki, P. R., Dickson, A. L., Anderson, H. N., & Jones, G. E. (1981). Relationship between contrived in vivo and role-play assertive behavior. *Journal of Clinical Psychology, 37,* 104–107.

Härtel, C. E. J., Smith, K. A., & Prince, C. (1991, April). *Searching mishaps for meaning.* Paper presented at the Fifth International Symposium on Aviation Psychology. Columbus, OH: The Ohio State University.

Helmreich, R. L., Foushee, H. C., Benson, R., & Russini, W. (1986). Cockpit management attitudes: Exploring the attitude–performance linkage. *Aviation, Space, and Environmental Medicine, 57,* 1198–1200.

Henderson, M., & Furnham, A. (1983). Dimensions of assertiveness: Factor analysis of five assertion inventories. *Journal of Behavior Theory and Experimental Psychiatry, 14*, 223–231.

Higgins, R. L., Alonso, R. R., & Pendleton, M. G. (1979). The validity of role-play assessments of assertiveness. *Behavior Therapy, 10*, 655–662.

Jentsch, F. (1997). *Metacognitive training for junior team members: Solving the "copilot's catch-22."* Unpublished doctoral dissertation, University of Central Florida.

Jentsch, F., Bowers, C., Martin, L., Barnett, J., & Prince, C. (1998). Identifying training needs for junior first officers. In R. S. Jensen & L. Rakovan (Eds.), *Proceedings of the Ninth International Symposium on Aviation Psychology* (pp. 1304–1309). Columbus, OH: The Ohio State University.

Kleinmann, M. (1993). Are rating dimensions in assessment centers transparent for participants? Consequences for criterion and construct validity. *Journal of Applied Psychology, 78*, 482–486.

Kolotkin, R. A. (1980). Situation specificity in the assessment of assertion: Considerations for the measurement of training and transfer. *Behavior Therapy, 11*, 651–661.

Komich, J. (1985). An analysis of the dearth of assertiveness by subordinate crew members. In R. S. Jensen & J. Adrion (Eds.), *Proceedings of the Third Symposium on Aviation Psychology* (pp. 431–436). Columbus, OH: The Ohio State University.

Lazarus, A. A. (1973). On assertive behavior: A brief note. *Behavior Therapy, 4*, 697–699.

Lorr, M., & More, W. (1980). Four dimensions of assertiveness. *Multivariate Behavioral Research, 14*, 127–138.

Milanovich, D. M., Driskell, J. M., Stout, R. J., & Salas, E. (1998). Status and cockpit dynamics: A review and empirical study. *Group Dynamics: Theory, Research, and Practice, 2*, 155–167.

Milgram, S. (1974). *Obedience to authority.* New York: Harper & Row.

Nance, J. J. (1984). *Splash of colors: The self-destruction of Braniff International.* New York: Morrow.

National Transportation Safety Board. (1969). *Braniff Airways, Inc., Lockheed L-188, N9707C, near Dawson, Texas, May 3, 1968.* Washington, DC: National Technical Information Service.

National Transportation Safety Board. (1980). *Downeast Airlines, Inc., DeHavilland DHC-6-200, N68DE, Rockland, Maine, May 30, 1979* (Rep. No. AAR-80-05). Washington, DC: National Technical Information Service.

National Transportation Safety Board. (1994). *Safety study: A review of flightcrew-involved, major accidents of U.S. carriers, 1978 through 1990* (Rep. No. NTSB/SS-94/01). Washington, DC: National Technical Information Service.

National Transportation Safety Board. (1997). *Wheels-up landing, Continental Airlines Flight 1943, Douglas DC-9, N10556, Houston, Texas, February 19, 1996* (Rep. No. FTW96FA118). Washington, DC: National Technical Information Service.

Nevid, J. S., & Rathus, S. A. (1979). Factor analysis of the Rathus Assertiveness Schedule with a college population. *Journal of Behaviour Therapy and Experimental Psychiatry, 10*, 21–24.

Northrop, L. M. E., & Edelstein, B. A. (1998). An assertive-behavior competence inventory for older adults. *Journal of Clinical Geropsychology, 4*, 315–331.

Rathus, S. A. (1973). A 30-item schedule for assessing assertive behavior. *Behavior Therapy, 4*, 398–406.

Rouiller, J. Z., & Goldstein, I. L. (1993). The relationship between organizational transfer climate and positive transfer training. *Human Resource Development Quarterly, 4*, 377–390.

Ruben, D. H., & Ruben, M. J. (1989). Why assertiveness training programs fail. *Small Group Behavior, 20*, 367–380.

Salas, E., Fowlkes, J., Stout, R., Milanovich, D., & Prince, C. (1999). Does CRM training improve teamwork skills in the cockpit?: Two evaluation studies. *Human Factors, 41*, 326–343.

Seamster, T. L., Boehm-Davis, D. A., Holt, R. W., & Schultz, K. (1998). *Developing advanced crew resource management (ACRM) training: A training manual.* Washington, DC: Federal Aviation Administration, Office of the Chief Scientific and Technical Advisor for Human Factors.

Sharp, S. (1993). *Improving aircrew performance through crew formation.* Unpublished manuscript.

Smith, K. A. (1994). *Narrowing the gap between performance and potential: The effects of team climate on the transfer of assertiveness training.* Unpublished doctoral dissertation, University of South Florida.

Smith, K. A., & Marion-Landais, C. A. (1993, August). *The effects of constraining motivational aspects of performance in a work-sample test.* Paper presented at the annual meeting of the Academy of Management, Atlanta, GA.

Smith, K. A., Marion-Landais, C. A., & Blume, B. J. (1993, April). *Development of a scale to predict assertiveness in team situations.* Paper presented at the 8th annual conference of the Society for Industrial/Organizational Psychology, San Francisco.

Smith-Jentsch, K. A. (1996, April). *Should rating dimensions in situational exercises be made transparent for participants? Empirical tests of the impact on convergent and predictive validity.* Paper presented at the 11th annual meeting of the Society of Industrial and Organizational Psychology, San Diego, CA.

Smith-Jentsch, K. A., Jentsch, F. G., Payne, S. C., & Salas, E. (1996). Can pretraining experiences explain individual differences in learning? *Journal of Applied Psychology, 81,* 110–116.

Smith-Jentsch, K. A., Johnston, J. H., & Payne, S. C. (1998). Measuring team-related expertise in complex environments. In J. A. Cannon-Bowers & E. Salas (Eds.), *Making decisions under stress: Implications for individual and team training* (pp. 61–87). Washington, DC: American Psychological Association.

Smith-Jentsch, K. A., Milanovich, D. M., & Salas, E. (1999, April). *The impact of pre-briefing style on crewmember assertiveness.* Paper presented at the Tenth International Symposium on Aviation Psychology. Columbus, OH: The Ohio State University.

Smith-Jentsch, K. A., Salas, E., & Baker, D. P. (1996). Training team performance-related assertiveness. *Personnel Psychology, 49,* 909–936.

Smith-Jentsch, K. A., Salas, E., & Brannick, M. (in press). To transfer or not to transfer: Investigating the combined effects of trainee characteristics and team transfer environment. *Journal of Applied Psychology.*

Stebbins, C. A., Kelly, B. R., Tolor, A., & Power, M. A. (1977). Sex differences in assertiveness in college students. *Journal of Psychology, 95,* 309–315.

Tannenbaum, S. I., Smith-Jentsch, K. A., & Behson, S. J. (1998). Training team leaders to facilitate team learning and performance. In J. A. Cannon-Bowers & E. Salas (Eds.), *Making decisions under stress: Implications for individual and team training* (pp. 247–270). Washington, DC: American Psychological Association.

Taylor, G. (1991). *The sky beyond.* New York: Bantam.

Thompson, K. R., & Luthans, F. (1990). Organizational culture: A behavioral perspective. In B. Schneider (Ed.), *Organizational climate and culture* (pp. 319–344). San Francisco: Jossey-Bass.

Torrance, E. P. (1955). Some consequences of power differences on decision making in permanent and temporary three-man groups. In A. P. Hare, E. F. Borgatta, & R. F. Bales (Eds.), *Small groups* (pp. 482–492). New York: Knopf.

Tracey, J. B., Tannenbaum, S. I., & Kavanaugh, M. J. (1995). Applying trained skills on the job: The importance of the work environment. *Journal of Applied Psychology, 80,* 239–252.

Webster, M., & Hysom, S. J. (1998). Creating status characteristics. *American Sociological Review, 63,* 351–378.

Wills, T. A., Baker, E., & Botvin, G. J. (1989). Dimensions of assertiveness: Differential relationships to substance use in early adolescence. *Journal of Consulting and Clinical Psychology, 57,* 473–478.

Wilson, L. K., & Gallois, C. (1985). Perceptions of assertive behavior: Sex combination, role appropriateness, and message type. *Sex Roles, 12,* 125–141.

6

Training Aviation
Communication Skills

Barbara G. Kanki
NASA Ames Research Center

Guy M. Smith
Embry-Riddle Aeronautical University

The greatest problem in communication is the illusion that it has been accomplished.
—George Bernard Shaw

This sage observation points out one of our main failures as communicators. It is ironic that we, as authors of this chapter, are attempting to write a chapter on communications when we ourselves must depend solely on communication (the tool) to discuss communication (the skill). Moreover, because our objective is to address training aviation communication skills, we must first acknowledge that, besides some acronyms and jargon, the essence of aviation communication is not exceedingly unique; it encompasses all of the nuances, subtleties, and complexities of human interaction.

Why are we attempting to distinguish communication (the tool) from communication (the skill)? The distinctions appear to be arbitrary, or at least subjective, and there is no apparent parallel differentiation in most aviation training courses. However, in the daily business of an aviation enterprise communication does have two distinguishable roles.

The most fundamental function of communication (the skill) is to deliver a message from one human being to another. Effective communication skills require these messages to be clearly transmitted and clearly understood. Since the advent of crew resource management (CRM) training, communication has been taught as a skill, typically analyzing the communications of another flight crew. One widely used example is the air traffic control (ATC) tower's conversation with Garuda Indonesia Airways (GIA) Flight 152, just before it crashed amid forest-fire smoke and haze on September 26, 1997:

ATC: GIA 152, turn right heading 046. Report established localizer.
152:    Turn right heading 040, GIA 152, check established.
ATC: Turning right, Sir?
152:    Roger 152.
ATC: Confirm you're making turning left now?
152:    We are turning right now.
ATC: OK. You continue turning left now.
152:    A (pause) confirm turning left? We are starting turning right now.
ATC: OK (pause) OK. ("Tower tape," 1997, p. 2)

Although both the air traffic controller and the Garuda crew were using English, not their native language, the first two lines demonstrate a fairly standard use of communications skills. The air traffic controller clearly transmitted a message and received proper acknowledgment from the crew of GIA 152. The next six lines show the opposite—the worsening breakdown in communication until neither the controller nor the Garuda crew was sure if they were supposed to turn left or right. Effective instruction will transcend the obvious language problems or castigating the crew and ATC for "poor communication procedures"—exercises that produce more righteousness than learning. A skilled training plan will lead the crew to discuss the double meaning of the phrase "turning right now" and how one interpretation of that phrase could transform the entire exchange into credible communications. Another powerful learning opportunity is the "Red Flag" communication in Line 3 (AMA Guides Red Flags, 1998). The query, "Turning right, sir?" is the first indication that something isn't proper. A training session in which pilots learn to recognize the inaccuracies of procedural communications and to be alert for "Red Flag" communications is an illustration of training communications as a skill.

In almost every aspect of aviation work, communication also fulfills a secondary role as an enabler (or tool) that makes it possible to accomplish a piece of work. An example of using communications as a tool is the briefing of exit-row passengers, a training event that must be accomplished on every flight segment. The goal of the training session is not communication; it is technical and analytical performance. Flight attendants must ensure that exit-row passengers can expeditiously perform certain functions, if they are required to do so. These functions can be surprisingly complex, including the following:

4. Operate the emergency exit;
5. Assess whether opening the emergency exit will increase the hazards to which passengers may be exposed;

6. Follow oral directions and hand signals given by crewmembers;

8. Assess the condition of the escape slide, activate the slide, and stabilize the slide after deployment to assist others in getting off the slide;

10. Assess, select, and follow a safe path away from the emergency exit and the aircraft. (Federal Aviation Administration [FAA], 1993, p. 121.585)

These skills, and five more, are vital to personal survival and to the rescue of passengers and crew members. Yet the training given to exit-row passengers rarely includes even the most essential information, such as the hand signals one may have to interpret or the criteria for assessing hazards, condition of the escape slide, and safe paths. Moreover, the training never addresses the built-in inconsistency that the exit-row passenger is both the first to depart the scene by the safe path and the last to depart after assisting others in getting off the slide. Most flight attendants give exit-row passengers a personal reminder that they are sitting in the exit row and should look at the safety information card in the seat back pocket. A few creative flight attendants use selective and cryptic communication (tools) to instruct the required evacuation skills and, more important, to assess whether the passenger might truly perform the required tasks in an emergency.

The examples above illustrate how safe and effective flight operations are critically dependent on communication, the skill, and communication, the enabling tool. Furthermore, to be successful, communication must fit the operational context and needs of the moment. For instance, communications that might be satisfactory during a normal, uneventful flight may be severely inadequate during an emergency situation. In other words, an important aspect of communication skill is to correctly identify the communication need dictated by the operational and interpersonal context. In addition to providing the pertinent information content needed (what is communicated), other aspects, such as when, to whom, and under what conditions, are equally important.

From an operational perspective communication serves as an enabling tool for achieving task objectives. The technical objectives of a flight require communication for conveying information on aircraft configuration, fuel, navigation, and so on. Communication is also a means for coordinating routine operations, such as flight control, checklists, briefings, and maintaining contact with ATC. Communication also supports CRM by providing a means for achieving team situation awareness, solving problems, distributing workload, and many other management functions.

From a training perspective it seems obvious that a certain level of communication skill is required by the flight task. However, to determine what skills need to be trained, we first must determine how communication supports aspects of task proficiency. We must then link communication training with the task objective it supports (e.g., training communications skills in conjunction with training emergency procedures). Furthermore, communication skills training should be integrated consistently throughout the training program. They need to fit with the structure and requirements of the program as well as its resources. In sum, communication skills (like any other skill to be trained) must be incorporated into the curriculum through clearly defined training objectives and into the evaluation process through equally clear performance standards.

But what are the requirements for communication skills, and how are they best trained and evaluated? To answer these questions, the next three sections identify the communication skills needed in flight operations by discussing basic communication principles, the communication process, and how they apply to an aviation setting. The chapter then focuses on how to design communication training by considering stand-alone communication skills as well as those needed for accomplishing technical, procedural, and CRM objectives. Some basic training principles and the advantageous use of various training media are discussed. The chapter concludes with a summary of principles and guidelines for communication training, what communication skills should be taught, and how they are trained most effectively.

COMMUNICATION PRINCIPLES

When we speak of communication as a "stand-alone" skill, we conjure up notes from communications theory: the sender, the receiver, the message, the medium, filters and barriers, feedback, and so on (Griffith, 1999). The skill of human communication is universal; it begins before birth with messages that are so faint that only mothers (or skilled practitioners) can discern; and it never ends, as messages of our ancestors reach us daily through our culture's media. An abbreviated description of communication theory encompasses messages, messengers, and media. The variety of messages (information, facts, emotions, feelings, questions, etc.), the variety of messengers (people, manuals, procedures, instruments, computers, etc.) and the variety of media (speech, text, video, audio, sensory, etc.) defy meaningful generalization.

Even when we narrow the scope of communication to the aviation context, it is too large to establish consistent principles and guidelines. Aviation communication, although somewhat unique, encompasses the entire realm of human communications: verbal, nonverbal, text, datalink, and so on. By limiting the subject to verbal communication, we avoid the lengthy and requisite discussions about electronic versus paper, readability, and writing skills. Even when we limit the discussion to verbal communications, we still encompass a totality of human communication—bartering, befriending, proclaiming, negotiating, apologizing, mediating, manipulating, and so on. So even within aviation we must further narrow the scope of this discussion to one significant and essential arena—real-time, interpersonal, verbal communications in the operational environment. The majority of this discussion will aggregate around flight operations and will be most applicable to pilots, flight attendants, air traffic controllers, dispatchers, and to those who instruct and evaluate them. However, these themes are a segment of universal communication; therefore, most of the principles and guidelines are relevant to both aviation and nonaviation domains.

The general principles described below are not principles people need to know in order to communicate, but they are fundamental to understanding how to define, train, and evaluate communication skills in action.

1. Communication takes place in a social, physical, and operational context. Information contained in these contexts contributes to the communication content, process, and interpretation.

Social Context

A cardinal principle of human communication is that it takes place in a social context. Even when speech is misdirected or misheard, there are an implied speaker and receiver and an intended message sent from a speaker to a receiver. Therefore, communication success is not merely a matter of using the correct words and grammar; it has more to do with whether the receiver appropriately understands the message intended by the speaker. In order to discuss a speaker's communication skill, we must understand the communication needs represented in the social context and assess whether they were met. In casual conversation, small misinterpretations are tolerated, sometimes even intentional (e.g., jokes). However, in high-risk, closely coordinated, complex tasks, small misinterpretations and communication inefficiencies may have dire consequences, as in the GIA 152 crash in September, 1997. The social context also includes characteristics of the communicators (speakers and receivers), such as the roles they perform and their purpose for communicating. One's age, status, task role, authority, experience level, and so on contribute to the meaning and impact of a statement. Settings in which roles are formally specified evoke communication styles that are more structured compared to informal settings. However, allowances are routinely made for switching between formal and informal styles. For instance, even in an informal setting (e.g., in the pilot's lounge) there are times when communication is formal (e.g., a messenger announces an unexpected crew change), and times

In a formal setting when communication can be informal (e.g., casual conversation between pilots after established in cruise). Organizational rules and policies, as well as the practices and norms of the constituents, dictate what is and what is not appropriate. At times, communications may be very restricted and predictable, and at other times they may be free to vary.

Physical Context

The *physical context* for communication refers to aspects of the location of the communication event. For example, some communicators are co-located and speak face to face; other communicators are remotely located and speak via interphone or radio. The physical context may not affect the information content of the message, but it often affects the impact, efficiency, and nature of the communication process. For example, face-to-face communication opens up the possibility of referring to elements in the immediate environment, using facial expression, gaze direction, gestures, and other nonverbal cues. In short, co-located communicators may take advantage of the shared situation, including the actions of themselves and others and objects and changes in the physical environment. In contrast, remote communication can utilize vocal cues only (e.g., intonation, phrasing, timing); shared visual information is largely absent, thus eliminating many sources of detail that can supplement or validate the communication message.

The potentially rich face-to-face context can also present problems. With the availability of numerous information sources, the possibility of conflict and ambiguity arises. For instance, communication and nonverbal cues are at odds with each other when words convey one meaning and nonverbal cues indicate the opposite (e.g., "Yes, Sir" is spoken, but facial expression and intonation indicates non-agreement). Communicators shortcut the process by assuming the expected or most probable response rather than checking for complete consistency across too many information sources. In this regard, remote, indirect communications may tend to be more precise even though there are fewer ways to verify what was conveyed. One way to distinguish the differences between face-to-face and remote communication is to consider the "noise" potential. In addition to physical noise in the communication location or on the communication "line," we can consider other aspects of the physical location that potentially distract the communication process. An effective communicator is able to draw on the strengths of the communication channel and be mindful of the limitations and potential for mishearing or misinterpreting the communication message. The use of feedback and clarifications (discussed in the next section) also come into play.

Operational Context

In addition to the social and physical characteristics of a work setting, there is an operational context that provides a constant reference point to the task at hand. Compared to some work settings, the aviation operational context is relatively structured by flight phase and standard operating procedures that organize task performance. However, when operational anomalies arise, the operational context may change abruptly, requiring accommodation to unexpected events, troubleshooting, and quick response. In either case, the operational environment defines situational factors that are relevant to the shared task, sets standards for acceptable performance, and creates a work environment in which participants become familiar with certain terminology and develop expectations for the way things work and the way to communicate with others.

2. Communication is a tool for accomplishing objectives, but it is a tool that can be used in unlimited ways. Consequently, there is more than one way to communicate the same message or to achieve the same objective.

Because communication is a tool that accomplishes objectives in a great variety of ways, the use of particular words, grammar, dialect, directness, emphasis, and so on, is relatively free to vary. This allows for individual creativity and style, but it also creates the potential for misunderstandings. In a high-risk work environment, some of the arbitrariness is removed in order to decrease human error potential and to support standardization (e.g. procedures for conducting checklists, or communication with ATC). Although this is an effective strategy when specific actions are well prescribed, it is sometimes less effective when the operational environment is abnormal or unpredictable. In some circumstances, the standardized response may be inadequate, and a higher level of communication skill may be needed. Unfortunately, information need and availability are also the most demanding in these situations.

Communication variations create a special problem for instructors and evaluators. On the one hand, some variations are merely stylistic and do not add to or detract from the purpose or intent of the communication. For instance, pilots maintain team situation awareness in a number of different ways: asking direct questions, making casual observations, providing additional information in briefings. Each method may promote team situation awareness equally well. On the other hand, some variations are highly detrimental to performance, and the essential communication elements must be standardized to strict performance criteria. Instructors and evaluators must be able to distinguish a range of situations in order to judge real differences in communication skill from mere differences in style. Understanding the performance consequences of inadequate communication is a first step, but understanding the communication process is important as well.

COMMUNICATION PROCESS

Operational aviation communications are unique in several ways. Most aviation communication is confined to small audiences—often, just two people—thus communications are tailored to a unique audience and circumstance. Furthermore, most aviation communication, within the operational scope, is time-sensitive and expeditious. This is true for both routine and nonroutine conditions with the possible exception of the routine cruise phase of long-haul flights. The clock, primarily the minute hand, is the overseer of most aviation activity, and operational communications reflect this subservience to time. A third unique aspect of aviation communication is that it is con-strained or limited in some way by the physical environment (as discussed previously). Thus, circumstantial factors (noise, static, vibration, weather, etc.) combine with barriers (cockpit doors, workstations, distances, etc.) to limit, restrict, and confound the channels used in everyday communications. Even pilots, sitting side by side, are often inattentive to the subtleties of com-munication because they do not face each other, they may be preoccupied, or their attentions may be diverted by different task concerns. These factors combine to make aviation speech somewhat more concentrated and imper-ative than the communication of other industries. It is noteworthy that the consequences of communicating unsuccessfully can be catastrophic, as sub-stantiated by numerous aircraft accident reports that list "failure to com-municate" or "poor communication" as the probable cause of the accident.

In an industry where it is often so imperative to "get it right," it is remark-able that there are numerous opportunities to "get it wrong." In aviation, as needs—particularly physiological and safety needs (Maslow, 1987)—escalate, messages become more urgent and the requirements for communications more exacting. These circumstances often manifest themselves as emer-gency or abnormal situations that develop without warning and transform the environment into a situation where it is essential to "get it right." In these time-critical and restricted circumstances, the chain of communication must transpire flawlessly.

First, the need must be transmitted, accurately and clearly; second, the need must be properly interpreted; third, a response must be formulated to accommodate the need; finally, the originator must perceive that the need was satisfied. As stated, there is one chain of communication to "get it right" and four opportunities to "get it wrong" (see Fig. 6.1). Fortunately, as in most aspects of aviation, there is built-in redundancy. If the need is not transmit-ted accurately and clearly, the receiver may guess, assume, or interpret the

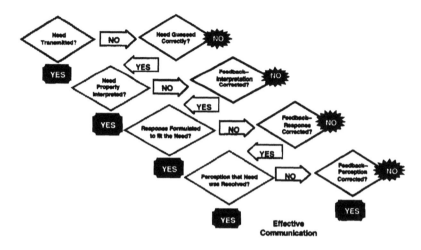

FIG. 6.1. Communication process with feedback opportunities.

need correctly. If the receiver does not understand the need, feedback to the originator may elicit a correction or clarification. If the answer does not fit the need, the originator may request a correction to the response. Finally, if feedback to the originator does not convey that the need was resolved, further communication may ensue to make the needed correction. In each case, feedback is the key to redundancy and correction. For this reason, feedback is considered a vital aspect of aviation communication, particularly in emergency and abnormal circumstances.

APPLYING COMMUNICATION PRINCIPLES AND PROCESS IN FLIGHT OPERATIONS

What Are the Communication Objectives in Flight Operations?

Communication is used for a seemingly endless list of diverse uses; from greeting passengers and requesting coffee, to talking to ATC and conducting briefings, to declaring an emergency and discussing evacuation options. Clearly we must identify the core communication skills that are so critical to flight operations that specific performance objectives must be met through training and evaluation. These skills are the competencies that enable communicators to achieve flight task objectives through the application of communication principles and process. For example, the communication skill that enables effective communication between pilot and ATC is more than the appropriate use of standard phraseology. It also reflects an understanding of the social context that includes the air traffic controller, who has different goals, procedures, and information access than the pilot does. Communications must be suitable to the physical context and accommodate the limitations of remote radio communication. Finally, it must be appropriate for the operational context, that is, adhering to standard operating procedures in the way controllers are contacted and information is exchanged. Skill in communication process means that ATC information needs are properly interpreted, communications are accurately formulated, and ambiguities in communication are resolved through active listening, inquiry, and clarification when needed. Verification that communications have been received is standard practice for guarding against error.

From a training perspective there are at least three kinds of flight task objectives: technical objectives, procedural objectives, and CRM objectives. The operational context provides a framework for considering how communication helps to achieve these objectives and for assessing how the communication skills that enable their accomplishment should be trained. We can further subdivide these objectives according to relevant condition sets (e.g., routine vs. nonroutine, normal vs. abnormal and emergency) as shown in Fig. 6.2.

Routine Operations

In routine operations we observe numerous communications that range from relatively informal transfer of information to highly structured procedural speech. Communication is an indispensable tool for accomplishing the technical, procedural, and CRM objectives of everyday operations.

Technical Objectives. In routine operations, task-related communications enable technical task performance in many ways. Commands are issued and acknowledged. Information related to the normal monitoring of systems, route, traffic, weather, and passengers are routinely communicated to one another in support of the flight task. Plans are made, verified, and carried out. Alternate options are discussed and agreed on in the event of a change from the expected. In addition to communication within the flightdeck, there are routine points of contact with personnel outside the flightdeck, such as cabin crews, passengers, ATC, dispatch, maintenance, and ramp. Normal, everyday operations require communication for coordinating with these groups.

Procedural Objectives. In routine operations, procedural communications are very structured and similar from one flight to another. Although parameters differ with respect to routes, airports, and aircraft, the general communication format—including checklists, briefings, and communications

Using Communication Enabling Skills
In Different Operational Contexts

	Routine Operations	Non-routine/Abnormal/ Emergency Operations
Technical Objectives	Flight Control: performing PF/PNF functions, standard commands, minor workload redistribution Navigation: clarifying and executing the flight plan, contingency plans, programming FMS Systems Management: adjustments, monitoring	Flight Control: time critical diagnosis, coordination with flight crew Navigation: complex planning with ATC and company Response to Hazard: enhanced, verbalized situation awareness Systems Management: responding to uncertain requirements and changes
	Preparedness and combating complacency are key	*Time critical, accurate information transfer, shared situation awareness*
Procedural Objectives	Checklists: normal Briefings: standard ATC: standard	Checklists: normal, non-normal Briefings: standard, and contingency Procedures: identification and performance of non-normal procedures ATC: standard, plus problem resolution
	Must guard against role performance, complacency, assumptions	*Anomalies may require nonstandard techniques for resolution, clarification & verification process critical*
CRM Objectives	Leadership: setting the tone, vigilance, team building Monitoring: normal systems, route, traffic, flight efficiency, weather Task Mgmt: minor adjustments, adherence to established routines and procedures	Leadership: setting the tone, utilizing resources, setting priorities Monitoring: normal, and non-normal, continual; dynamic event consequences, time management Task Mgmt: major adjustments, coordination Problem Solving: time limited inquiry, assertiveness, advocacy of changes Decision Making: time critical, explicit online planning, risk sensitive
	Closely associated with technical objectives, minor task management, resolution of minor problem	*Driven by non-normal events, no tolerance for inaccuracies, reliance on all resources*

FIG. 6.2. Communication, an enabling skill for achieving flight operational objectives. PF = pilot flying; PNF = pilot not flying; FMS = flight management system; ATC = air traffic control; CRM = crew resource management; Mgmt = management.

with ATC—are almost identical because of their conformity to regulations and procedures. It is relatively easy to hear whether pilot–ATC communications are correct and efficient and when checklists, briefings, and callouts are accomplished in compliance with company policies. The timing and content of procedural communications may show some variation, but the actual communication content is almost formulaic. A danger of proceduralized speech is evidenced by pilots who perform procedural communications through rote memory without actually performing the procedure they are articulating. It is critical for instructors and evaluators to judge communication effectiveness according to the procedural objective, not merely the performance of the communication "script." If the procedure is not accomplished to a proficient level, the communication has failed to serve its function as well.

CRM Objectives.   In routine operations, most CRM objectives are closely associated with technical objectives of the normal flight task. For example, team situation awareness is achieved through routine monitoring of systems, route, traffic, weather, and so on. Within the cockpit, communication facilitates crew coordination involving minor changes in workload distribution or the transition between pilot-flying and pilot-not-flying roles. Communication supports coordination with personnel outside the cockpit, such as ramp or cabin crew, in getting passengers on and off the aircraft or waiting out a delay in the gate area. If conditions are running normally, good leadership may require little more than conducting briefings in a conscientious manner and setting the tone for working cooperatively. By definition, problem solving is minimal, although noncritical decision making takes place routinely regarding weather, turbulence, passenger handling, interpretation of procedures, and so on. Although it is always possible for miscommunications to escalate into a nonroutine situation, many elements of CRM are incorporated into standard operating procedures and accomplished through structured communications as discussed previously.

Nonroutine, Abnormal, and Emergency Operations

Nonroutine, abnormal, and emergency operations are characterized by a significant shift in the way communications are used. First, crew members develop a shared recognition of the change or difference in conditions and determine the impact on their flight task. Often, they must carry out their tasks in the face of unfamiliar or difficult conditions and procedures; increased workload; and any number of unexpected, cascading events that create the need for additional deliberation. Communications that facilitate this process become increasingly critical as time pressure mounts and safety of operations becomes an issue.

Technical Objectives.   Achieving technical objectives in nonroutine, abnormal, or emergency conditions relies heavily on the role of communications, but some patterns are now more likely to be observed than others. For example, non-task-related communications may drop out completely as questions and problem solving increase and become more focused. Nonnormal procedures become substituted for normal ones and, depending on flight conditions, the Captain (CA) may fully exercise leadership authority in redistributing workload, soliciting input for problem solving, verbalizing contingency plans, and making command decisions. The First Officer (FO) may engage ATC and dispatch with information requests, provide input to the problem solving, acknowledge and verify plans, and provide continual verbal monitoring on aircraft and environmental conditions.

Procedural Objectives. The structured format of procedural communications provides a safety net against errors. Thus, they are critical in conducting nonroutine, abnormal, or emergency procedures. The communications must be unambiguous so that proper sequences, conditional steps, and appropriate warning information are communicated and understood. As in routine operations, communication enables procedural objectives, but the primary objective is to accomplish the procedure, not simply read a script accurately.

Unfortunately, there are times when flight conditions become so unusual that non-normal procedures and standard ATC phraseology fail to accommodate the situation completely. For instance, the pilot may be forced to use nonstandard terminology when a problem is not yet diagnosed. ATC may also lack "standard techniques" for helping pilots resolve some anomalies. In short, the clarification and verification process of communication may go far beyond routine readback and hearback. Standard techniques cannot anticipate every set of circumstances, and the stand-alone communication skills, such as active listening, advocacy, and assertiveness, come fully into play. All steps in the communication process must be completed quickly and accurately.

CRM Objectives. In nonroutine, abnormal, and emergency conditions, many types of CRM behaviors are triggered. Team situation awareness, monitoring, and planning are raised to a critical level because of the additional requirements imposed on the crew. When operations are routine, crew members can rely on their expectations and SOP reminders. When operations cease to be routine, monitoring and planning must be completely explicit through accurate and timely communications. Leadership actions may involve extensive task, automation, and workload management and resetting of task priorities. Communication skills involved in achieving effective CRM outcomes (e.g., solving problems, making decisions) involve the application of communication principles and following the communication process to effective completion.

Who Are the Communicators? Delineating Four Communication Arenas

Within Flightdeck. Most of the objectives described above are achieved through communication between pilots on the flightdeck. As such, the communication senders and receivers occupy the roles of captain, first or second officer, as well as pilot-flying or pilot-not-flying roles. With these roles are associated differences in responsibilities, skills, and authority. Effective communication relies on a shared understanding of these roles and their respective tasks and how these roles relate to communications in the operational context. This does not lessen the importance of individual styles, but it does acknowledge the standard level of competence and behavior expectations associated with work roles.

Flightdeck and Cabin Crew. The flightdeck and cabin crews have markedly different task objectives and procedures; however, they share many aspects of the physical and operational context and follow established policies and practices that help them to coordinate their tasks. In routine operations, there are times when communication is limited (sterile cockpit), times when it is relatively informal in a face-to-face context (cruise), and times when communications are remotely accomplished via interphone. In nonroutine operations, communications are often less structured, yet precise coordination among crew members is critical for passenger safety and safety of flight. In some circumstances, the cabin crew are in a good position to obtain critical aircraft status information; thus they become a crucial partner in problem-solving and emergency response.

Flightdeck and ATC. As mentioned earlier, communications between the flightdeck and ATC are noted for their high degree of structure and standardization. Only in extremely anomalous conditions do communications lapse into problem-solving discourse. Typically, procedural speech is the rule, in spite of pilot and ATC differences in goals and work roles. Occasionally, ambiguities in responsibilities emerge, particularly when new technologies (e.g., Traffic Alert and Collision Avoidance System [TCAS]) or new procedures (precision monitoring on final approach) create coordination issues. Even in these situations, however, there is usually a protocol for communications to follow.

Unlike flightdeck–cabin communications, which take place in a variety of communication contexts (including both face-to-face and remote channels), flightdeck–ATC communications are always remote via radio transmissions. As such, it is important to remember that elements in the airspace environment may or may not be perceived in exactly the same way, because they do not share the same physical context. For example, when traffic is the topic of communication, multiple information sources (radar vs. TCAS vs. visual) are available to pilots and ATC; thus creating potential confusion in communication. In addition, there are many other well-known sources of pilot–ATC miscommunications, including "stuck mike," callsign confusion, international inconsistencies in terminology and phraseology, and the need for improved readback and hearback practices (Connell, 1994). In spite of crowded airwaves and workload issues, adequate feedback, clarifications, and verifications can serve to maintain the communication process.

Flightdeck and Ground (Maintenance, Dispatch, Ramp). This final communication arena is difficult to characterize, because interactions between flightdeck and ground vary from company to company and airport to airport. Nevertheless, this arena represents the ground support organizations (usually within the company) that coordinate flight operations prior to and following every flight. Typically, the communications are minimal and take place on the ground in a variety of ways. For example, maintenance communicates primarily through the maintenance log and occasionally face to face with pilots in the cockpit. Dispatch is likely to communicate over phone or radio and through "datalinked" messages, whereas ramp may communicate over headset or by pre-established hand signals. Although standard procedures usually exist, the primary communication channel is often supplemented by other cues in the environment (e.g., watching the de-icer as well as communicating over headset). This can help or hinder the communication process depending on whether information is consistent or inconsistent. Again, it is critical that the procedures underlying communications are clearly understood to both senders and receivers.

Where, When, and How Communications Take Place In the Aviation Workplace

We have touched on the fact that the interactions of the primary communicators tend to occur at particular points in flight operations. To describe this further, flight operations can be broken down into flight phase segments, such as the following: (a) preflight and start, (b) taxi and takeoff, (c) climb, (d) cruise, (e) descent, (f) approach, (g) landing, and (h) taxi and shutdown. Each flight phase has its own technical objectives; follows its own standardized procedures; and is characterized by its own sets of routine, nonroutine, and abnormal conditions. Thus, when communications occur within particular phases of flight, they are automatically grounded in an interpretive context. Figure 6.3 shows the flight phases typically associated with the four communication domains. The chart also indicates the type of physical context in which the communication takes place (e.g., face to face, remote). Both flight phase and physical context influence the types of messages communicated, the communicators involved, and the communication channels available. These factors comprise more than a setting for communications, however; they directly bear on how the communication process can be successfully completed.

Flight phases delineate sequential segments in the operational context that dictate the informational needs of the moment. The physical context dictates the limitations as well as the resources on which communicators can draw. As

KANKI AND SMITH

Flight Phase

COM ARENA	Prelight & Start	Taxi & Takeoff	Climb	Cruise	Descent	Approach	Landing	Taxi & Shutdown
within Fltdeck	F-F	F-F	F-F	F-F	F-F	F-F	F-F	F-F
Fltdeck-Cabin	F-F & REM	REM	REM	F-F & REM	F-F & REM	REM	REM	REM & F-F
Fltdeck-ATC		REM (GRD/TWR)	REM (DPT)	REM (CTR)	REM (CTR/APP)	REM (APP)	REM (TWR)	REM (GRD)
Fltdeck-ground	F-F & REM							F-F & REM

[begin...Sterile Cockpit...end]   [begin................Sterile Cockpit................end]

Physical Context of Communication: Face-to-Face (F-F) Remote (REM)

FIG. 6.3. Typical flight phase and communication context for four communication arenas. ATC Ground = (GRD); ATC TWR = (TWR), ATC Departure = (DPT), ATC Center = (CTR), ATC Approach = (APP).

described earlier, face-to-face communication provides the communicators a visually shared environment in which persons, objects, and events can be used as a common reference. Remote communications must rely on vocal cues and explicit verbal clarifications. Because the methods used for seeking and providing feedback incorporate different communication channels, communicators use different methods for completing the communication process. Table 6.1 provides a list of characteristics associated with the face-to-face versus remote physical context of communication. Typically, speakers do not consciously orchestrate these factors; however, it is important for instructors and evaluators to understand that these factors intentionally or unintentionally influence the communication message and process.

DESIGNING COMMUNICATION TRAINING

Because effective communication is essential for achieving success in every aviation undertaking, it is logical that the associated communication should be an integral part of every aviation training curriculum. Yet in reality communication often is relegated to a module in the human factors or CRM syllabus or to a topic for recurrent training or captain upgrade training. When trainers isolate communications as a stand-alone "soft" CRM skill they overlook the limitless potential of communication—the mechanism for achieving

TABLE 6.1
Typical Physical Contexts and Characteristics for Four Communication Arenas

Communication Arena	Within Flightdeck	Flightdeck–Cabin	Flightdeck–ATC	Flightdeck–Ground		
				Mx	Ramp	Dispatch
Face-to-Face						
Verbal speech content	✓	✓				
Nonverbal intonation, inflection	✓	✓		✓		
Silence, often visually explained	✓	✓		✓		
Pauses, often visually explained	✓	✓		✓	✓	
Visual access: objects & events	✓	✓		✓	✓	
Actions observed, task behavior, signals, other communications	✓	✓		✓	✓	
Remote						
Verbal speech content	✓	✓	✓		✓	✓
Nonverbal intonation, etc.	✓	✓	✓		✓	✓
Silence unexplained		✓	✓			✓
Pauses unexplained, Timelag potential		✓	✓			✓
Visual access: none		✓	✓			✓
Actions observed: none, indirect indicators only		✓	✓			✓

Note. ATC = air traffic control; Mx = maintenance.

proficiency in technical, procedural, and CRM skills. The implications of communication as a tool for achieving objectives permeates all aspects of instruction and evaluation, just as it does in actual flight operations.

Prerequisites of Training Program Development

The primary component of an effective airline training program is instructional systems design (ISD). Even if the airline has multimillion dollar simulators, dynamic and resourceful instructors, and attentive and eager learners, the training has little chance of succeeding without a fully developed, comprehensive training program. The airlines with blue-ribbon training programs put extensive resources and personnel at the front end—into training development efforts that include expertise from both line pilots and ISD experts. Particularly in airlines that use off-line pilots as the primary instructors, there is a strong cadre of experienced pilot–instructors who work cooperatively with the instructional development team, producing meaningful training programs, evaluating the effectiveness and sufficiency of the programs and resources, and instructing and standardizing the fleet instructors. The instructional development team is the essential element for integrating communications skills into technical, procedural, and CRM training. If communication skills are uniformly included in approved training programs, there is a good prospect of instilling communications skills across the spectrum of airline operations. If communication skills are not included in technical, procedural, and CRM training at the instructional development stage, implementation is left to individuality and chance, and the probability of achieving integration diminishes considerably.

The second component of an effective airline training program is the instructor cadre. In *Instructor Excellence*, Bob Powers (1992) wrote, "The majority of employer-sponsored training programs were delivered by instructors who had received little or no formal training on how to instruct, who do not consider themselves training professionals, and who will, in fact, leave their instructing assignment within the next two to three years" (p. xiii). If this quote fits the instructor profile of an airline, there is trouble ahead. Instructors are the key pilots of an airline; if they are meticulous, thorough, and standardized, the pilot group will reflect their professionalism. Likewise, if instructors are indifferent, unwatchful, or discontented, the fleet climate will begin to acerbate. Thus, the key to integrating communication into pilot training is to prevail on instructors that communication training is essential to their success as teachers of technical, procedural, and CRM skills. Communication issues should figure prominently in the initial orientation programs or foundation courses for new instructors. Then, communication principles should be reinforced during instructor standardization sessions and instructor recurrent training.

The third component of an effective airline training program is quality learning, which implies motivated learners. For adults, quality learning is not the mere acquisition of knowledge, it is developing knowledge while interacting meaningfully with others in a substantial learning activity (Telfer & Moore, 1997). For adult learning to be substantial, it must have a social aspect that allows the adult to relate the current learning to prior experience. Unquestionably, communication training can integrate and elaborate technical and procedural training, thus providing the motivation to learn material that otherwise may be tedious.

Training Content

Recognizing that verbal communication is an intricate and detailed process, we categorized aviation communications into four objectives: communication as a process, communication to achieve technical objectives, procedural communication, and communication to achieve CRM objectives.

Training Communication: The Process. The skills of effective communication, or lack of them, came to the forefront in the late 1970s as the industry struggled with the cause of the tragic Christmastime crash of United 173 in Portland, Oregon on December 28, 1978 (National Transportation Safety Board [NTSB], 1979). It became apparent to the NTSB investigators that mechanical difficulties did not bring the Boeing 707 down into the forests of Portland, but fuel exhaustion while the crew of three considered a minor malfunction. From the wreckage of 173, United built the Command, Leadership, & Resources (C/L/R) training program, which featured communications as "the exchange of ideas, information and instruction in a clear and timely manner so that the message is received and understood with a minimum of confusion and misunderstanding" (United Airlines Flight Operating Manual, 1995, p. C/L/R-24). After the C/L/R model came many variations and transformations as companies created their own CRM models featuring skills that were tailored to accommodate their own company and cultural identities. Communications skills are eminently featured in each one of them. From a few of these models we extracted a list of some principal communications skills, also known as *behavioral markers* or *crew performance indicators*:

- Advocacy, negotiation, and persistence
- Appropriate language and style
- Assertiveness
- Automation impacts
- Overcoming barriers
- Critiquing (analyzing, scrutinizing, and praising)
- Discernment, insight, and perceptions
- Attitude, disposition (nondefensive, responsive, agreeable)
- Feedback, acknowledgment, and verification
- Inquiry and questioning
- Listening skills
- Recognizing and acknowledging inconsistency or disharmony
- Appropriate terminology, phraseology, and jargon
- Time management
- Establishing tone (trust, openness, candor, and confidence)
- Use of voice characteristics (volume, clarity, enunciation, pauses)

These skills are the subject matter of most airline and military training modules in communications. It is odd to think of them in isolation, because they seem to beg for context. For instance, without context, teaching the skill of verification is only an academic exercise, too theoretical for effective learning. If a pilot learns to verify the flight management system (FMS) mode, it is a technical communication; if a pilot learns to verify checklist responses, it is procedural communication; and if a pilot learns to verify the captain's decision, it is CRM communication. It is reasonably useful to teach communication theory as a CRM awareness course, particularly as a part of new-hire indoctrination; it is imperative to teach it to instructors during instructor qualification courses; and it is necessary to regularly review it for instructors during recurrent training. However, training becomes more meaningful when the communication skills in the above list eventually migrate into technical, procedural, and CRM training.

Training Communications to Achieve Technical Objectives. On February, 19, 1989, Tiger 66, a Flying Tigers Boeing 747 on a short repositioning flight from Singapore, was cleared for a predawn non-directional beacon (NDB) approach into Kuala Lumpur Airport in Malaysia. It was clear from the cockpit communication among the three experienced crewmembers that they were not properly set up or configured for the approach, yet they commenced the procedure, failed to verify the descent clearance, and crashed. The crew did not lack the technical skills to perform an NDB approach or read the approach charts. The problem was clearly stated in two critical communications from the first officer (FO). Neither communication elicited a reasonable response from the captain (CA):

FO: # this #, let's go over and do an ILS.
CA: We can do . . .
FO: I haven't even got the # plate in front of me (NTSB, 1989, p. 6)

The inclusion of communication skills into technical training and checking is not a natural tendency for designers and deliverers of technical training. Instead there is a widespread tendency to "finish up the CRM stuff" to get down to the "real stuff" of flying an airliner. Yet the successful performance of most technical maneuvers is as much dependent on communication as it is on flying skill. When Air Florida Flight 90 (NTSB, 1982) crashed into Washington, DC's, Fourteenth Street bridge, it was in a full stall, yet both pilots, qualified in stall recovery procedures since primary flight training, failed to verbalize anything—"Stall!" "Stall recovery!" "Nose down, full power!" "Power!"—that would have triggered an automatic stall recovery response, that is, applying full power. Instead, the inadequate communication that did occur, "We're (falling) . . . we're going down" (NTSB, 1982, p. 133) was more an acceptance of fate than a last-ditch effort to correct the problem.

    If integrating communication into technical training is not intuitive or natural, the training staff must proceduralize it or it will be excluded. The

Advanced Qualification Program (AQP) was envisioned to merge CRM with technical training and checking (FAA, 1996). This is often implemented by constructing scenarios with technical tasks accompanied by CRM problems. For example, the technical skill of managing an in-flight engine fire can be encumbered by a weather forecast showing that the nearest suitable divert airport is below minimums. The crew is expected to use their technical skills to combat the engine fire and their CRM skills to decide whether to shoot the approach below minimums or to search for a more suitable divert airport. However, continually complicating the technical tasks to include CRM objectives can become too elaborate and sometimes unrealistic and frustrating to the crew. Furthermore, separating technical objectives from CRM objectives is often artificial and therefore difficult for instructors and evaluators to distinguish reliably. Training multiple objectives can be effectively accomplished in a simpler and more integrated way. The expected and observable crew communications that are inherent to effectively managing an in-flight engine fire include proper terminology and phraseology, verification, clarity of intentions, accuracy and completeness, automation impacts, and so on. One or several of these communication enablers can be scripted into the AQP "expected and observed behaviors" without intensifying the task with a CRM add-on.

Designers of technical training should have an extensive list of communication behaviors that they consult as an integral part of designing the technical training. The experts involved with the daily business of teaching and evaluating technical skills should be just as proficient in teaching and evaluating communications processes. Usually when pilots are selected to be instructors, their technical abilities have been demonstrated, and their CRM skills have been proven. However, it is just as remiss to presume that every good CRM practitioner can teach and evaluate communication skills as it is to conclude that every good pilot can teach and evaluate technical flying skills. Instructors need extensive formal training in both the technical and communications concepts involved in the technical skills they are expected to teach and evaluate. The duality of technical and communication skills should be designed into instructor qualification so that everyone who designs technical training, teaches technical skills, or evaluates technical performance is qualified as both a technical and communications expert.

Training Communication to Achieve Procedural Objectives. In Flushing, New York, at 11:21 pm on September 20, 1989, USAir 5050 came to a stop after a rejected takeoff (NTSB, 1990). The Boeing 737-400 was partially submerged in the waters of Bowery Bay off the departure end of Runway 31 at LaGuardia Airport. Less than 2 minutes earlier, the airplane, controlled by the first officer, had drifted to the left of the runway, and a "bang" was heard, followed by a continued rumbling noise:

2320:53.6 CA: Got the steering
2320:54.4 FO: Okay
2320:56.2 FO: Watch it then.
2320:58.1 CA: Let's take it back (NTSB, 1990, p. 90).

The procedural sequence of a rejected takeoff is very familiar to every airline pilot, including the crew of USAir 5050; however, the communication sequence that initiated this particular rejected takeoff is unconventional. In fact, the phrase "Let's take it back" could have several meanings in the off-centerline condition of 5050. The most obvious response is not a rejected takeoff but taking it back to the runway centerline. "Let's take it back" could also mean to take it back to the gate, implying a rejected takeoff. The few seconds that could have been gained by the standard "Abort" command might have been enough to keep the airplane on dry land.

Procedural communication is the norm for commercial pilots, perhaps even more than for highly standardized military aviators. The complexities of crew pairing in the major airlines leaves no safe alternative but to resolutely train standardized procedures and to conscientiously evaluate their compliance. Even the suspicion that flight crews are becoming lax in procedural compliance is a serious concern for an airline.

New-hire pilots are sometimes stunned by the requirement to memorize endless pages of "script" that comprise the standardized procedural speech they are required to use on every flight. Just as daunting is the procedural speech that air traffic controllers must memorize from the "Pilot–Controllers Glossary." The major advantage of procedural speech over spontaneous or unconstrained speech is its ability to convey an elaborate concept with very few words. When the captain declares, "The cockpit is sterile;" the entire crew knows exactly what is expected, for how long, and what types of communication are allowed. Procedural speech is carefully constructed, trained, and rehearsed. Most important, procedural speech fulfills expectations, so that every flight with a different crew member is not an adventure or a disaster. On October 31, 1979, Western 2605, a DC-10-10 flight from Los Angeles, California, to Mexico City, Mexico, touched down on Runway 23 Left, which was closed for repairs. It struck a truck on the runway, killed 73 people, and destroyed the aircraft. Western 2605 was directed to make the ILS approach to Runway 23 Left but to land on Runway 23 Right. There is one procedural word that describes this complex maneuver, but neither the first officer nor the controllers ever spoke it: *sidestep* (NTSB, 1980).

Because procedural communication has such an impact on commercial aviation, it rightly deserves major emphasis in airline training. The training principle is simple: Make them memorize it, make them use it, evaluate their compliance with it. That training rubric works well in elementary school, but adults justifiably balk at unexplained restraints to their expertise and creativity. Nevertheless, procedural speech must be memorized, rehearsed, evaluated, and used on every flight. It is a monumental task for trainers to accomplish unless the trainers themselves are absolutely convinced of the wisdom and the benefits of procedural speech. Trainers must be well versed in the background and logic of procedures and should be able to refute arguments such as "I can think of a better way to say that." To accept procedural speech as the norm, pilots must be convinced that it simplifies their work and adds to their professionalism.

One of the greatest challenges for training procedural speech is to create a healthy balance between blind compliance with procedures and judgment about the situational impacts of a procedure. This is a hotly contested subject among airline training departments, and there is no universal doctrine concerning procedural compliance. The important ingredient is that procedural compliance doctrine should be standardized within an airline and not left to the discretion of individual instructors. Too often, procedures are coordinated with technical training modules but the company's doctrine on procedural compliance is discussed elsewhere or not at all. Essentially, procedural compliance should become the expected and automatic response of every crew member; deviations and noncompliance with procedures should be purposeful, deliberate, and justifiable. Thus, procedures training and evaluation of procedural compliance must minimally include procedural speech, time management skills, and decision-making skills. The policies underlying procedural speech should be explicit and a part of everyone's shared understanding.

Training Communication to Achieve CRM Objectives. Most adults have been exposed to some version of the exercise in which two people sit back to back and try to fit puzzle pieces together while their communications are limited in various ways. If nothing else, participants usually report that the team task was hampered by the deficiencies in two-way communication. Effective communication, as a process, is an essential element of all the other major CRM skills: command and authority, conflict management, crew climate, decision making, leadership and followership, problem definition, resource management, situation awareness, team building, team maintenance, and workload management.

CRM training programs usually contain a module on conflict management, which is frequently divided into management of operational conflicts and management of interpersonal conflicts. Imagine how laborious it would be to resolve an operational conflict when the communications are limited. On February 3, 1988, American 132 was on approach to Nashville, Tennessee. The captain was flying the approach, and the first officer was speaking on the interphone to the flight attendants and to a deadheading pilot about fumes in the cabin and a section of the cabin floorboard that was getting soft. Acting sometimes as a communications filter and sometimes as a barrier, the first officer did not effectively communicate the urgency of the cabin situation to the captain; thus the operational conflict was not resolved until several minutes after the aircraft had landed at Nashville.

Interpersonal conflicts are even more difficult to resolve with limited communications. In the case of Western 2605, which crashed in Mexico City, Mexico, the first officer never properly briefed the captain on the sidestep maneuver required. However, on numerous other recent trips to Mexico City, this first officer had thoroughly briefed his captains on the required maneuver. Why did he fail to do so on that fateful morning? Was it a mind lapse or an intentionally vague and confusing briefing? Why would a first officer want to make his captain look bad by landing on a closed runway? Earlier that day, the first officer had been reprimanded by the airline's chief pilot for failure to adhere to dress codes and standards. Imagine the icy atmosphere that permeated the flight deck when the first officer realized he was flying to Mexico City with the same captain who had initiated the report to the chief pilot.

In a CRM training program, communication loses much of its significance if it is taught only as an isolated, stand-alone CRM skill. As each of the CRM skills or clusters of behavior markers are considered in the training session, there should be a discussion of the elementary communications skills that enable the CRM skill to be accomplished. When developing CRM skill training, it is not difficult for trainers to select a few applicable communication skills from the list above. For example, in decision making it is always desirable for the flight crew to assess the situation if time allows. Thus, decision-making training should include instruction in time management, negotiation, persistence, analyzing, questioning and listening skills, and so on. When learning situation awareness, the flight crews should learn, practice, and be evaluated on advocacy, assertiveness, analyzing, scrutinizing, insight and perceptions, and so on. When learning leadership and followership, crewmembers should become proficient in the communication skills of inquiry and questioning, listening skills, openness, and so on.

An imaginative training developer can blend the role of communication skills with each CRM skill. In fact, the best material for pilot training does not come from the imagination of training developers; it comes from the true stories and incidents related by line pilots. In an appropriately unthreatening environment, pilots are usually eager to tell about their experiences flying the line. Course developers can benefit greatly from encouraging them to provide examples of positive, maybe even heroic experiences and, most of all, from listening to the role that communication played in the outcome of the story. Basing communication training on real-world factual situations makes the training program more authentic and compelling. Moreover, there will always be new stories about flying the line, thus allowing training managers to feel the pulse of their airline and to adjust the training accordingly. Pilots will respond more enthusiastically if they perceive that the CRM training program is credible and in touch with their world.

Training Media

Most airlines use a combination of the following training media: full-flight simulators, part-task trainers, computer-based training (CBT) and desktop computers, traditional classrooms with instructors, videos, modeling, on-the-job training, and operational experience. Whenever the topic of media is broached, we confront the issue of the "haves" versus the "have nots." However, the gap can be considerably narrowed by focusing on the end product: learning.

What type of medium is best suited to deliver the desired learning? Media only facilitates learning; it does not ensure its attainment. As we struggled to determine which training media are most suitable for communications training and evaluation, it became more apparent that the three most important ingredients of learning are the training developer, the trainer, and the trainee. Surprisingly missing from the top three are the training media used to deliver the training. With a well-developed and researched lesson plan, a competent and imaginative instructor, and an intelligent and eager learner, more learning could occur in a 1-hour session using two chairs and a broomstick than in a 4-hour period in a level D simulator. In other words, all training media have strengths that can be exploited and weaknesses that should be avoided, but conscientious implementation (including an understanding of media tradeoffs) is the key to getting the best value from the media.

Without doubt, resources often drive the choice of media; it is not unusual for a full-flight simulator to be used for systems familiarization or flight management system (FMS) orientation when a part-task trainer would suffice. On the other end of the spectrum, instructors may be forced to use aircraft mockups for line-oriented flight training (LOFT) because all other resources are expended. Armed with creativity, a good developer or instructor can design any training medium into a communication-training program. It is better to be creative and resourceful than to expunge the training because the ideal resources are not available. Is it possible that there is "best" equipment for training communication skills under the technical, procedural, and CRM objectives? To be "best," it must be suitable for the learners and for the airline. For adult learners, the primary requirement is that the training should drive learners toward blending their experiences with the new material. For the airline, the training media must be the lowest cost to achieve the objective. Using these as criteria, full-flight simulation is probably the best way to teach the communication aspects of technical objectives, part-task trainers are probably best for teaching procedural communication, and classroom exercises with practice and feedback are probably

best for teaching stand-alone communication skills. Communication aspects
of CRM objectives may be effectively taught in the classroom, but the practice of the skills may require full-flight simulation in order to present a more
realistic operational context. However, every organization must consider its
own resources and training priorities in order to makes its media choices.
Table 6.2 indicates the most suitable and the most limited training media
(based on interaction and cost) for each of the communications objectives—
stand-alone, technical, procedural, and CRM training. Note that all media are
useful, but some may not be appropriate as the only training medium used.
For instance, video is not an ideal standalone medium, because it lacks the
interactive element. On-the-job training and experience do not appear to be
the best stand-alone training media, because they are unstructured and haphazard, often leaving learning to chance. Nevertheless, both categories of
training media may be useful in conjunction with other training methods.

PRINCIPLES AND GUIDELINES

In this chapter we addressed the training of aviation communication skills
first by defining communication skills and then discussing how they can best
be trained. Communication is a large, complicated topic, and we have tried
to capture both universal and aviation-specific perspectives. Universal elements, represented by the communication principles and process, are
essential because they provide the training developer, instructor, and evaluator some basic tools for understanding and assessing all types of communication. This is important because communication serves many objectives
in a great variety of conditions. The aviation-specific view of communication

TABLE 6.2
Suitability of Media to Training Objective

	Communication Enabling Skills			
Training Media	Stand-alone Skill	Technical Objectives	Procedural Objectives	CRM Objectives
Full-flight simulation	xxx	xxx	xxx	xxx
Part task trainer	xxx	xx	xxx	x
CBT/desktop	x	x	xxx	xxx
Classroom/instructor	xxx	x	x	xxx
Modeling	xxx	xxx	x	xxx
Video	xxx	x	x	x
OJT/experience	x	x	x	x

Note. xxx = most suitable, x = most limited. CRM = crew resource management; CBT =
computer-based training; OJT = on-the-job-traiing.

provides the social, physical, and operational contexts in which aviation communications occur and focuses on the technical, procedural, and CRM objectives that communications serve. In a nonaviation domain, a similar strategy could be easily followed by identifying the relevant communication contexts and defining the task objectives.

It is tempting to assume that the training staff needs only domain-specific knowledge, but this provides only a partial understanding of how communication is used. It describes the aviation contexts for communication, and it outlines the basic task objectives that communications serve. However, domain knowledge alone fails to provide the principles and tools for instructing and assessing the actual communication process and level of skill. Communication is a flexible and powerful tool that serves multiple objectives in a variety of ways. It is not a trivial matter to distinguish among differences in communication skill level as opposed to differences in communication style or differences due to aspects of the communication context.

The following principles and guidelines do not provide step-by-step instructions, and they do not represent all possible recommendations. Rather, they are meant to broaden the scope of what is usually defined and trained as communication skills and to suggest new opportunities for integrating communication skills training into the overall flight training program. Following the general structure of this chapter, we will begin with communication and training principles followed by general guidelines for training program development. We will continue with guidelines that specifically pertain to standalone communication skills and, finally, guidelines pertaining to communication skills that enable technical, procedural, and CRM objectives.

Principles: Communication

C1. Communication takes place in a social, physical, and operational context. Information contained in these contexts contributes to the communication content, process, and interpretation.

C1.1 Communication routinely involves different interactants within the flightdeck, flightdeck to cabin, and flightdeck to ATC and other ground support personnel. In each of these communication domains there are different requirements specified by the roles and tasks these people bring to the situation. These differences have implications for training content, choice of training media, and performance standards.

C1.2 Communication requirements differ significantly depending on operational conditions (e.g., routine, nonroutine, abnormal). Training content, instruction, and evaluation must reflect these requirement differences in a realistic way.

C1.3   In physical contexts that permit face-to-face communication, aspects of the shared situation, including the speaker's own actions, objects, and changes in the physical environment may be incorporated. However, the availability of numerous information sources may introduce opportunities for information conflict and ambiguity.

C1.4   In physical contexts that permit only remote communications, a shared context among communicators is not available. Therefore, aspects of the verification and feedback process must be accomplished verbally and explicitly. In some conditions there will be no standardized procedures for accomplishing this.

C2.   Communication is a tool for accomplishing objectives, but it is a tool that can be used in an unlimited number of ways. Consequently, there is more than one way to communicate the same message or to achieve the same objective.

C2.1   Communication can be taught as a stand-alone skill and as enabling skills; that is, as a means by which technical, procedural, and CRM objectives are achieved.

C2.2   The primary objective of the communication (e.g., technical, procedural, CRM) is the primary indicator of communication effectiveness (e.g., the technical goal must be accomplished for the communication to be effective). Even though communication is a necessary tool for achieving objectives, it does not in itself guarantee proficiency.

C2.3   Instructors and evaluators must recognize differences between communication skill level as opposed to communication variations that are merely stylistic (allowable) and when they are due to aspects of the communication context.

Principles: Training

T1.   Designers should incorporate adult learning theory into all aspects of communication training. Emphasizing the communications topics within technical, procedural, and CRM training introduces a social, interactive aspect to the training, allowing adult learners to relate their prior experience to the current topic.

T2.   Adult learning is improved if the learners have an opportunity to relate the subject matter to their relevant experiences. Exercises that encourage learners to say the words and to contemplate their meaning and impact can increase their learning about communication skills. We have learned from LOFT that videotaped interactions and facilitated debriefs promote self-analysis of communication and intensify the associated learning.

T3. Another adult learning principal is to maintain an atmosphere of respect for both learners and facilitators. Never embarrass a person or create an unsafe environment. The media for training should be selected to meet this criteria.

Guidelines I: Training Program Development

G1.1 Protect your investment in professionalism by budgeting abundant resources and superior talent for the company's training development program. Considering that both line pilots and ISD personnel have unique experiences to contribute to the development process, include the best expertise available from both sources in development efforts.

G1.2 Require that the performance and evaluation of communication skills be an objective and activity in each phase of technical, procedural, and CRM training. This requirement should be reflected in your company's training philosophy and proceduralized by the training staff.

G1.3 Teach communication theory as a stand-alone course for instructors. Review it regularly during instructor recurrent training. Make sure the instructors understand how to integrate communication into technical, procedural, and CRM training, and give them concrete parameters for evaluating communication performance across the spectrum of pilot activities.

G1.4 Define performance objectives for communication as an enabling skill. They should incorporate the standard for the primary objective (i.e., technical, procedural, or CRM objective) as well as standards for completing the communication process. It is critical that developers, instructors, and evaluators interpret these objectives consistently.

G1.5 Select equipment or media for training not only for their compatibility with technical, procedural, and CRM objective, but also for their ability to facilitate instruction of the associated communication objectives. The selected media should be interactive, tapping into the experiences of the participants; for the airline, the selection criteria should be the lowest costing media that achieve the learning objective.

G1.6 Balance your use of media to maximize the strengths and minimize the weaknesses of each. Some media (e.g., video, on-the-job training) may not be ideal as a stand-alone medium but may serve well to augment the use of other media.

G1.7 When constructing training scenarios with technical, procedural, or CRM tasks, avoid confounding the exercise with additional, nonrelated communication tasks. Instead, evaluate the expected and observable communications skills that are already intrinsic to the technical, procedural, or CRM duties.

G1.8 Create training examples and scenarios from true stories and incidents related by line pilots. Encourage pilots to describe examples of positive experiences and replicate the details of the situation in the training program to emphasize the pertinent communications. This will support the development of communication training programs that are authentic and relevant to line operations.

G1.9 Integrate communications training with technical, procedural, and CRM training associated with the introduction of new technologies or new procedures. It is a prime opportunity to better understand the implications of the "new" process. This is particularly helpful when the task involves coordination within the flightdeck (e.g., heads-up display) and across domains (e.g., de-icing procedures, TCAS).

Guidelines 2: Stand-Alone Communication Skills

G2.1 When teaching communication concepts as a stand-alone "awareness course," include information about communication principles and the communication process. Its relevance to safety could easily be made by considering aviation incidents and accidents.

G2.2 Recognize the different types of communication failures that can occur within the communication process in order to conduct more effective training and debriefing.

G2.3 Instruct on how aspects of the physical and operational environment create different opportunities and barriers to the communication process. Build awareness of these conditions into training content and provide opportunities to practice "repairing" the process under different conditions.

G2.4 When training communication skills across different communication domains (within the flightdeck, flightdeck to cabin, flightdeck to ATC, etc.), recognize the difference in requirements specified by the social, physical, and operational contexts. These differences have implications for training content and choice of training media.

G2.5   Teach communication theory as a stand-alone course primarily to instructors. Review it regularly during instructor recurrent training. Make sure the instructors understand how to integrate communication into technical, procedural, and CRM training, and give them concrete parameters for evaluating communication performance across the spectrum of pilot activities.

Guidelines 3: Communication Skills Enabling Technical Objectives

G3.1   Integrate communication skills for enabling technical objectives into training wherever those skills are most relevant.

G3.2   Provide formal training in communication aspects of technical skills to technical instructors and evaluators. Expertise in teaching and evaluating communication skills should not be taken for granted; it should be designed into instructor qualification courses and regularly reinforced in instructor recurrent training.

G3.3   Recognize that communication skills enabling technical objectives may appear unremarkable under routine conditions; nevertheless, they are needed to guard against fatigue, complacency, and unwarranted reliance on expectations. Furthermore, communication needs may escalate dramatically if operations become non-normal.

G3.4   Recognize that numerous technical objectives are achieved through communications with personnel outside the flightdeck. The communication process may be achieved differently when the interactants serve different operational roles and when communications take place in different physical contexts (e.g., face to face vs. remote channels).

Guidelines 4: Communications Skills Enabling Procedural Objectives

G4.1   Integrate communication skills for enabling procedural objectives into training wherever those skills are most relevant.

G4.2   Teach pilots the rationale for procedural speech by explaining the company position on procedures and by elaborating on the advantages of procedural speech over spontaneous or unconstrained speech.

G4.3   Ensure that trainers who teach procedures are well versed in the company philosophy concerning procedural compliance and that they are informed about the background, logic, and safety implications of company procedures.

G4.4   Create a healthy balance between blind, unthinking compliance with procedures and communications involved in deliberate decisions to forsake established procedures. Procedural-compliance doctrine should be standardized within an airline and not left to the discretion of individual instructors and evaluators.

G4.5   When evaluating procedural speech, recognize differences between incidents of deviation from procedures and procedural noncompliance. Ensure that such deviations are purposeful, deliberate, and justifiable.

Guidelines 5: Communication Skills
Enabling CRM Objectives

G5.1   Teach communication skills for enabling CRM in awareness, classroom training, practical simulator training, and wherever they are relevant to CRM skills training.

G5.2   Avoid teaching communication as a separate and detached skill; instead, whenever there is a training session on any CRM skill, include some discussion of the communications skills that enable the CRM skill to be accomplished.

G5.3   Create an extensive inventory of communication behaviors as an integral part of the instruction design process, so it is readily available to be incorporated into CRM training.

G5.4   Differentiate CRM objectives that are needed under different operational conditions (e.g., routine, nonroutine, emergency). Communication requirements supporting CRM must be appropriate to the demands and constraints of those conditions.

REFERENCES

American Medical Association Guides Red Flags. (1998). *AMA guides to the evaluation of permanent impairment* (4th ed.). American Board of Independent Medical Examiner. Beverly Farms, MA: OEM Press.

Connell, L. J. (1994, October). Pilot and controller communications: Incidents reported to the NASA Aviation Safety Reporting System (Society of Automobile Engineers Tech. Paper 942137). Paper presented at Aerotech '94, Los Angeles.

Federal Aviation Administration. (1993). *FAR Part 121* (14CFR). Washington, DC: Federal Register.

Federal Aviation Administration. (1996). *Special Federal Aviation Regulation No. 58—Advanced Qualification Program Amendment 121–259*. Washington, DC: Author.

Griffith, E. (1999). *A first look at communication theory* (4th ed.). Boston: McGraw-Hill.

Maslow, A. (1987). *Motivation and personality* (3rd ed.). New York: Addison-Wesley.

National Transportation Safety Board. (1979). *Aircraft accident report: United Airlines, Inc., McDonnell Douglas DC-8-61, N8082U, Portland, Oregon, December 28, 1978* (NTSB-AAR-79-7). Washington, DC: Author.

National Transportation Safety Board. (1980). *Aircraft accident report: Western Airlines, DC-10-10, Mexico City International Airport, Mexico City, Mexico, October 31, 1979* (NTSB/AAR-80/03). Washington, DC: Author.

National Transportation Safety Board. (1982). *Aircraft accident report: Air Florida, Inc., Boeing 737-222, N62AF, collision with 14th Street Bridge, near Washington National Airport, Washington, DC, January 13, 1982* (NTSB-AAR-82-8). Washington, DC: Author.

National Transportation Safety Board. (1989). *Aircraft accident report: Flying Tiger Line, Inc., Boeing 747-200, N807FT, Subang, Malaysia, February 18, 1989* (NTSB-AAR-89-1). Washington, DC: Author.

National Transportation Safety Board. (1990). *Aircraft accident report: USAir, Inc., Boeing 737-400, LaGuardia Airport, Flushing, New York, September 20, 1989* (NTSB/AAR-90/03). Washington, DC: Author.

Powers, B. (1992). Instructor excellence: Mastering the delivery of training. San Francisco, CA: Jossey-Bass.

Tower tape transcript of Garuda Flight 152. (1997, October 5). *Avweb, the Internet's aviation magazine and news service* [On-line]. Available: http://www.avweb.com/other/garuda.html

Telfer, R. A., & Moore, P. J. (1997). Aviation training: Learners, instruction and organization. Brookfield, VT: Avebury Technical.

United Airlines Flight Operating Manual (UAFOM). (1995). p. C/L/R-24. Denver, CO: Jeppesen-Sanderson.

TOOLS FOR RESOURCE MANAGEMENT TRAINING

7

Training Raters to Assess Resource Management Skills

David P. Baker
Casey Mulqueen
American Institutes for Research

R. Key Dismukes
National Aeronautics and Space Administration Ames Research Center

Training work teams in resource management is becoming increasingly important in a wide variety of organizations and industries. For example, within commercial aviation, effective resource management is critical on the flightdeck. In this industry, where the consequences of error are extreme, the vast majority of incidents and accidents have been attributed to breakdowns in the resource management skills of crew members (Helmreich, Foushee, Benson, & Russini, 1986; Helmreich, Weiner, & Kanki, 1993; Prince & Salas, 1993; Ruffell-Smith, 1979). As a result, commercial aviation has been a leading contributor to the development of effective resource management training, or *crew resource management* (CRM), as it is known in the airline industry. This training has continually evolved over the last 20 years from short lecture and discussion-based classes focused on aircrew members' attitudes toward teamwork to a fully integrated performance-based training curriculum known as the *Advanced Qualification Program* (AQP; Birnbach & Longridge, 1993).

An important feature of AQP is the fact that aircrew members must complete a line operation evaluation (LOE) scenario at the end of initial and recurrent training. This type of training event is similar to other resource management training programs in which trainees are provided with practice and feedback or are evaluated at the end of training on their resource management skills (Baker & Salas, 1997; Brannick, Salas, & Prince, 1997). Essentially, an LOE is a job simulation that includes identifiable scenario events that are designed to elicit technical and CRM behaviors by the crew (Air Transport Association, AQP Subcommittee, 1994). A pilot instructor observes a crew's performance during the LOE and rates the crew on specific technical and CRM skills. These ratings are used to determine whether the pilots composing the crew should be certified to fly the line or require additional training.

A critical factor in the evaluation a flight crew's resource management skills is the pilot instructor. Inevitably, the reliability and validity of the process rest on the ability of the instructor to observe relevant crew behavior and make an accurate evaluation. As Birnbach and Longridge (1993) noted, LOEs will be valid only if pilot instructor ratings are accurate and reliable. The most direct and efficient method for ensuring that pilot instructors will be capable of evaluating a crew's resource management is to provide them with rater training. Formal rater training can familiarize pilot instructors with the scenario events, the rating forms, and the CRM skills to be assessed.

The primary purpose of this chapter is to examine the relevant research literature on rater training in order to develop a series of guidelines for training raters to evaluate resource management skills. To do this, we first examine how rater training is conducted in airlines and review the available empirical literature on its effectiveness. As mentioned earlier, the commercial airline industry has been one of the leaders in all aspects of resource management training. Second, we review four strategies that have been traditionally used to train supervisors who conduct performance appraisals. In addition, we present research on each strategy's effectiveness and discuss the relative merits of these approaches. Finally, we combine the results from the literature review and summarize them into a set of guidelines for developing rater training in the future. These guidelines delineate what we believe are the best practices for training raters to assess resource management skills.

RATER TRAINING IN THE AIRLINES

Prior to reviewing specific strategies for training raters to evaluate CRM in the airlines, we first present a brief review of what we believe are the key criteria for determining the effectiveness of rater-training programs. This information is presented here because it is useful for examining the utility of the different rater-training strategies presented in this section and the next.

Criteria for Determining Rater-Training Effectiveness

The primary purpose of rater training is to improve rater accuracy and agreement (Borman, 1979). With respect to accuracy, two forms of accuracy are important. The first is *observation accuracy*, which is defined as the

extent to which raters can correctly identify and record behavioral information. This form of accuracy is critical, because the assessment of resource
management skills typically requires raters to observe and rate a ratee's
(e.g., an aircrew, pilot, team, team member, etc.) performance on a simulated or actual task (Baker & Salas, 1997; Brannick et al., 1997). The second is *rating accuracy*, which is defined as the extent to which raters assign the correct rating (i.e., on a defined rating scale) to the particular level of
performance that was observed. Here, accuracy is critical in order to provide a valid assessment of the ratee's resource management skills (Baker,
Swezey, & Dismukes, 1998; Gaugler & Thornton, 1989).

With respect to rater agreement, trained raters should agree regarding
both their behavioral observations and their performance ratings. Agreement is important because trained raters should be interchangeable; the
evaluation should not be dependent on any particular rater. Typically, agreement is assessed using the within-group agreement index (i.e., r_{wg}; James,
Demaree, & Wolf, 1984), which ranges from a low of .00 (no agreement) to a
high of 1.0 (perfect agreement).

Rater Calibration Training

In the airline industry, rater calibration training has been used to train raters
(i.e., pilot instructors) to be accurate and reliable when evaluating an aircrew's technical and CRM skills. In AQP, this training typically consists of a
1-day workshop in which pilot instructors receive information and discuss
aspects of the LOE grading process. Once the grading process is reviewed,
pilot instructors spend the remainder of the training session rating the
videotaped performance of several crews flying LOE scenarios. This practice phase usually consists of the following sequence of events. First, several videotapes of different crews flying a specific LOE scenario, or one of the
scenario's component events, is shown to the class. For each videotape,
pilot instructor trainees independently rate the crew's technical and CRM
performance using the LOE grade sheet. Next, during a class break, ratings
are analyzed to determine the current level of interrater agreement that
exists among the instructors in the class and the areas where significant rating discrepancies exist. On reconvening the class, the results of these analyses are fed back to the pilot instructor trainees, and rating discrepancies are
discussed to reach consensus. Finally, a videotape of a different crew flying
the same LOE or events is then rated by the trainees to determine the level
of postfeedback agreement (Williams, Holt, & Boehm-Davis, 1997).

To date, only a handful of studies have examined the effectiveness of rater
calibration training. The vast majority of these studies have focused on the
extent to which rater calibration training has improved rater agreement. Thus
far, the results from these limited investigations have been mixed. In some

cases rater calibration training has been found to be effective in enhancing interrater agreement, whereas in other cases no noticeable effects have been found (Williams et al., 1997). For example, Holt and his colleagues (George Mason University, 1996; Williams et al., 1997) have reported levels of interrater agreement that are within what these researchers identified as a "desired range" of .70 to .80, but rater calibration training has not been shown to significantly improve interrater agreement. Essentially, raters in these investigations were calibrated (i.e., demonstrated high levels of agreement) before training, and therefore no training effect was observed.

Given the limited empirical evidence, one can examine the effectiveness of rater calibration training by reviewing the primary features of this training and hypothesizing about its likely effectiveness. A number of researchers have suggested that effective training should include the following phases: *information, demonstration, practice*, and *feedback* (Prince & Salas, 1993; Salas, Dickinson, Converse, & Tannenbaum, 1992; Swezey & Salas, 1992). Regarding the *information* phase, rater calibration training characteristically includes discussion of the technical and CRM skills to be rated and the technical and CRM standards associated with each LOE scenario. Furthermore, this discussion typically includes a review of the grade sheets and the LOE scenario scripts (i.e., the *demonstration* phase). Regarding the *practice* and *feedback* phases, rater calibration training includes significant opportunities for pilot instructors to watch and rate the videotaped performance of aircrews flying LOE scenarios and/or their component events. However, *feedback* is primarily based on the extent to which instructors agree with each other (i.e., rater agreement) and in most cases does not include feedback regarding observation or rating accuracy, which are important for improving validity. Therefore, although rater calibration training possesses many highly desirable features, we hypothesize that this approach to rater training may produce raters who are calibrated within a particular training class (high levels of agreement) but not necessarily across separate rater-training classes. For example, high levels of interrater agreement would be observed within each of two training classes, but low levels of agreement would be observed between these training classes. Data from an ongoing rater-training program at a major U.S. airline has provided some support for this argument. Essentially, high levels of rater agreement were observed within several classes, but several significant differences were found when comparing ratings for the same practice videotapes across these classes (Baker & Mulqueen, 1999).

Summary

A review of the research on rater training in the airlines suggests that rater calibration training is the primary strategy used for training pilot instructors to evaluate CRM skills. This approach possesses many desirable features, including significant opportunities for practice and feedback with the rating task, which is viewed as essential for effective rater training. Of the few empirical investigations that have been reported, the results suggest that rater calibration training likely leads to reasonable levels of rater agreement. However, feedback has primarily focused on the extent to which pilot instructors agree with each other. We hypothesize that this approach to feedback creates the potential problem that different groups of instructors trained in separate classes might all establish different norms for rating CRM skills. Therefore, it is possible that even within a single organization, although pilot instructors all receive the same rater training they may possess many different rating standards at the end of training depending on the particular class in which they were trained. In such a scenario, the evaluation of CRM depends on the particular instructor conducting the evaluation, not the performance of the crew being assessed; instructors are not interchangeable.

In the next section we review a series of specific rater-training strategies that have been presented in the performance appraisal literature to offset the potential problem described in the preceding paragraph. On the basis of research on each strategy's effectiveness and what we believe are the best practices for effective rater training, we derive a set of specific guidelines for training raters to evaluate resource management skills.

RATER TRAINING FOR PERFORMANCE APPRAISAL

The accuracy of performance appraisal has been an ongoing area of study within the field of industrial and organizational (I/O) psychology. Initial attempts to increase the accuracy of supervisors evaluating the performance of their subordinates concentrated on the development of rating scales that were intended to reduce ambiguity in the rating process. Although the advent of behavioral anchored rating scales and behavioral oriented scales have provided for more behaviorally based and meaningful performance appraisal formats, such scale modification has done little to improve the accuracy of supervisory performance ratings (Landy & Farr, 1980). More recent efforts on improving this process, with noticeably greater success, have focused on rater training.

Strategies for Training Raters

Historically, four strategies have been advocated in the performance-appraisal literature for training supervisors (i.e., raters) to be more accurate and reliable when making performance assessments. These are: rater error training (RET), performance dimension training (PDT), behavioral observation training (BOT), and frame-of-reference (FOR) training. With the exception of BOT, each has been widely studied. In this section we review each of these strategies in detail and then present empirical evidence pertaining to their effectiveness. Table 7.1 presents an overview of the specific goals and methods of each training approach.

RET. The purpose of RET is to familiarize raters with common rating errors in the hope that such knowledge will reduce these errors and produce more accurate ratings. This is typically accomplished through a detailed lecture in which rating errors are described and discussed. Errors covered in RET characteristically include *halo error* (i.e., the tendency of a rater's global impression of a ratee to dictate ratings on all the performance dimensions assessed), *leniency error* (i.e., the tendency of a rater to give ratees high performance ratings), *severity error* (i.e., the tendency of a rater to give ratees low performance ratings), and *central tendency error* (i.e., the tendency of a rater to give ratees performance ratings only near the middle of the performance scale). In all cases, the rater making the performance assessment fails to correctly distinguish among different performance levels and typically clusters ratings within one part of the rating scale. Therefore, the desired outcome of RET is ratings that are more normally distributed.

PDT. The purpose of PDT is to familiarize raters with the rating scales that will be used to evaluate different dimensions of performance. This training is usually accomplished by having raters review and discuss the rating scales or involving raters in the actual development of the scales. PDT was

TABLE 7.1
Strategies for Training Raters

Strategy	Goals	Method
Rater error training (RET)	Reduce rating error, produce more normal distributed ratings	Familiarize raters with common rating errors (e.g., halo, leniency)
Performance dimension training (PDT)	Increase rating accuracy by facilitating dimension-relevant evaluations	Familiarize raters with performance dimension and rating scales
Behavioral observation training (BOT)	Increase rating accuracy by focusing on the observation of behavior	Use strategies that focus on observing and recording behavior (e.g., note-taking)
Fame-of-reference training (FOR)	Increase rating accuracy by focusing on the different levels of performance	Provide raters with different standards of performance on dimensions; include rating practice and feedback

developed on the basis of research that suggests that people tend to form evaluative judgments at the time when behavior is observed rather than at a later time when making performance ratings (Woehr & Feldman, 1993). Therefore, PDT trains raters to recognize and use the appropriate performance dimensions and to rely on these dimensions when making observations. As a result, performance ratings should be based on behavior that was observed and organized by job-related dimensions, producing more accurate ratings (DeNisi, Cafferty, & Meglino, 1984; Woehr & Huffcutt, 1994).

BOT. As might be inferred from its title, BOT focuses on the observation of behavior rather than the actual evaluation of behavior. It is based on the premise that there is a critical distinction between the processes involved in observation (e.g., detection, perception, and recall) and those involved in evaluation (e.g., categorizing, integrating, evaluating; Thornton & Zorich, 1980; Woehr & Huffcutt, 1994). Inadequacy of behavioral observation processes is viewed as the primary reason for inaccurate ratings. Typically, BOT encompasses any strategy that focuses on the observation or recording of behavioral events as opposed to information integration and evaluation. A common method used with this approach is training raters to take detailed notes during the evaluation. These notes should reflect only the behaviors that are observed during the evaluation and not the rater's premature evaluations of performance. Actual evaluation occurs after observations are completed, when raters review their notes. Discussion that focuses on recognizing and avoiding systematic errors of observation, contamination from prior information, and overreliance on a single source of information may also be included (Hedge & Kavanagh, 1988).

FOR Training. Finally, the purpose of FOR training, as the name implies, is to train raters to adopt a common frame of reference (Bernardin & Buckley, 1981). Raters are presented with information about the rating task and the relevant performance dimensions to be assessed. Raters are given samples of varying levels of performance on behaviors that represent each dimension, along with practice and feedback in the use of these performance standards (Woehr, 1994). The defining characteristic of FOR training is the nature of practice and feedback provided to rater trainees. Practice usually involves rating a series of training videotapes that present varying levels of ratee performance, and feedback usually compares a rater trainee's practice ratings with a set of previously defined "true scores." True scores are assigned to each practice videotape by task experts who review the tape, independently rate the performance, and discuss their ratings to reach consensus. The resulting ratings are believed to reflect the actual performance level displayed on the videotape and are critical for developing a common performance standards among raters (Baker et al., 1998; Sulsky & Balzer, 1988).

Effectiveness of Rater Training

Woehr and Huffcutt (1994) conducted a review of the literature that included a total of 29 studies pertaining to the four rater-training strategies just outlined. They evaluated each of these methodologies against two measures of rater-training effectiveness: rating accuracy and observational accuracy. The meta-analysis reports results using an effect-size statistic d, which in this context represents the effectiveness of rater training. A positive d value indicates that the rater-training strategy was effective, and a negative d value indicates the opposite (Woehr & Huffcutt, 1994). Cohen (1977) proposed that a d of 0.2 indicates a small effect, a d of 0.5 represents a medium effect, and a d of 0.8 represents a large effect.

Woehr and Huffcutt (1994) found that FOR training was the most effective single strategy for training raters to make accurate ratings, with a mean effect size of 0.83. FOR training had small to moderate effect (d = 0.37) on observational accuracy. BOT was found to have a moderate to large effect on both rating and observational accuracy (ds = 0.77 and 0.49, respectively). However, caution is warranted regarding findings for BOT, because only four data points were available for analysis. BOT and, to an even greater extent, behavioral accuracy as an outcome variable, have not been widely studied. PDT was found to have a weak positive effect on rating accuracy (d = 0.13), but no data were available to test effects on observational accuracy. Finally, RET had a slight positive effect on rating accuracy (d = 0.26) and a slight negative effect on observational accuracy (d = –0.17). In summarizing these results, Woehr and Huffcutt noted that the supremacy of FOR training for increasing rating accuracy is consistent with the purposes of the FOR methodology. Raters who are "trained to evaluate performance using the same standards as 'expert raters' will produce ratings more like the 'expert ratings' (the operationalization of rating accuracy)." (Woehr & Huffcutt, 1994, p. 200).

In addition to examining the effectiveness of each rater-training strategy, Woehr and Huffcutt (1994) reported results from a small number of studies that included combinations of rater-training strategies. Because the number of these studies is small, the results need to be interpreted with caution. However, Woehr and Huffcutt pointed out that the results of the combined training strategies are consistent with the goals of their respective individual training methodologies. For instance, the combination of BOT and PDT resulted in strong effects on both rating and observational accuracy (ds = 1.14 and 1.10, respectively). No combinations of BOT and FOR were reported, but theoretically such a combination should be quite effective for increasing both types of accuracy.

Smith (1986) conducted another comprehensive review of the effectiveness of different rater-training strategies (RET, PDT, and FOR) as well as the effectiveness of different rater-training methods (lecture, group discussion, and practice and feedback). Smith referred to FOR training as *performance standards training*; however, for the sake of clarity we refer to it here as FOR. With respect to the effectiveness of the four rater-training strategies, Smith's findings generally paralleled those found by Woehr and Huffcutt (1994). He found that RET, when used alone, did not lead to increases in accuracy, whereas PDT and the combination of PDT and FOR training did result in higher levels of rating accuracy.

Regarding rater-training methods, Smith (1986) found that the lecture method was ineffective at increasing rating accuracy. Five out of the eight studies he reviewed that used a lecture format failed to improve accuracy with this method. Two of the studies that reported an increase in rating accuracy either combined lecture with practice and feedback or combined lecture with discussion, practice, and feedback. A third study reported an increase in rating accuracy with lecture alone but did not define how rating accuracy was measured.

Smith (1986) reported results indicating that the inclusion of practice and feedback is critical for improving the accuracy of ratings. Increases in accuracy were reported in five out of the six studies he reviewed that included practice and feedback. He reviewed three studies that combined discussion with practice and feedback, and each of these studies reported increases in rating accuracy. Only one study used discussion alone, and it failed to result in an increase in rating accuracy.

Given the importance of practice and feedback, a critical question that was not addressed by Smith's (1986) review is: How much practice and feedback is necessary for optimal rater training? In their formulation of FOR training, Bernardin and Buckley (1981) proposed using three practice ratees representing three levels of performance (i.e., outstanding, average, unsatisfactory). Other researchers have endorsed this methodology, claiming that it is critical to provide behavioral incidents across the spectrum of performance, particularly for the middle ranges, where discrimination among performance levels is most difficult (Stout, Prince, Salas, & Brannick, 1995). However, McIntyre, Smith, and Hassett (1984) found significant increases in accuracy for an FOR-trained group that used only one practice ratee. Although it may be preferable to include a spectrum of behavioral incidents for practice, it may not always be entirely necessary.

Summary

Several conclusions can be drawn from the literature cited above that have direct relevance for training raters to assess resource management skills. First, of the various training strategies reviewed, FOR training produced the greatest increase in rating accuracy, and this strategy would likely be effective at increasing rating accuracy for resource management skills as well. Essentially, raters who are trained to evaluate resource management skills using expert standards should produce ratings more like resource management experts. Second, the literature review indicated that PDT can produce increases in rating accuracy but, on the basis of Woehr and Huffcutt's (1994) more empirically rigorous meta-analytic review, these gains may be slight. Therefore, although we would argue that resource management rater training should not solely be structured after PDT, desirable features of PDT, such as detailed discussion of the resource management skills to be assessed, appear to be valuable and should be included. Third, although BOT represents a relatively new and unstudied methodology, it appears to be effective for increasing both observational and rating accuracy. Because the evaluation of resource management skills typically requires raters to make behavioral observations during job simulations, we recommend that training in observational skills be included in any resource management rater-training program. Fourth, the literature reviewed indicated that combinations of strategies that are individually effective lead to even higher gains in accuracy. On the basis of this finding we recommend that resource management rater training combine the best practices of PDT, BOT, and FOR training. Finally, the literature suggested that the combination of group discussion with significant opportunities for practice and feedback represents the most effective training methods. Although the amount of practice and feedback required has yet to be determined, and may in fact vary by the rating task, we recommend that rater-training programs for evaluating resource management skills include several opportunities to practice the rating task that span a wide range of performance.

GUIDELINES FOR TRAINING RATERS

On the basis of the literature review and the conclusions drawn earlier, we present five guidelines for the structuring and development of resource management rater training. These guidelines combine what we believe are the best practices for rater training and extend these practices to training raters in the evaluation of resource management skills. Each guideline is presented, followed by a discussion of the rater-training best practices on which it was based and a description of how it might be operationalized in the development of training for raters who evaluate resource management skills. We have attempted to structure these guidelines so that they are sufficiently detailed to facilitate the development of rater training yet also sufficiently generic so that they could be applied in a wide variety of organizations (e.g., airlines, military, nuclear power plants, hospitals, etc.).

Guideline 1: Training Raters to Assess Resource Management Skills Should Include a Detailed Discussion of the Skills to Be Evaluated

Guideline 1 is based on the best practices of PDT. Consistent with PDT, we recommend that rater-training programs include a detailed discussion of the resource management skills to be assessed. This discussion should review the definitions of each skill and any corresponding behavioral examples. Furthermore, when possible, raters should be involved in the development of all aspects of the resource management exercise they will evaluate. This includes development of the grade sheets that will be used. The research on PDT suggests that a detailed understanding of the resource management skills to be rated should enhance behavioral observation and rating accuracy by causing raters to use these skills to organize their observations.

Guideline 2: Training Raters to Assess Resource Management Skills Should Include a Review of the Performance Standards Associated With Each Skill

In addition to a discussion of the resource management skills to be assessed, rater training should also include information on the standards of performance associated with each resource management skill to be rated. Guideline 2 is based the best practices of PDT and FOR training. We feel that including information on performance standards is critical in rater training, because these standards provide rater trainees with useful information on what constitutes acceptable and unacceptable performance regarding resource management within an organization. Furthermore, we would advocate that specific behavioral examples be developed for different performance levels on the grading sheets. This information can then serve as a referent when practicing the rating task and, as a result, should enhance rating accuracy.

Guideline 3: Training Raters to Assess Resource Management Skills Should Include Training on Observational Skills

Guideline 3 is based on the best practices of BOT. In the case of assessing resource management, raters need to possess strong observational skills, because the assessment of resource management typically relies heavily on the observation of ratees performing a simulated task (Baker & Salas, 1997; Brannick et al., 1997). For example, in the airlines, pilot instructors observe and evaluate an aircrew's resource management skills during high-fidelity simulations that mirror the task of flying. Therefore, accurate assessments cannot occur without accurate observations. We believe that, to address this rater-training need, observation training should be included when training raters to assess resource management. This training should include discussion about the characteristics of a good observation and opportunities for practice and feedback.

Guideline 4: Training Raters to Assess Resource Management Skills Should Provide Trainees Opportunities to Practice and Receive Feedback on the Rating Task

Relative to the other guidelines, the requirement for practice and feedback with the rating task is most critical. Guideline 4 is based on the best practices associated with PDT and FOR training. Practice should ideally include videotapes that display actual ratee performance, as opposed to scripted performance, because actual performance typically contains more subtle variations that are harder for raters to observe and distinguish. Rater training should ideally consist of practice videos that display a range of performance levels (e.g., excellent performance, good performance, and poor performance) on the resource management skills to be assessed (Bernardin & Buckley, 1981). Although there is some debate in the literature regarding the amount of practice required, we recommend that raters practice and receive feedback on at least six ratees: two representing poor performance, two representing average performance, and two representing outstanding performance.

Guideline 5: Training Raters to Assess Resource Management Skills Should Include Feedback That Compares Trainee Ratings to Standards That Are Established by Resource Management Experts

On the basis of the strong empirical support for FOR training (Woehr & Huffcutt, 1994), Guideline 5 advocates that rater training should include feedback that is based on expert scores that were previously developed for the practice videotapes. In addition, feedback should include information on expert rationales for the scores that are assigned. In the airline industry, these expert scores have been referred to as *gold standards* (Baker et al., 1998).

The research suggests that gold standards are imperative for training raters to assess resource management skills like task experts. Furthermore, by using the same gold standards across rater-training classes, as opposed to norming rater trainees to the group's standard (i.e., as done in rater-calibration training), greater reliability and accuracy should be observed across classes, and common performance standards for resource management should develop within an organization.

SUMMARY AND CONCLUSIONS

The guidelines proposed here for training raters to evaluate resource management skills are based on strategies that have been shown to be effective in other similar performance assessment contexts. These guidelines are meant to provide practitioners with strategies for designing and developing rater training. Although we are confident that these guidelines will be effective at increasing the overall effectiveness of rater training, several questions remain to be answered, including: (a) how much practice and feedback are required for training raters?, (b) to what extent does rater training generalize from one resource management evaluation to the next?, and (c) how often and at what intervals is recurrent rater training required? These are some of the obvious theoretical and practical questions that require study, although other questions of interest will undoubtedly arise as rater-training programs are developed, applied, and tested for the purposes of assessing and evaluating resource management skills.

ACKNOWLEDGMENT

This research was supported by a grant from the National Aeronautics and Space Administration (NASA) Ames Research Center. The views presented in this chapter are those of the authors and should not be construed as an official NASA position, policy, or decision, unless so designated by other official document.

REFERENCES

ATA, AQP Subcommittee. (1994). *Line operational simulation: LOFT scenario design and validation.* Washington, DC: Author.

Baker, D. P., & Mulqueen, C. (1999). Pilot instructor rater training: Guidelines for development. In *Proceedings of the Tenth International Symposium on Aviation Psychology, 1,* 332–337.

Baker, D. P., & Salas, E. (1997). Principles for measuring teamwork: A summary and look toward the future. In M. T. Brannick, E. Salas, & C. Prince (Eds.), *Team performance assessment and measurement* (pp. 331–355). Mahwah, NJ: Lawrence Erlbaum Associates.

Baker, D. P., Swezey, R. W., & Dismukes, R. K. (1998). *A methodology for developing gold standards for rater training video tapes.* Washington, DC: Federal Aviation Administration, Office of the Chief Scientific and Technical Advisor for Human Factors.

144

BAKER, MULQUEEN, DISMUKES

Bernardin, H. J., & Buckley, M. R. (1981). Strategies in rater training. *Academy of Management Review, 6*(2), 205–212.

Birnbach, R. A., & Longridge, T. M. (1993). The regulatory perspective. In E. L. Wiener, B. G. Kanki, & R. L. Helmreich (Eds.), *Cockpit resource management* (pp. 263–281). New York: Academic Press.

Borman, W. C. (1979). Format and training effects on rating accuracy and rater errors. *Journal of Applied Psychology, 64*, 410–421.

Brannick, M. T., Salas, E., & Prince, C. (1997). *Team performance assessment and measurement.* Mahwah, NJ: Lawrence Erlbaum Associates.

Cohen, J. (1977). *Statistical power analysis for the behavioral sciences* (rev. ed.). New York: Academic Press.

DeNisi, A. S., Cafferty, T. P., & Meglino, B. M. (1984). A cognitive view of the performance appraisal process: A model and research propositions. *Organizational Behavior and Human Performance, 33*, 360–396.

Gaugler, B. B., & Thornton, G. C. (1989). Number of assessment center dimensions as a determinant of assessor accuracy. *Journal of Applied Psychology, 74*, 611–618.

George Mason University. (1996). *Developing and evaluating CRM procedures for a regional air carrier, Phase I report.* Washington, DC: Federal Aviation Administration.

Hedge, J. W., & Kavanagh, M. J. (1988). Improving the accuracy of performance evaluations: Comparison of three methods of performance appraiser training. *Journal of Applied Psychology, 73*, 68–73.

Helmreich, R. L., Foushee, H. C., Benson, R., & Russini, W. (1986). Cockpit management attitudes: Exploring the attitude–behavior linkage. *Aviation, Space, and Environmental Medicine, 57*, 1198–1200.

Helmreich, R. L., Wiener, E. L., & Kanki, B. G. (1993). The future of crew resource management in the cockpit and elsewhere. In E. L. Wiener, B. G. Kanki, & R. L. Helmreich (Eds.), *Cockpit resource management* (pp. 479–501). New York: Academic Press.

James, L. R., Demaree, R. G., & Wolf, G. (1984). Estimating within-group interrater reliability with and without response bias. *Journal of Applied Psychology, 69*, 85–98.

Landy, F. J., & Farr, J. L. (1980). Performance rating. *Psychological Bulletin, 87*, 72–107.

McIntyre, R. M., Smith, D. E., & Hassett, C. E. (1984). Accuracy of performance ratings as affected by rater training and perceived purpose of rating. *Journal of Applied Psychology, 69*, 147–156.

Prince, C., & Salas, E. (1993). Training and research for teamwork in the military aircrew. In E. L. Wiener, B. G. Kanki, & R. L. Helmreich (Eds.), *Cockpit resource management* (pp. 337–366). New York: Academic Press.

Ruffell-Smith, H. P. (1979). *A simulator study of the interaction of pilot workload with errors* (NASA Tech. Rep. No. TM-78482). Moffett Field, CA: National Aeronautics and Space Administration, Ames Research Center.

Salas, E., Dickinson, T., Converse, S. A., & Tannenbaum, S. I. (1992). Toward an understanding of team performance and training. In R. W. Swezey & E. Salas (Eds.), *Teams: Their training and performance* (pp. 3–29). Norwood, NJ: Ablex.

Smith, D. E. (1986). Training programs for performance appraisal: A review. *Academy of Management Journal, 11*, 22–40.

Stout, R., Prince, C., Salas, E., & Brannick, M. (1995). Beyond reliability: Using crew resource management (CRM) measures for training. In *Proceedings of the Eighth International Symposium on Aviation Psychology, 1*, 619–624.

Sulsky, L. M., & Balzer, W. K. (1988). Meaning and measurement of performance rating accuracy: Some methodological and theoretical concerns. *Journal of Applied Psychology, 73*, 497–506.

Swezey, R. W., & Salas, E. (1992). *Teams: Their training and performance.* Norwood, NJ: Ablex.

Thornton, G. C., & Zorich, S. (1980). Training to improve observer accuracy. *Journal of Applied Psychology, 65*, 351–354.

Williams, D., Holt, R., & Boehm-Davis, D. (1997). Training for inter-rater reliability: Baselines and benchmarks. *Proceedings of the Ninth International Symposium on Aviation Psychology, 1,* 514–520.

Woehr, D. J. (1994). Understanding frame-of-reference training: The impact of training on the recall of performance information. *Journal of Applied Psychology, 79,* 525–534.

Woehr, D. J., & Feldman, J. M. (1993). Processing objective and question order effects on the causal relation between memory and judgment in performance appraisal: The tip of the iceberg. *Journal of Applied Psychology, 78,* 232–241.

Woehr, D. J., & Huffcutt, A. I. (1994). Rater training for performance appraisal: A quantitative review. *Journal of Occupational and Organizational Psychology, 67,* 189–205.

8

Aviation Crew Resource Management Training With Low-Fidelity Devices

Carolyn Prince
Florian Jentsch
University of Central Florida

If "experience is the best teacher," the question for training developers is "what experience teaches best?" For skill training, there is evidence that practice and feedback are important and should be a part of training (Salas et al., 1999; Thornton & Cleveland, 1990). To enhance this experience, training designers attempt to replicate the job context as faithfully as possible. Thus for training crew resource management (CRM) skills (e.g., decision making, communication, workload management), realistic scenarios combined with the full-mission–high-fidelity simulator is the preferred method. This method, referred to as *line-oriented flight training* (LOFT), is well accepted by crew members and instructors and is believed to provide good training (Lauber & Foushee, 1981). Unfortunately, the cost of using sophisticated flight simulators limits their availability and their use in providing a range of practice for CRM skill training. At present, the most commonly used alternative to the high-fidelity simulator for CRM practice and feedback is role play, but effective role plays are difficult to design, and they may not be well accepted by the people being trained. To address the need for more available CRM practice and feedback opportunities, research has been conducted using low-cost trainers that only partially replicate the cockpit environment. In this chapter we present the possibilities of these devices for use in CRM skill training. We give a rationale for their use, discuss their capabilities and limitations, cover some of the research for CRM training with the devices, and outline training guidance. Our purposes in this chapter are to acquaint readers who are unfamiliar with low-fidelity training devices with some of the research showing their usefulness for CRM and to give guidance to those who are interested in integrating the systems into their training programs. Although our emphasis in this chapter is on the use of low-fidelity devices for CRM training in aviation, we have drawn some of our rationale for the use of these systems from management training; we also acknowledge a similarity between the skills defined for CRM and general teamwork skills. This similarity suggests that the concept of low-fidelity simulation may prove useful for other professional groups who must be trained in similar skills and for whom a simple job simulation would be possible.

DESCRIBING LOW-FIDELITY TRAINERS AND THEIR ROLE IN CRM TRAINING

Before proceeding, it is important to clarify what we are talking about when we say *low-fidelity* and to depict our approach to CRM training. These descriptions are found in the following two sections.

Training Device Description

When people refer to *low-fidelity simulation* they may mean anything that ranges from role-playing exercises in a classroom (the low end of the range) to scenarios conducted in simulators that represent much of what is in the airplane (the high end). Between these two extremes are the training devices we have used. By adding *low cost* (i.e., hardware and software costs do not exceed $10,000) to the description of the training device we exclude simulators that closely replicate the job environment. At the same time, requiring the crew members to carry out many of the flight-related activities (e.g., maintaining attitude and heading, communicating with outside agencies, looking for traffic, completing checklists at appropriate times), rules out simple role-playing exercises.

The devices referred to here are generally nothing more than a desktop computer, a video splitter, separate monitors for the instructor and each crew member, and peripherals such as joysticks or yokes, headsets, recorded voices on tape, and a dummy box for dialing frequencies. The software used with this hardware is an off-the-shelf program that is readily available in computer stores. Clearly, the physical fidelity is low; the simulated cockpit does not resemble the cockpit environment except for the existence of certain dials and gauges. However, pilots have to fly and navigate, following the same approach plates and charts they would use in the airplane. Beyond this, the careful design of scenarios used with the device, including the requirements for flight planning and briefing, raises both the functional and cognitive fidelity of the device well above the level of its physical fidelity. Although the system cannot simulate highly technical equipment or complex technological environments, it presents a more realistic task environment than simple, traditional role play.

Matching the Training Device
to the Type of CRM Training

There are a number of different approaches to CRM and its training. The specific type of CRM that can be enhanced and increased through training with the low-fidelity systems is defined by observable skills, behaviors, or actions that have been linked to effective teamwork in the cockpit (Prince & Salas, 1993, 1999). Although there is evidence that attitudes can be important contributors to CRM, training for attitude change has not been specifically investigated with these systems.

The training for CRM that we address in this chapter is based on the formula that has proved effective in training managerial skills (Thornton & Cleveland, 1990) and CRM skills (Salas et al., 1999), that is: give training participants information about the skills, demonstrate the skill behaviors, let the participants practice the skills in realistic situations, and provide feedback on participants' use of the skills.

RATIONALE FOR THE USE OF LOW-FIDELITY
SYSTEMS FOR CRM TRAINING

Despite the evidence that skill training requires practice, CRM practice is virtually limited to the realistic LOFT scenarios conducted in the high-fidelity simulator, with feedback in the debrief that follows the LOFT scenario. Role play is attempted in some classroom training but, as we have pointed out, it can be difficult to engage crew members, particularly when makeshift props that do not demand task-related skills are used.

When we began working with PC-based simulations, there were three main reasons that suggested they would be useful in CRM skill practice: (a) the underlying theories that suggest that realistic practice and observation (e.g., Bandura, 1977; Knowles, 1970) are important, (b) the research demonstrating the potential of low-fidelity simulation for other kinds of training, and (c) the need for practice solutions in organizations with limited access to high-fidelity simulators. Our research subsequently confirmed the usefulness of the systems for CRM training.

Low-Fidelity Simulation Training
in Management and Aviation

Two diverse approaches to low-fidelity simulation used in training different skills resulted in training success. One is described by Thornton and Cleveland (1990) in training general management skills. The other approach is that used by Gopher and his associates (Gopher, 1990; Gopher, Weil, & Baraket,

1994; Gopher, Weil, Baraket, & Caspi, 1988; Gopher, Weil & Siegel, 1989) to train attention control to pilots. In the former, the simulation is described as a simplified version of the real situation (e.g., although physically similar, it represents only one task in a multitask, dynamic situation); in the latter, training was done with a computer game that is considered to be context relevant to flight.

In describing and evaluating simulation as it is used in management training, Thornton and Cleveland (1990) recommended the use of simulation with simplified tasks as a way to help trainees practice and receive feedback on basic job skill elements. By separating out different tasks beforehand and training on each of these separately, the trainee can subsequently combine the learned skills in a complex situation; if the initial training is in an elaborate simulation, the development of basic skill elements may be overlooked. Although low-fidelity simulation represents only part of a task, Thornton and Cleveland maintained it must have stimulus and response elements that are representative of the job situation (p. 191). They recommended the training itself include some lecture or other form of classroom training, followed by simulations that are sequenced from simple to complex. Sequencing allows trainees to focus on skills in basic tasks and then advance to the practice of multiple skills in the complex, high-fidelity simulation. They cited social learning theory (Bandura, 1977) and adult learning theory (Knowles, 1970) as the basis for their recommendations. Both Bandura and Knowles suggested that interactions with others—including observing others, modeling their behaviors, and thinking about one's own behavior—are crucial in the training situation.

Gopher et al. (1994) pointed out that the traditional physical fidelity approach to training environment design relies on surface properties of the real job. As an alternative, they recommended using a "skill oriented task analysis" to define the "deep structure of the task in terms of its processing, response, and resource demands" (p. 404). For training, Gopher et al. used a computer game that has no physical resemblance to the flying task but has skill requirements similar to those found in the skill-oriented analysis of the task. These requirements include high attention demands and information processing. Both the flying task and the computer game on the tabletop system require players to use manual control, discrete motor responses, visual scanning, memory, decision making, and to manage cognitive resources, all under high time pressure (Gopher, 1990).

In contrast to Thornton and Cleveland (1990), Gopher et al. (1989) originally suggested that a part-task approach for training would not be effective. They explained that teaching skills separately would prevent the building of important connections between different skill elements. However, when Gopher et al. tested two different training strategies with the computer game (one in which participants are exposed to the full task as they practice on individual aspects, the other in which they are trained with a hierarchical part-task method) they found a transfer effect for both. They gave two possible explanations for training transfer. One was that transfer of strategies learned in the game occurred because the game and the task had similar functional and demand characteristics. Their second proposed explanation was that the individual being trained learned to explore alternative response modes and to develop attention strategies in a relevant training context and transferred these to the job environment.

In a review of the use of low-fidelity training devices (computer-based simulations) in aviation, Koonce and Bramble (1998) found that these systems had been used in basic flight training and instrument familiarization with some success. They concluded that an important reason for transfer of training may be that the systems are able to adequately provide training in the underlying cognitive principles of the task rather than the transfer of proprioceptive cues.

This was reinforced by Dennis and Harris (1998), who demonstrated the value of a tabletop system for training some technical flight skills. They found that increased fidelity of the control interface of the system did not affect transfer to the performance of the participants in the flight situation, suggesting that the system was not training psychomotor skills used in the task. Rather, Dennis and Harris observed that the system practice appeared to be helping participants become familiar with the pace of the task, with what effective task performance looks like, and with some of the control interactions.

The examples just cited show that low-fidelity simulation is effective for training a variety of skills. The relevant job task demands made by the system, not those demands related to physical manipulation, appear to be the key to their usefulness. These results encouraged CRM specific research on the systems.

CRM Behaviors and Low-Fidelity Simulation

To illustrate the type of events that are included in scenarios, we offer the following excerpt, which is from a scenario used in a CRM research study (Prince, Brannick, Prince, & Salas, 1997):

> As a crew of two pilots prepare to land their small transport plane they are informed that one of their two passengers appears to be experiencing a heart attack. They have already been cleared to land, but the airfield is in a rural area, with no nearby hospital and the availability of a physician is unknown. Because they are about 15 minutes flying time from a large metropolitan area, they are faced with making a decision: either they will land at the destination airfield or they will divert to the location closer to qualified medical assistance. If they choose to divert, they must communicate with air traffic control (ATC),

then plan and prepare for their landing at an unfamiliar airport. In either case, they need to make arrangements for an ambulance to meet the plane.

Crew members were observed as they communicated their awareness of the relative merits of the two choices: to land immediately or to divert. Their decision-making skills were shown in the manner in which they sought information about the patient's condition, medical facilities, and timing and distance to the alternate airport and the way they used that information to make a decision. Planning skills were evident as they prepared to change routes and fly to an airport that had not been part of their original flight plan. An instructor could observe communication skills as the crew members sought information from, and gave information to, the passengers, air traffic controllers, and one another. Finally, the crew's handling of the multitasking (e.g., maintaining assigned altitude and heading, completing checklists, completing required communications, and following instructions while dealing with the decisions required by the circumstances) could also be observed.

This scenario did not occur in a high-fidelity simulator but was designed and carried out as part of CRM training research using a tabletop trainer (Prince et al., 1997). In this experiment, two scenarios were used. They were as realistic as possible, given the constraints of the system (e.g., no aircraft system problems were included, because the systems differed from those on the aircraft that the pilots flew). Fifty crews (two pilots in each) flew the scenarios on the training device. Videotapes of each flight allowed multiple observers to view the crews' CRM behaviors. Observers found that crew members acted as they would be expected to act in the cockpit, and their actions were found to be related to the separately made ratings of technical performance. In addition, cockpit experience was found to be relevant to performance in the scenario. More experienced crew members (those with more flight hours) performed more CRM actions than did those with less experience. Furthermore, observed crew actions in early parts of the scenario were related logically to subsequent performance in the scenario (Jentsch, Bowers, Sellin-Wolters, & Salas, 1995). For example, behaviors related to situation awareness (e.g., assessing the current situation, keeping one another aware of the flight status) that were observed in the 5-minute period before the beginning of each scenario decision event were related to the speed with which the crew members made the decision. This training situation, conducted on a low-fidelity device, illustrates the ability of these devices to support scenarios that can be complex and challenging enough to elicit a range of CRM skills.

Experiments have been conducted in which specific manipulations of workload, cockpit climate, decision type, and requirements for situation awareness (Jentsch & Bowers, 1998) have all been tested. In one experiment (Jentsch et al., 1995), when changes in workload were made they were related to differences in performance and in subjective assessments of workload made by those flying the scenarios. Smith (1994) varied cockpit climate in a training experiment, in which it was shown that a captain could develop a "tone" in the device's cockpit sufficient to either discourage or encourage copilots to use newly trained assertive behaviors. In another experiment (Prince et al., 1997), pilots discriminated between a scenario with a demanding and unusual decision and one with a familiar type of decision, rating the former more difficult. Measures of crew situation awareness actions were found to be related to scenario difficulty; that is, in the less difficult scenario, both experienced and inexperienced pilots demonstrated similar situation awareness actions. In the more difficult scenario, experienced pilots demonstrated more situation awareness actions than less experienced pilots.

To summarize, experiences with low-fidelity training devices have consistently shown that they are flexible enough to support a wide variety of scenario manipulations. Design choices have been shown to be related to different levels of CRM skill behaviors.

Pilot Acceptance of the Training Device

Pilots who were part of four different research studies with a low-fidelity training system were asked questions on the relevance of the experience on the training device to the cockpit environment (Jentsch & Bowers, 1998). Their responses to questions such as: "Was the experience useful for CRM?" and "Would you recommend participation in the scenario on the training device?" were positive. Across experiments, spontaneous comments about the system also were positive, with many pilots volunteering to take part in more research scenarios.

Training Evidence for CRM

The most important concern with these devices (as with any training device) is whether they offer any actual training value. To determine this, several experiments were conducted (Brannick et al., 2000; Jentsch & Bowers, 1998) to look at skill transfer both to other scenarios and to the high-fidelity simulator.

Scenario-to-Scenario Training. In the first training investigation, crews of two flew a scenario on the low-fidelity device, took a short break, briefed a second flight, and flew the second scenario (Prince et al., 1997). No feedback from an observer was given, and few crews made any attempt to debrief their flight. Measurement of crew behaviors showed that practice through participation in the first scenario alone did not result in improved performance in the second scenario.

In a follow-up experiment (Prince et al., 1997), all crew members were given informal feedback during the scenario, and half of the participants were formally debriefed at the end of the scenario to reinforce the feedback they had received. Unlike in the first experiment, all participants increased their use of CRM behaviors in the second scenario.

Training-Device-to-Simulator Training Transfer. To determine if training on a low-fidelity device could transfer to performance in the high-fidelity simulator, crews were trained with the low-fidelity device and tested in the high-fidelity simulator (Brannick et al., 2000). Fifty crews of volunteer pilots received classroom CRM training that included lecture, videotape observation, and discussion. Half of the crews then flew a scenario on the training device and were given feedback on their use of the CRM skills that were part of the classroom training. This was followed by a LOFT in the high-fidelity simulator. The other half of the pilots "played" several management development games, similar to those included in a number of CRM-awareness phase programs. After the games, they flew the same LOFT scenario in the high-fidelity simulator as the first group of pilots had flown. Analysis of instructor ratings of the crews showed that those who flew the low-fidelity simulation showed improved performance in the LOFT for events that were the same as those in the scenario flown on the low-fidelity system. In addition, their CRM performance in these events was superior to the performance of those who had merely played games before the LOFT.

These studies demonstrate training potential for the low-fidelity systems that is directly relevant to CRM. It is also clear from the research that it is not the device that is responsible for training but the device combined with the selection of specific training goals, scenarios designed around those goals, and feedback on CRM skills.

SUGGESTIONS AND CAUTIONS
FOR DEVICE USE

Experience with the low fidelity device suggests that it may be used in a variety of ways in CRM training. The strengths and limitations of the systems, however, must be recognized to get the best results from the training.

Fitting Low-Fidelity Training
Into Existing CRM Training

These training devices can be placed easily in a number of locations (e.g., the classroom) and can be adapted to various uses. They can introduce simple scenarios for building skills that can be transferred to a more complex situation (as recommended by Thornton & Cleveland, 1990) while providing much of the multitasking of the cockpit (as espoused by Gopher, 1990). There are three ways that low-fidelity systems can be used in existing CRM training: to supplement training, to enhance training, and to replace less effective elements in training programs.

Supplement Training. Because skill practice is important for skill training (Salas et al., 1999), CRM training programs need to offer crew members opportunities to practice CRM skills in a controlled situation where crew members can receive feedback. It is in bridging the gap of needed practice between receiving information about CRM on the one side, and the high-fidelity simulator experience on the other side, that low-fidelity systems offer much promise. Thus, to add to existing training, practice with the table-top system should be scheduled for crew members after they have received classroom information and demonstration of CRM skills and before they fly a CRM scenario in the simulator. If these two existing training sessions (i.e., classroom and simulator) are not scheduled close in time, practice with the low-fidelity system could be used just before the simulator training to refresh the skills introduced in the classroom. If the simulator session closely follows the classroom training, the system can be used to demonstrate the classroom training and begin the active practice.

Enhance Current Training. The training device can be included in existing CRM classroom training without increasing training time or changing scheduling. Instead of using a role-play exercise with props that are unrelated to the skills of flying, the role play can include the low-fidelity system. The demands that the system makes at even the simplest level (e.g., maintaining altitude and heading) increase the similarity of the skills required in the role-play situation to those necessary for the cockpit and will help engage crew members in the situations. This makes them more likely to accept the role play, to participate, and to recognize the relevance of the training. Having the system in the classroom gives crew members the opportunity not only to fly a scenario but also to watch another crew fly it and provide them with feedback. This, according to adult learning theory (Knowles, 1970) and to behavior modeling (Bandura, 1977), is important for training. The training value of the high-fidelity simulator is then increased by eliminating the more basic CRM skill training from its LOFT scenarios and allowing the instructors to concentrate on the complex combinations of actions that can be called for only in a system that closely mimics the plane.

Replace Current Training Elements. If existing programs are examined for training value, it may be found that part of the training can be substituted with active training. In programs where management games are still a mainstay, replacing games with scenarios on low-fidelity devices could make the program both more effective and more acceptable. Although some information and demonstration are necessary for CRM training, there are some programs for which this part of the training may be too basic, or too long, to be very effective. CRM represents high-level skills; that is, every aviator enters CRM training knowing how to talk and listen, how to make a plan and follow it, and how to come to a decision in the cockpit. Rather than basic information on these skills, they need more opportunities to learn to recognize when these skills are needed and appropriate. Imagine that a crew has been diverted. One crew member, after descending to a new altitude, is not certain where they are. The other recognizes the area and feels comfortable about knowing their location. Neither pilot is likely to need to learn how to communicate; instead, and more important, they may need to learn to recognize when the other crew member needs information and when it is appropriate and even necessary to ask a question in the cockpit.

Limitations of Training With Low-Fidelity Devices

First, despite the positive results from research with the use of the low-fidelity training systems (Brannick et al., 2000; Jentsch & Bowers, 1998; Prince et al., 1997) and the theoretical basis for their use (Gopher, 1990; Thornton & Cleveland, 1990), there is no evidence to suggest that these systems can, or should, replace high-fidelity simulators in high-level training or evaluation. The CRM training transfer research on the tabletop systems has been exclusively for transfer to a simulator and not to the aircraft. For this reason, the recommendations arising from that research can be applied only to the system's use as an adjunct to simulator training.

Second, training designers must take into account the restrictions in situation complexity with low-fidelity simulations. A comparison of training in the simulator and in the training device illustrates both a use and a limitation of the device. Crew members at all levels of experience need to be ready to respond to a situation in which they lose communications with their controlling agency. They must recognize their situation, troubleshoot to find the cause, follow the lost communications procedure that is appropriate for their situation, and attempt to establish communications, always maintaining safe flight. To do this, the crew members need to communicate with one another about the problem and their plan for handling the problem, maintain awareness of their position, manage the workload that is created by the abnormal situation (when added to their normal workload), and may need to make some decisions based on the unexpected situation. This event can be practiced in a tabletop trainer as well as a high-fidelity simulator. With both, the communication, division of workload (i.e., one person flying, one person troubleshooting), assumption of appropriate cockpit roles, and the recognition of what the situation requires—all relevant to CRM—can be done. However, in the high-fidelity simulator the situation can be made more demanding by using instruments and conditions not found in the low-fidelity systems. For example, the problem can be made more complex by combining it with a systems failure that increases workload, increases the range of resource management skills required, and changes the possible courses of action that could be taken by the crew.

USING THE SYSTEM: GUIDELINES AND POTENTIALS

In this section we discuss how the system may be put to use within a training program. Each crucial consideration has been condensed into a guideline.

Making the System Work

It is a common misconception that aircraft, high-fidelity simulators, or devices such as PC-based simulations become effective training tools solely through their purchase. Instead, they are merely pieces of equipment that become effective training tools through their structured, and sometimes imaginative, use in a training program. All too often, however, organizations seem to overlook this point; examples abound where organizations budget substantial amounts of money to purchase simulators of all kinds, without allotting the funds needed to incorporate the systems into training programs. In these cases, organizations may find that well-meaning acquisitions of simulators turn into expenses that do not create the expected return on investment. At worst, simulators sit idle; at best, the crews scheduled to use them engage in a free play where skills may be practiced—but not necessarily the skills that are the actual goal of the simulator practice session.

Unfortunately, tabletop systems such as the ones discussed here are not immune to this problem. Their low cost may even suggest to the organization that additional expenses associated with making the device effective (e.g., development of training goals, methods, and strategies) are not warranted. The following guidelines for their use should therefore be followed to create effective and efficient training opportunities. We have based these recommendations in part on an earlier article by our group (i.e., Prince, Oser, Salas, & Woodruff, 1993), with additional guidelines from the subsequent experiences we have had with these devices.

Scenarios. By far the most important step in turning a piece of training equipment into an effective training tool is the use of appropriate scenarios. Scenarios situate the practice opportunity within the context of the real world in which the learned skills will be used. They provide the background needed for skill development and allow for the integration of various skills into a meaningful whole. In this way they create a direct link between the skills needed by the pilot, their introduction in the classroom, and their eventual use on the job. This suggests the first guideline for the use of low-fidelity devices in CRM skill training: *1. Use structured and scripted scenarios that link training objectives to realistic flight tasks.* In this sense, scenarios for low-fidelity simulations are not different from those for high-fidelity simulators. Consequently, many of the guidelines for scenario development in high-fidelity simulators also apply in the low-fidelity environment.

One structured way to design training scenarios for low-fidelity simulations is the methodology developed for the line-oriented simulations, used in the Federal Aviation Administration's (FAA's) Advanced Qualification Program. The value of this methodology is so clear that it suggests our second guideline: *2. Develop and script scenarios in accordance with accepted guidelines for the design of line-oriented simulations.* According to this methodology, scenario design begins with the identification of required skills from a task analysis and then links these skills to tasks that are part of the job description. The combination of skills and tasks should then be augmented by the selection of realistic conditions (e.g., weather, geography) from the job environment that support the need to use the identified skills. Several practice opportunities should then be strung together in the form of a realistic flight, as is used in LOFT and line-oriented evaluation. Route data should match published charts and procedures, and communications to the crew from air traffic control or other units (e.g., ground personnel, dispatch) should be scripted to ensure realism and reduce instructor workload. Finally, instructors (or the crew members themselves) should be provided with a listing and description of the targeted skills, when they should be practiced, and other important details. This allows the instructor to use the briefing, simulation, and debriefing cycle to focus the crew's attention on those skills, observe their performance, and give meaningful feedback that links back to the training goals for the simulator session (Naval Air Warfare Center Training Systems Division, 1998; Smith-Jentsch, Zeisig, Acton, & McPherson, 1998). By following this methodology, the developer can ensure that he or she will adhere to our third guideline: *3. Give instructors and crew members a framework for the training session through the use of a structured briefing, simulation, and debriefing cycle.* As our research has shown, debriefing the scenario actions in relation to the CRM skills is an essential part of the training. We feel that it is so important that we have developed a separate guideline: *4. Give participants structured opportunities for feedback and critique.*

At present, the method of scenario generation depends on the organization, including its commitment to CRM training, the expertise of its personnel in developing scenarios, and its collection of information on existing scenarios. As CRM training and knowledge about the appropriateness of scenarios for training specific skills evolve, the process of scenario design can become easier and, at the same time, result in more effective training scenarios. This continual growth in the field is the basis for another guideline: 5. *Keep informed of changes in scenario development; in particular, be on the lookout for methods and tools that will improve scenario design.* There is a promising tool for the creation of valid and reliable simulator scenarios for all sorts of simulations currently under development. This tool, the *rapidly reconfigurable line-oriented evaluation* (RRLOE) generator (Bowers, Jentsch, Baker, Prince, & Salas, 1997) will ease the process of scenario design and will help the designer to rely on chance factors to a lesser extent. The RRLOE allows users to create, catalogue, and combine event sets that are linked to training objectives into a full scenario. The generator selects event sets on the basis of the training goals, checks the continuity from event set to event set, and makes sure that the generated scenarios are neither too difficult nor too easy. It then identifies a fitting route for the scenario, generates a flight plan and all the associated paperwork, and prints a full scenario script for the instructor. Together, the material generated by the RRLOE software allows an instructor to brief, administer, grade, and debrief an entire simulation scenario while taking the limitations of the simulator into consideration. This last item—the simulator's limitations—is another important concern in designing training with a low-fidelity device.

Training on the device will be hampered if the scenario makes crew members take important actions (especially actions that require handling equipment as part of emergency procedures) that are dissimilar from those they would take in their own aircraft, or if they are prevented from taking any action they normally would take, because the equipment is missing from the training device. For example, adding a simulated engine fire to the scenario without the equipment that would normally be available and necessary to the procedure used in the aircraft would not be recommended. Our guidance here is the sixth guideline: 6. *Select scenarios for use in low-fidelity simulations that are supported by the simulation software and hardware.*

Aircraft Model. The second major consideration in making low-fidelity simulations work for training has to do with the selection of an adequate aircraft model for the simulation. This selection should be based on (a) the scenarios required to achieve the training objectives and (b) the available software and hardware for the planned expenditure. Both of these considerations interact with and affect each other.

Traditional wisdom tells us that pilots should practice in a simulator that approximates as closely as possible the aircraft they are currently flying or training to fly. For CRM training, however, we would make the case that there are other (and sometimes more important) considerations that should determine the selection of an aircraft model for simulation. The first of these is whether all crew members who are attending a CRM training course are flying or planning to fly the same aircraft. In some organizations (e.g., corporate aviation, the military, or organizations such as the U.S. Forest Service) this may not be the case. Here, it might be better to use a somewhat simplified and generic aircraft model for the training rather than trying to train everyone on the intricacies of specific aircraft models.

Similarly, some crew members may need to practice CRM skills at a point in their training where they are not yet familiar with the very complex aircraft simulated in a high-fidelity simulator. Attending to the unfamiliar aspects of the equipment prevents them from focusing attention adequately on the CRM training objectives (see Jentsch, 1997). When the most difficult part of the simulation becomes figuring out how to start the engines in a full-mission simulator, crew members cannot be expected to appropriately crystallize the expected CRM experience. Also, a somewhat simplified aircraft model, such as those found in the typical PC-based flight simulators, can help experienced crew members focus on the CRM aspects of training, because it reduces the temptation to use specific tricks that work in their aircraft alone in solving a CRM problem or to focus on technical or systems issues rather than CRM issues, during the debrief. This is a problem that has been observed at times in air carrier CRM training. Our specific guidance on this is: *7. Select software and hardware on the basis of training objectives and factors such as previous experience of the crew members to be trained and the homogeneity of the group to be trained; keep in mind that a simplified and generic aircraft model may be better for CRM training than a very specific and detailed aircraft model.*

Although a generic aircraft model can be advantageous for CRM training in low-fidelity simulations, the simulations cannot be completely devoid of fidelity or realism. For example, the simulated aircraft should move at approximately the same speed as the category and class of aircraft in which the crew members are to use the CRM skills. Maintaining the temporal realism of simulations is one characteristic that we have found to be important for pilot acceptance, and therefore we feel that it merits a separate guideline: *8. Select a device such that the temporal aspects of the flight (speeds, time to climb) are similar to those in the aircraft flown by the crew members in the CRM training.* The simulation should provide pilots with an adequate and realistic navigation database. It is in this area that PC-based flight programs are exceptional in that they allow crews to conduct a flight using realistic charts and flight plans, thereby immensely increasing the usefulness and acceptability of the devices for CRM training. The accurate replication of navigation gives the crew members the chance to be involved in an activity that is an integral part of their job. If there is a choice in device capabilities, follow Guideline *9.: Choose accurate replication of navigation systems over software that can simulate system fail-*

ures. Although "simple is better" in many respects, there are certain requirements that should not be ignored: *10. At a minimum, select an aircraft model that will allow simulation of all basic flight maneuvers.*

Training Accessories, or Props. In addition to valid and structured scenarios and the fitting selection of an appropriate simulation system, the third most important step an organization can take to make effective use of low-fidelity simulations is to select props that adequately support the training objectives. The use of real aeronautical charts, real checklists (as far as the simulation allows) and other realistic flight paperwork is not difficult to realize but has, in our experience, a tremendous impact on crew members' acceptance of low-fidelity simulations. Using headsets and intercoms is beneficial for the instructor as well in that it facilitates communication. This constitutes another important guideline: *11. Use realistic props such as real charts and air traffic control communications that replicate the communications carried on when flying.* Finally, the controls used by the pilots should have some semblance of realism. Although for CRM training it is not necessary to provide the very realistic (and consequently, rather expensive) controls used in FAA-approved PC-based flight training devices, a good computer game yoke or joystick with throttle capability should be used. This saves pilots from having to control the aircraft with a mouse or keyboard. Thus, Guideline *12.: Use a simple yoke or joystick for most simulations when CRM training is the objective.* The selection of all controls should be determined by the criterion of relearning–unlearning. For example, because most PC-based aircraft models can be flown easily without rudder inputs, providing rudder pedals, in our experience, is largely unnecessary. In fact, we believe it pays only if the maneuvers required for the scenario cannot be conducted adequately without a separate rudder control system (such as aerobatics maneuvers, helicopter flight, and certain engine-out conditions in a multiengine airplane). Thus, the selection of the right control system interacts with scenario design and the selection of simulated aircraft.

Embedding Low-Fidelity Simulation. In our view, the final important step in making low-fidelity simulations work for CRM training and practice is to embed these simulations within the greater context of the aviation training system. As we indicated earlier, low-fidelity simulation is not a panacea; it should not be used to reduce training in high-fidelity simulations. Instead, it should be used to supplement that practice. Specifically, low-fidelity simulation can be used effectively to bridge the gap between the introduction of CRM principles in the classroom and their eventual practice in the full-mission simulator. It is in this way that organizations such as the U.S. Navy and air carriers such as Aer Lingus have used these devices.

Embedding the simulations in the training system means more than merely specifying a time at which these simulations should be used. It also means that the list of training objectives for such simulations must flow directly from the training task analysis to the classroom training and from there to the simulations of varying fidelity. Thus, supporting paperwork for a low-fidelity simulation should tie the simulation's objectives directly to the overall training goals and to the related lectures, classroom exercises, and high-fidelity simulator sessions. It should be possible for the instructor, as well as the crew members who are being trained, to treat a session in the low-fidelity simulator in the same way as any other part of the training curriculum. In this sense, low fidelity simulations fit easily into a structured system of aviation training, such as the FAA's Advanced Qualification Program training. Our final guideline is derived from this argument: *13. Use low-fidelity simulation as a practice tool to link classroom instruction on CRM principles and the practice of those principles in the high-fidelity simulator or the actual aircraft.*

THE FUTURE OF LOW-FIDELITY TRAINING DEVICES FOR CRM IN AVIATION AND BEYOND

In the few years since the research into low-fidelity systems for CRM training began, the speed and capacity of personal computers have increased, and software programs have been upgraded. This means that, in replicating a complex system, a higher level of physical fidelity is possible. This helps increase the range of scenarios that can be developed for training use, while improved reliability makes the systems easier for instructors to use. More important, and potentially more valuable for CRM than developments in the physical systems, are advances in training research and knowledge about CRM. As we gain understanding about important elements for skill learning, such as amount of practice needed and when and how to provide feedback, these systems will increase in training value.

We have addressed some issues in this chapter about low-fidelity devices for CRM training in aviation. On the basis of our investigations, we feel confident in stating:

1. Active practice for job-relevant interaction skills must be done in a situation where the skills and knowledge relevant to job tasks are required.
2. There is a poor fit between the needs for active practice and the availability of high-fidelity simulators.
3. Low-fidelity systems offer valid training solutions that increase the usefulness of high-level, complex training environments.

Pilots must navigate, complete routine procedures, keep in contact with controllers, abide by regulations for speed and altitudes, and handle unusual or unexpected situations, all while physically controlling the aircraft. Other jobs that require interactions with others, completing multiple tasks, and seeking information and making decisions, while never diverting attention from a crucial primary task (e.g., medical emergency teams, firefighters, process analysts for computer chip manufacture) are candidates for similar low-fidelity training innovations. Aviation has been fortunate in that it can take advantage of commercially developed, readily available software programs for flying. Few (if any) other professions are represented in the same way, but many have PC-based procedure trainers that could be used for training interaction skills. There is much in the low-fidelity training research literature that is applicable. Management and teacher education have both used part-task simulations to train people in necessary interaction skills (Allen & Ryan, 1969; Thornton & Cleveland, 1990). Aviation research and guidelines can be used to update this already-existing training. In medicine, high-fidelity simulators are being developed to train physicians (Gaba, Howard, & Small, 1995). Principles from the aviation training devices can be used to develop low-fidelity simulations that are easier to implement and less expensive than high-fidelity simulators.

As a final word, it is important to remember that at any level of training it is not the low-fidelity device by itself that is recommended for training. The device's worth can be realized only through its use with the elements that have been defined as important to active practice training: (a) a scenario design based on training objectives, (b) feedback, (c) job/task skills that are defined with observable behaviors, (d) trained instructors, and (e) integration into a planned training program. More than 25 years ago Smode, Hall, and Meyer (1966) determined that it is how a training system is used, not its level of fidelity, that determines its value for training. Despite amazing advances in technology, this statement remains true.

REFERENCES

Allen, D., & Ryan, D. (1969). *Microteaching.* Reading, MA: Addison Wesley.

Bandura, A. (1977). *Social learning theory.* Englewood Cliffs, NJ: Prentice Hall.

Bowers, C., Jentsch, F., Baker, D., Prince, C., & Salas, E. (1997). Rapidly reconfigurable line-oriented evaluation (RRLOE). In *Proceedings of the Human Factors and Ergonomics Society 41st annual meeting* (pp. 912–915). Santa Monica, CA: Human Factors and Ergonomics Society.

Brannick, M., Prince, C., & Salas, E. (2000). *Reliability of instructor evaluations of crew performance: Good news and not so good news.* Manuscript submitted for review.

Dennis, K., & Harris, D. (1998). Computer-based simulation as an adjunct to *ab initio* flight training. *International Journal of Aviation Psychology, 8,* 261–276.

Gaba, D., Howard, S., & Small, S. (1995). Situation awareness in anesthesiology. *Human Factors, 37,* 20–31.

Gopher, D. (1990, July). *The skill of attention control: Acquisition and execution of attention strategies.* Paper presented at the 14th conference of the International Society for the Study of Attention and Performance, Ann Arbor, MI.

Gopher, D., Weil, M., & Baraket, T. (1994). Transfer of skill from a computer game trainer to flight. *Human Factors, 36*, 387-405.

Gopher, D., Weil, M., Baraket, T., & Caspi, S. (1988). Fidelity of task structure as a guiding principle in the development of skill trainers based upon complex computer games. In *Proceedings of the 32nd annual meeting of the Human Factors Society* (pp. 1266-1270). Anaheim, CA: Human Factors and Ergonomics Society.

Gopher, D., Weil, M., & Siegel, D. (1989). Practice under changing priorities: An approach to training of complex skills. *Acta Psychologica, 71*, 147-179.

Jentsch, F. (1997). *Metacognitive training for junior team members: Solving the copilot's catch-22.* Unpublished doctoral dissertation, University of Central Florida.

Jentsch, F., & Bowers, C. (1998). Evidence for the validity of PC-based simulations in studying aircrew coordination. *International Journal of Aviation Psychology, 8*, 243-260.

Jentsch, F., Bowers, C., Sellin-Wolters, S., & Salas, E. (1995). Crew coordination behaviors as predictors of problem detection and decision making times. In *Proceedings of the Human Factors and Ergonomics Society 39th annual meeting* (pp. 1350-1354). Santa Monica, CA: Human Factors and Ergonomics Society.

Knowles, M . (1970). *The modern practice of adult education: Androgogy versus pedagogy.* New York: Association Press.

Koonce, J., & Bramble, W. (1998). Personal computer-based flight training devices. *International Journal of Aviation Psychology, 8*, 277-292.

Lauber, J., & Foushee, C. (1981). *Guidelines for line oriented flight training* (NASA Conference Pub. No. 2184). Moffett Field, CA: National Aeronautics and Space Administration.

Naval Air Warfare Center Training Systems Division. (1998). *The handbook of team dimensional training.* Orlando, FL: Author.

Prince, A., Brannick, M., Prince, C., & Salas, E. (1997). The measurement of team process behaviors in the cockpit: Lessons learned. In M. Brannick, E. Salas, & C. Prince (Eds.), *Team performance assessment and measurement: Theory, methods and applications* (pp. 289-310). Mahwah, NJ: Lawrence Erlbaum Associates.

Prince, C., Oser, R., Salas, E., Woodruff, W. (1993). Increasing hits, reducing misses in CRM LOS scenarios: Guidelines for simulator scenario development. *International Journal of Aviation Psychology, 3*(1), 67-82.

Prince, C., & Salas, E. (1993). Training and research for teamwork in the military aircrew. In E. L. Wiener, B. G. Kanki, & R. L. Helmreich (Eds.), *Cockpit resource management* (pp. 337-366). San Diego, CA: Academic Press.

Prince, C., & Salas, E. (1999). Team processes and their training in aviation. In D. Garland, J. Wise, & V. Hopkin (Eds.), *Handbook of aviation human factors* (pp. 183-214). Mahwah, NJ: Lawrence Erlbaum Associates.

Salas, E., Prince, C., Bowers, C., Stout, R., Oser, R., & Cannon-Bowers, J. (1999). A methodology to enhance crew resource management training. *Human Factors, 41*, 161-172.

Smith, K. A. (1994). *Narrowing the gap between performance and potential: The effects of team climate on the transfer of assertiveness training.* Unpublished doctoral dissertation, University of South Florida.

Smith-Jentsch, K., Zeisig, R., Acton, B., & McPherson, J. (1998). Team dimensional training: A strategy for guided team self-correction. In E. Salas & J. Cannon-Bowers (Eds.), *Making decisions under stress* (pp. 271-297). Washington, DC: American Psychological Association.

Smode, A., Hall, E., & Meyer, D. (1966). *An assessment of research relevant to pilot training* (Vol. 11, Tech. Rep. AMRL-TR-66-196). Dayton, OH: Wright Patterson Air Force Base, Aerospace Medical Research Laboratory.

Thornton, G., & Cleveland, J. (1990). Developing managerial talent through simulation. *American Psychologist, 43*, 190-199.

9

Evaluating Resource Management Training

Robert W. Holt
Deborah A. Boehm-Davis
J. Matthew Beaubien
George Mason University

Resource management is a critical component of effective group perform-ance in a number of domains, including aviation, medicine, and the military. Although a fair amount of research has been devoted to the development of resource management training programs (Helmreich & Foushee, 1993; Wiener, Kanki, & Helmreich, 1993), much less effort has been devoted to their evaluation. The evaluation of a training program is important for a number of reasons, not the least of which is to determine whether the organization's investment pays off in terms of demonstrable performance improvements. In many domains, however, changes in performance are difficult to measure because of uncontrollable factors that exist within the larger organizational context.

This chapter outlines the steps required to evaluate the effectiveness of a resource management training program and highlights the various practi-cal and theoretical issues that arise during this process. We first cover gen-eral requirements for defining, implementing, and evaluating resource man-agement training; we then illustrate these principles by applying them to crew resource management (CRM) in the aviation domain. Although this chapter emphasizes the application of statistical techniques and research design, page constraints limit our discussion of these topics. Interested read-ers should refer to more comprehensive expositions provided by Campbell and Stanley (1963), Cook and Campbell (1979), Howell (1997), and Pedhazur and Pedhazur Schmelkin (1991).

DEVELOPING A RESOURCE MANAGEMENT
EVALUATION PLAN

Principles of Evaluation

Although several different approaches are available for evaluating the effectiveness of a resource management training program (Goldstein, 1993; Guttentag & Struening, 1975; Joint Committee on Standards for Education Evaluation, 1994), certain principles remain invariant. For example, the primary objective is to determine if resource management training makes a noticeable difference in the critical variables and the magnitude of the training program's effect.

At a minimum, training should make a difference that is noticeable. A noticeable difference has two components. First, it is a difference that statistical methods determine to be nonchance (above a background level of noise due to measurement error). Second, the difference should have practical value to the organization. If it is determined that training made a noticeable difference, then the size of the training effect should be estimated so that cost–benefit analyses can be performed. If multiple training programs have been developed, the data can be used to assess the relative effectiveness of the different training methods.

When evaluating a training program, it is critical to collect measures of performance at the appropriate time. If performance is evaluated before training has "sunk in," a training effect may not be observed (Kraiger, Ford, & Salas, 1993). Similarly, evaluating performance after too long an interval may contaminate the data with uncontrolled intervening events that obscure the effects of training. Thus, the right time interval must be chosen to accurately evaluate training effects. If a valid theory of performance is available for the training domain, the time interval can be based on this theory. Alternatively, if the right time interval is unknown, evaluation should be repeated over a reasonable period of time to check for both immediate and delayed training effects.

It is also important to determine where to look for changes in performance. For example, Kirkpatrick (1976) suggested that training effectiveness can be manifest at several levels of analysis: the individual, the team or crew, and the organization. A majority of the resource management literature focuses exclusively on the *immediate* transfer of trained material at the individual or team level. This is not unreasonable, as individual/team behaviors are most directly under the control of trainees. However, *long-term* aggregate performance data—for example, at the department or organizational level—are also important to the organization. Unfortunately, performance data, unlike measures of behavior, are frequently beyond the control of the individuals or team (J. P. Campbell, 1990). For example, an aircrew may manage a crisis situation perfectly, yet factors beyond their control, such as faulty equipment, can nonetheless lead to a disaster.

Therefore, it is important to remember that any measured effect can have multiple causes. Although training is one such cause, a systematic evaluation should attempt to rule out as many plausible alternatives as possible so that the training program can be isolated as the primary source of the observed differences (D. T. Campbell & Stanley, 1963; Cook & Campbell, 1979). For these reasons, the effects of resource management training should be evaluated in a systematic, step-by-step fashion. This requires developing a list of *targeted* changes in knowledge, skills, and attitudes that are expected to occur after training and investigating them in a *systematic* fashion (Kraiger et al., 1993).

Selecting an Evaluation Design

All evaluations of resource management training programs rely on some form of comparison. The simplest type of evaluation involves comparing groups with different degrees or methods of training with one another using the same set of criteria. Still another form of comparison is to compare pretraining performance with posttraining performance. The various approaches differ on the type of comparison emphasized and on the amount of control over confounds. Regardless of which approach is chosen, the goal is to develop the fairest and least confounded comparison of the effects of training (D. T. Campbell & Stanley, 1963; Cook & Campbell, 1976).

Evaluation approaches range along a continuum from extremely controlled studies modeled on the experimental method, with people randomly assigned to separate trained and untrained groups, to relatively uncontrolled field studies, in which the training is done *in vivo* and the effects are measured in the natural environment. There are costs and benefits associated with each approach. The sections that follow highlight these trade-offs.

Experimental Designs

A traditional experimental design requires the ability to randomly assign people to trained and untrained groups (or different levels or types of training). The trained group is then compared with the untrained group on each possible criterion variable. This is the most precise evaluation of training effectiveness but probably the least practical, as most organizations will usually want to train all job incumbents. One variation of the traditional experiment is a "waiting list" control group. In this variation, all people ultimately receive the training, but the people designated to receive the training first versus last are *randomly* determined. In the window of time where the first groups are trained and the last groups are not, the effects of training can be measured on what are essentially randomly assigned trained and untrained groups.

Quasi-Experimental Designs

If naturally occurring groups are available but cannot be randomly assigned, a quasi-experiment can be performed in which one group is trained and the other group is not. As in a traditional experiment, both groups are evaluated for the effects of resource management training. The major disadvantage of this design, however, is that the groups may not be equivalent on other relevant variables, such as ability, experience, and so forth.

In commercial aviation the naturally occurring groups are fleets, and fleets typically differ in the average age and experience of the pilots therein. Therefore, the possibility exists that some characteristics unique to the trained group may interact with the training to produce the measured effects. This makes it essential to measure possible confounds (e.g., differences in experience across fleets) and assess their effects on the evaluation criteria; for example, by means of hierarchical regression or analysis of covariance.

Pre–Post Evaluation

If everyone must receive training at the same time, evaluation studies can be set up to address changes in the trainees' performance. For example, after resource management training, trainees should have higher levels of efficiency and productivity while simultaneously having lower levels of errors and other undesirable outcomes. This is one of the easiest methods of evaluation and, at the very minimum, some form of pre–post design should be used to evaluate the effects of training.

Unfortunately, this evaluation method is also one of the weakest because it is subject to many confounds such as contextual effects and maturation. If these confounds occur between the pretraining and posttraining measurements, they can artificially cause the observed changes in performance. Therefore, the pretraining measurement should be taken early enough to be unaffected by the knowledge of, or anticipation of, training but not so early that the baseline performance could change a great deal prior to training.

Time Series Evaluation

A time series design extends the time where performance is measured before and after training. Extending these intervals of measurement provides the advantage of being able to rule out potential confounds, such as a general increase in performance that is due to maturation. However, it does so at the cost of additional measurements.

When making multiple measurements, one must consider the effect of the measurement process itself. For example, if supervisors are simply rating subordinates on naturally observed performance, the subordinates may not react negatively to the measurement process (although effects of multiple assessments being made by the supervisor should still be considered). However, if subordinates are put in a specially designed evaluation scenario for each measurement, then practice effects, learning of test-relevant knowledge and skills, and changes in performance motivation may very well occur. Any situation in which the subordinate is strongly aware of the testing and evaluation process is open to these types of confounds.

DEVELOPING MEASURES OF RESOURCE MANAGEMENT PERFORMANCE

Once the research design is selected, measures must be developed that address the constructs of interest. Accurate performance assessment requires several critical steps: defining the construct, developing appropriate measurement instruments, and objectively confirming the psychometric properties of these instruments. Although these steps are highly interdependent, they will be discussed separately for clarity of exposition.

To accurately assess resource management, it must first be defined. Without a specific operational definition, appropriate assessments of resource management cannot be developed. If the construct is multidimensional, then *multiple* measures need to be developed. Once developed, these measures must be evaluated for acceptable levels of statistical sensitivity, reliability, and validity. After the quality of these measures has been established, they may be confidently used to obtain a full and accurate evaluation of the resource management training program.

Defining Resource Management

Resource management is potentially difficult to define and measure, because it is complex, multidimensional, and process oriented (see Lauber, 1984, for more information). Given this complexity, it may be necessary to create several operational definitions, one for each of the various resource management dimensions and processes.

An operational definition is a precise, focused definition that is used for a specific purpose such as evaluation. Any operational definition must be complete and specific enough to clearly imply appropriate measurement strategies and techniques. As a general rule, the operational definition should specify (a) the core knowledge, skills, and behaviors required for effective resource management and (b) relevant situational factors that describe the context in which performance is measured.

Developing Appropriate Measures

Effective resource management should affect both task- and relationship-oriented aspects of performance (Borman & Motowidlo, 1993). The performance changes may occur at the individual, team, and organizational levels (Kirkpatrick, 1976), but their form and interrelationships may vary across levels (Chan, 1998). However, practical limitations generally require the evaluation process to focus on a selected *subset* of these possible effects. At a minimum, this subset should include process and outcome measures at both the individual and team levels.

Performance changes at the individual, team, or organization level may occur at different time frames. Kirkpatrick (1976) proposed a model that suggests that training results are manifested at multiple stages: initial reactions to the training program, changes in knowledge and behavior during the training, transfer of trained behaviors to the workplace, and changes in organizational effectiveness. According to Kirkpatrick's model, each stage is a necessary but insufficient precursor to the following stages. Despite previous criticisms and caveats (Alliger & Janak, 1989; Alliger, Tannenbaum, Bennett, Traver, & Shotland, 1997; Goodman, Lerch, & Mukhopadhyay, 1994), this model provides a useful framework for considering the effect of training interventions at different organizational levels. In general, individual effects of training appear first, followed by team changes and then organizational changes. Therefore, the appropriate time to measure individual, team, and organizational effects may vary considerably.

Unfortunately, measurement of resource management performance is more difficult than measuring the output of an assembly-line worker. When a physical object is being produced, productivity can be indexed in terms of output quality or quantity. In contrast, the evaluation of process variables such as resource management requires evaluating the interaction of a team within a complex system. For example, evaluating resource management in aviation crews depends on the interaction between the Captain and the First Officer as well as their interactions with flight attendants, air traffic control personnel, and the physical aircraft systems (see chap. 10). Therefore, it may be desirable to measure each construct by means of a number of different methods. The principle of converging operations (Campbell & Fiske, 1959) suggests that if different measurement methods provide the same result, confidence in that result is increased. Whenever possible, multiple measures of resource management performance should be included in the evaluation process.

At the same time, it is also wise to measure more than just one possible effect of training (Kraiger et al., 1993). For example, relevant outcomes of resource management training at the individual level may include attitudes toward resource management, declarative knowledge of resource management procedures, and changes in trainees' knowledge structures (Schvaneveldt, 1990). Relevant outcomes for the team may include increased task and social cohesion, a perception of more collective competence, an increase in shared knowledge structures, and better group interaction processes. In aviation, relevant crew outcomes would include improved communication, coordination, situation awareness, planning, and decision making. Relevant outcomes at the organizational level would depend on the domain. In the aviation domain, relevant outcomes may include on-time performance, decreased fuel consumption, fewer incidents, and decreased insurance costs.

Measuring Performance

Because of the complexity of resource management performance, the evaluation method of choice is often a performance *rating* regarding the quality of resource management behavior at the individual or crew level. This evaluation should be guided by appropriate tools and materials that help the evaluator make an accurate assessment. For example, carefully designed rater training programs and evaluation worksheets developed according to the principles of human factors have the potential to reduce the rater's cognitive workload. This may simplify the evaluation process and give more reliable results (see chap. 10). Other materials required for evaluation will depend on the evaluation context.

The context for evaluation can be either job performance in a normal context or performance measured in a special evaluation context. One common method is to have evaluators make an overall assessment of typical performance, which is often done annually. This particular evaluation has the advantage of reflecting the person's resource management in diverse job-related situations over an extended period of time. Nevertheless, there are disadvantages to this evaluation technique. These include incomplete or distorted recall for relevant events, recency bias, memory priming caused by the phrasing of evaluation questions, and the influence of pre-existing knowledge about the individual being evaluated (DeNisi, Cafferty, & Meglino, 1984). Other evaluation problems depend on the number of persons evaluated. If each evaluator rates only a few individuals, their evaluations may be poor because of limited practice with the rating system and exposure to a limited range of performance. At the same time, if each evaluator assesses multiple persons, carryover or contrast effects may adversely influence individual performance ratings.

Special evaluation contexts can be designed to avoid or minimize these errors but may have the disadvantage that performance in the special context is at a maximal rather than a typical level and thus may not generalize to the job (Dubois, Sackett, Zedeck, & Fogli, 1993; Sackett, Zedeck, & Fogli, 1988). In the aviation domain, the work sample of a normal flight is typically combined with realistic simulations of normal working conditions to increase generalization and obtain more typical levels of resource management behaviors. Special evaluation requires the preparation of extra materials, including the work sample itself and guides or scripts to standardize evaluators' behavior during the assessment. Furthermore, evaluators must be appropriately trained in the use and administration of these materials (Prince, Oser, Salas, & Woodruff, 1993).

Measuring Knowledge

One option for evaluating resource management is to evaluate the components that contribute to performance, such as the information that individuals have acquired as a result of the training program. Training may change two types of knowledge: declarative knowledge and procedural knowledge. *Declarative knowledge* refers to the static information about a domain that is represented in memory. It can be thought of as the definitions for constructs in the domain and rules for when this knowledge can (or should) be applied. *Procedural knowledge*, on the other hand, typically refers to rules regarding the *execution* of specific behaviors (Anderson, 1985). Although procedural knowledge is based in part on declarative knowledge, it is considered to be a "higher order" form of knowledge, because it involves the integration of multiple sources of information as well as the automation of specific behaviors.

For example, before one can turn an aircraft by coordinating aileron and rudder movements, one must have the appropriate foundation of declarative knowledge about adverse yaw caused by moving the ailerons. Because procedural and declarative knowledge are manifested in different forms, they must be assessed differently. Typically, *elements* of declarative knowledge are assessed by means of paper-and-pencil measures, whereas the *organization* of declarative knowledge is assessed by means of techniques such as Pathfinder (Schvaneveldt, 1990). Procedural knowledge, on the other hand, is typically assessed with some form of work sample test (Kraiger et al., 1993).

Effectiveness Criteria

Another issue to consider when developing appropriate measures is the different ways training can affect performance. The goal of training may be to change the mean (average) level of performance or to change the distribution (variability) of performance.

Traditionally, training is evaluated in terms of mean differences. For example, the mean performance of trained crews is often compared with that of untrained crews.

However, some researchers (e.g., Alliger & Katzman, 1997) have argued that certain training interventions can influence both the mean and the variability of performance data. For example, group consensus training or instructor calibration training are often used to decrease the random variability in people's response patterns while simultaneously having little or no effect on mean ratings. Conversely, training may attempt to increase the variance of ratings. For example, training in creativity may seek to increase the variability of ideas generated by a group. Therefore, it is essential that researchers avoid the temptation to assess training performance solely in terms of mean change.

The specific outcomes of training should be guided by an overall theory of resource management in the domain of interest. This theory of performance should, in turn, be used to develop a systematic measurement plan (Kraiger et al., 1993) that specifies which type and level of performance to be expected, the time at which this performance is expected to occur, and the appropriate measurement strategy for each facet of performance.

Multifaceted Approaches and Multiple Constituencies

In recent years a number of researchers have heeded Kraiger et al.'s (1993) call for a multifaceted approach to the evaluation of training programs (Leedom & Simon, 1995; Salas, Fowlkes, Stout, Milanovich, & Prince, in press; Stout, Salas, & Fowlkes, 1997; Stout, Salas, & Kraiger, 1997). In general, these studies have included a variety of individual-level (e.g., reactions to training, declarative knowledge, knowledge organization) and group-level (crew processes, crew outcomes) criteria as indices of the effectiveness of CRM training programs. Unfortunately, even these well-designed and well-intentioned studies attest to the difficulties of performing systematic training evaluation in organizational contexts. For example, several studies were limited by psychometrically deficient measures of declarative knowledge; small sample sizes; or the measurement of immediate, maximal performance to the exclusion of long-term, typical performance.

Although these groundbreaking efforts were more complete and multifaceted than previous evaluations, their weaknesses illustrate two basic principles that still jeopardize the usefulness of a training evaluation. First, no matter how many criterion variables are measured, the information that they provide is only as good as the measurement instrument. For example, a given study may measure both reactions to training and declarative knowledge. However, to the extent that the measures of declarative knowledge are psychometrically deficient (e.g., the lack of item difficulty results in ceiling effects), they provide little additional information regarding the effectiveness of the training program (Crocker & Algina, 1986).

Second, virtually every training program is going to have *some* effect on immediate performance, but these could be transitory effects. Commercial air carriers invest tens of millions of dollars every year with the implicit understanding that training programs will result in performance increases that carry over to typical performance in line operations over the long term, with the ultimate criteria being increased safety and efficiency in line operations. Therefore, training professionals must conduct studies that assess the *long-term effects* of CRM training programs. For example, if the effect of a training program wears off after 1 month of line performance, it would probably not be considered an effective program from the airline's perspective. Different constituencies, such as the researcher, the carrier, the union, the Federal Aviation Administration (FAA), and the general public may have different criteria for success, and these success criteria are often at odds with one another (Austin, Klimoski, & Hunt, 1996). Therefore, carriers and researchers alike need to be more considerate of the needs of these other constituencies. We believe that long-term measures of CRM training program effectiveness will address at least some of these needs.

Ensuring the Quality of Measurement

The third step in the evaluation process is to objectively confirm the psychometric quality of these assessment measures. Because evaluations are performed by individuals, the quality of evaluations is decreased by inaccuracy, subjectivity, or personal biases on the part of the evaluator. Objectively confirming the quality of measurement instruments involves three basic facets. A good measure of resource management must be *sensitive* enough to discriminate good from poor resource management, *reliable* enough to consistently provide the same estimate of resource management, and *valid* enough to ensure that the measure involves only resource management rather than other extraneous factors. We cover each facet of measurement in turn.

Sensitivity

Sensitivity refers to the extent to which a measure can detect changes in the construct being assessed. Specifically, a sensitive measure of resource management should show higher scores when resource management is above average and lower scores when resource management is below average. Although extreme examples of good or bad performance are usually easy to detect, sensitivity must also be established for subtle differences in resource management behaviors, such as *marginally* safe versus unsafe performance.

Sensitivity is influenced by the granularity of the measurement instrument. More specifically, the evaluation scale must be sufficiently fine grained to capture important differences in the quality of resource management that is observed yet still be accurately used by the evaluator. For example, a dichotomous "satisfactory versus unsatisfactory" scale might be accurately used by evaluators but would not be sensitive to varying degrees of good or bad resource management. Conversely, a 100-point scale might be extremely fine grained, but evaluators may not be able to use it accurately. A compromise for measurements based on human evaluations is often a 5- or 7-point scale with meaningful definitions assigned to each scale point (Likert, 1932).

To objectively index the sensitivity of measurement, it is necessary to compare the judgments made by evaluators with pre-established levels of resource management. One method for indexing the sensitivity of evaluation is to have evaluators rate "test" cases of varying levels of resource management proficiency (as determined by subject matter experts [SMEs]). For example, average evaluator ratings for "good" test cases ought to be higher than ratings for "average" test cases, which in turn should be higher than ratings for "poor" test cases. One way to index sensitivity for each evaluator is to use Hays's (1988) omega-squared index for strength of effect (Holt, Johnson, & Goldsmith, 1997; Williams, Holt, & Boehm-Davis, 1997). This index reflects how different an evaluator's ratings are for different categories of test cases and has a range from 0 (no discrimination among levels) to 1 (perfect discrimination among levels).

Reliability

Reliability can be informally defined as the consistency or stability of measurement. Formally, reliability is defined as the lack of random error in the measurement instrument (Nunnally, 1967). Although different traditional methods of estimating reliability have been developed, we cover only two: test–retest reliability and internal consistency reliability (see Nunnally [1967] or Pedhazur & Pedhazur Schmelkin [1991] for more information). Because each method makes different assumptions about the main source of error in measurement, each has its own advantages and disadvantages.

Test–retest reliability is used to assess the stability of measurement over time. One method of assessing this form of reliability consists of having evaluators assess the same set of performances at two different times and correlating these two sets of evaluations. The calculation is based on the Pearson product–moment correlation and results in an index r that reflects reliability. In this case, a value of r near 0 indicates a lack of test–retest reliability, whereas values near 1 indicate near-perfect test–retest reliability. However, test–retest reliability assumes that the only important source of random error is spontaneous changes over time. Unfortunately, systematic evaluator differences are common in evaluating resource management in the aviation domain (Williams et al., 1997). To the extent that these differences are stable over time, the test–retest reliability is inflated. Therefore, although simple to execute, the test–retest reliability method addresses only one potential source of error and may be positively biased.

Internal consistency reliability refers to the internal coherence of a set of items that are all measuring the same thing (Nunnally, 1967). For evaluation of resource management, this type of reliability requires a set of multiple items that all reflect resource management. If resource management has distinct components, each distinct component must have its own set of multiple items. The intercorrelations among items in a set are summarized into a coefficient alpha index, which ranges from 0 (no internal consistency reliability) to 1 (perfect internal consistency reliability). Several factors influence coefficient alpha, such as the number of items included in the scale (Cortina, 1993; Cronbach, 1951), as well as *systematic* judgment errors made by evaluators (e.g., halo rating errors). To the extent that these systematic errors occur across items, internal consistency reliability will be inflated.

When used in isolation, both test–retest and internal consistency reliability estimates can provide misleading results. To check and correct such rater errors, we have developed an alternative approach for training and checking evaluator reliability that uses multiple statistical indexes for evaluating rater performance and giving training feedback. This multicomponent approach was labeled *interrater reliability* (IRR) training (Holt et al., 1997).

During the IRR process each evaluator's ratings of the test cases are compared with the group's judgments by using four indexes, each of which provides information on one aspect of reliability. In addition, an index of the sensitivity of judgment is also included if SMEs have evaluated the test cases. First, the overall distribution of each evaluator's ratings is compared with the group's distribution to ascertain its level of *congruency*. Low congruency suggests that the evaluator gives a different mix of ratings on the scale compared to the group. Second, *systematic differences* of harsher or more lenient grading among the evaluators are identified. Third, the interrater correlation is calculated to see if the raters shift in a consistent manner up and down in their ratings across evaluated items (*consistency*). Finally, if the test cases have been externally scored by SMEs, raters can also be assessed regarding the *sensitivity* of their evaluations. Rater-specific estimates of congruency, systematic differences, consistency, and sensitivity results are provided to each individual, and the aggregate results for all raters are provided to the group for discussion.

The group of raters is also provided with information concerning their level of group *agreement* on each item (James, Demaree, & Wolf, 1993). This feedback is critical because every item with low agreement should be discussed until a reasonable group consensus is reached. In summary, the IRR method compares each rater with the group using indexes that give the rater information about the congruency, systematic differences, consistency, and sensitivity of his or her evaluations. The information from these indexes and group agreement is then used to train and improve subsequent ratings (Williams et al., 1997).

Validity

Validity refers to the extent to which a measure really measures its intended construct (Landy, 1986; Nunnally, 1967). More specifically, validity is the proportion of variance in a measure that reflects real variation in the measured construct. From a resource management perspective, validity refers to the amount of variability in evaluator ratings that accurately reflects real differences in the resource management performance of the persons being evaluated. Assessing validity requires checking measurement items, the measurement process, and the results of the measurement process.

The items used for evaluating resource management should be checked for face and content validity (Nunnally, 1967). *Face validity* refers to the judgment of a group of experts that the items are plausibly measuring the desired construct. Such judgments are easy and convenient, but unfortunately they are also somewhat subjective. A more objective item analysis will often indicate that items designed by experts to measure a given construct do not in fact predict that construct. Face validity is, therefore, easy to establish but is only weak evidence that the construct in question is being assessed.

Content validity first requires a careful specification of the domain of all possible relevant items. Content validity can then be demonstrated by showing that the evaluation items are a fair, unbiased, and representative sample of items from this larger domain. Techniques for specifying relevant content items for training programs have been developed (Lawshe, 1975). However, because resource management typically requires an individual or team to interact in a complex system, the set of possible items is very large and ill defined. Therefore, the specification of the domain of all possible relevant items for this type of complex domain may be difficult or impossible.

The validity of measurement generally is established by empirically examining the relationship to other measures that should be related to the construct. Two basic principles apply. The first principle is *convergent* and *discriminant validity* (Campbell & Fiske, 1959). In convergent validity, measures that ought to be related to a construct should converge or correlate with the proposed measure. For resource management, measures that ought to positively relate to it, such as measures of teamwork, knowledge, skills, and abilities (Stevens & Campion, 1994), ought to positively correlate with the resource management measure. A valid measure of a construct should show the expected relationships with plausible criteria (criterion validity) and predict the expected outcomes of changing resource management (predictive validity). In divergent validity, measures that ought to be independent or distinct from resource management should diverge or not correlate with the proposed measure. For example, if resource management can be done equally well by men and women, then gender should not correlate with resource management measures. Divergent validity is particularly important if potential confounds such as popularity or appearance could influence a measure of resource management effectiveness; they must be shown not to do so.

The second principle is *network validity* (Pedhazur & Pedhazur Schmelkin, 1991). For network validity the nomological network of constructs that should be theoretically associated with the construct is empirically assessed to determine whether it demonstrates the expected pattern of relationships. For example, a valid measure of resource management ought to show a plausible set of relationships with antecedents, correlates, and consequences that one would expect for resource management. If the expected network of relationships is generally found, network validity is established.

EXAMPLE OF AN EVALUATION
OF PROGRAM DEVELOPMENT

We recently worked with a regional air carrier to develop and evaluate a resource management training program for pilots. This training program focused on improved crew briefings and communication during normal operations, as well as problem diagnosis, situation assessment, and planning and decision making during abnormal or emergency operations. This program was unique in that the resource management principles were translated into step-by-step operational procedures. Furthermore, these procedures were formally required as part of standard operating procedure (SOP) for one fleet and added to the operating manuals and handbooks for that particular aircraft.

Selecting the Level at Which Resource Management
Would be Evaluated

We evaluated the effectiveness of the training program by measuring performance at both the individual and crew levels. Clearly, the performance of individual pilots is important. First, individual pilots must be qualified to continue to legally operate an aircraft. Second, the performance of an individual can directly affect the performance of his or her team or crew. Third, some issues, such as the effects of ability on performance, were more sensibly addressed by comparing the assessed ability of individuals with their performance (Boehm-Davis, Holt, & Hansberger, 1997).

Although individual performance is important, commercial aircraft are always operated by crews. The performance of a team or crew may be quite distinct from the performance of individual team members, especially for highly complex, interdependent tasks (Steiner, 1972). Furthermore, evidence from aviation accidents and safety reports suggests that a lack of coordination among crew members has been the cause of a substantial portion of problems on the flightdeck (National Transportation Safety Board, 1994). As a result, this project also focused on crew-level performance.

Developing the Evaluation Plan

Selecting an Evaluation Design

This particular carrier was composed primarily of two fleets. We decided to provide the resource management training program to one fleet, while the other fleet continued to use existing procedures and management techniques. In this quasi-experimental design, the fleet with extra training and new procedures acted as the experimental group, and the fleet with normal training and procedures acted as the control group. One focus of the evaluation design was to compare pilots and crews in the two fleets.

To allow for gradual learning on the part of the pilots of the new procedures and processes, we also incorporated aspects of a time series design. Specifically, we collected pilot and crew performance measures over a 3-year period. During the first year, the pilots had additional resource management training, but the new procedures had not been formally implemented. This was our baseline performance year. In the second year, the new procedures were formally implemented and required as SOP for that fleet. Performance measured in that year would reflect the immediate impact of the resource management training and SOP changes. The third year was the final follow-up assessment that would either confirm or disconfirm long-term effects of the training, including a gradual acceptance of and accommodation to the new methods of cockpit interaction and coordination. In addition, during the final year of evaluation three auxiliary measures of resource training were developed. These additional methods allowed converging measures of the effects of this training with different samples of evaluators and performance situations.

Developing Measures of Resource Management Performance

Once the evaluation design had been selected, the next steps were to develop an operational definition of resource management, develop appropriate measures given that operational definition, and to ensure the quality of the measures that were developed.

Defining Resource Management

For this project, effective CRM was defined for two qualitatively distinct contexts: normal operations and abnormal/emergency situations. For normal situations, effective CRM was defined as the effective communication and coordination of crew members before, during, and after flying a typical flight. The operational definition of normal performance included quality of briefings and other communication; quality of workload management and avoiding overload; maintaining situation awareness of the aircraft and external traffic and weather situation; and preserving effective coordination on checklists, flows, and other sequential tasks during the flight.

For abnormal/emergency situations, effective CRM was defined as effective workload management and communication while performing normal flight tasks plus problem diagnosis, situation assessment, planning, and monitoring of plan execution. The operational definition of abnormal/emergency CRM was quite extensive and included, for example, the establishment of explicit "bottom lines" and "backup plans" during the planning task, plus clearly communicating these plan components to other crew members.

Developing Appropriate Measures

In carrying out this project, we developed a variety of measures to capture both individual- and crew-level performance. Furthermore, we realized that these metrics would be applied by a number of different evaluators (pilot instructor/evaluators). Thus, we felt that it also was important to develop a structured method for collecting assessments of pilot and crew performance.

Measuring Performance. We designed a structured evaluation process to achieve systematic and reliable observations and ratings of performance. The multiyear evaluations consisted of line operational evaluations (LOEs) and line checks. The LOE, conducted during the pilots' annual evaluation for flight certification, consisted of a work sample performance evaluation in which the crews performed a typical flight scenario in a full-motion simulator. The evaluator followed an LOE script to consistently introduce specific problems and distracting conditions into the flight. In this way, crew reactions to abnormal and emergency situations could be assessed in a standardized manner. The evaluation forms emphasized specific crew reactions for these events, including both technical and CRM performance items and related skills. The basis for the evaluation forms was the specification of a set of observable behaviors. These observable behaviors were carefully identified by SMEs as being central to successful performance on a specific event set. These behaviors and skills provided a point of focus for the instructor/evaluators during the observation and evaluation of the LOE and during the crew debriefing after the LOE.

The line check assessed pilot and crew performance during normal flight operations. Typically, instructor/evaluators would board a routine flight without prior announcement and evaluate the crew on a spectrum of technical and CRM items. For this carrier, crew performance ratings, both technical and CRM, were based on a standardized 4-point scale covering the full range of possible crew performance, from *unsatisfactory performance* (observed crew behavior does not meet minimal requirements) to *above standard performance* (observed crew behavior is markedly better than the standard performance).

To provide converging measurement of crew performance, we designed two auxiliary performance measures. First, the cadre of instructor/evaluators who had evaluated pilots from both the experimental and control fleets completed a detailed performance questionnaire regarding the *relative* performance of pilots from both fleets during upgrade or transition training from the control fleet to the experimental fleet. Second, a separate cadre of five evaluators assessed pilots from both fleets during normal flights using a direct observation form. This cadre was completely different from the carrier's line check evaluators or FAA evaluators, and the assessments were strictly voluntary. By using different sets of evaluators and different evaluation formats (e.g., the LOE and line check), these data provide converging information regarding the hypothesized performance differences between the two fleets.

To ensure a broader measurement of possible training effects besides performance, we also used a pilot survey to measure knowledge and attitudes as suggested by Kraiger et al. (1993). Knowledge acquired by individual pilots from the training program was measured by a survey of all carrier pilots in the final year of the project. Knowledge was measured only at post-training because the training introduced completely new procedures developed for this project that pilots could not have known about previously. Therefore, the focus of the knowledge evaluation was on the extent to which individual pilots were able to describe the new set of procedures and the appropriate context for enacting each procedure. The focus of attitude measurement concerned attitudes toward CRM in general and more specifically toward the trained resource management procedures. The survey also measured how often pilots performed the new procedures and briefings and the perceived effects of the new procedures and briefings.

Focus on Mean Changes.   The major focus in this project was on mean differences between the two fleets; that is, we were interested in demonstrating that the crews in the trained fleet would perform at a higher level on measured CRM skills than would crews that had not received the training.

Mean differences were the focus, because narrowing the range or variability of performance would not have been a useful outcome.

For attitudes, we compared the mean attitudes of pilots in the two fleets with one another as well as with a neutral baseline. For assessing knowledge, we measured the relative extent of relevant knowledge and tested whether the trained pilots could answer knowledge questions at an above-chance level (representing *some* knowledge). We similarly analyzed the pilots' perceptions of the frequency of performance and effects of the new procedures.

Ensuring the Quality of Measurement

Sensitivity

Instructor/evaluators were presented with videotapes showing different levels of resource management behavior, derived from simulation sessions conducted by the airline. They were asked to rate the level of resource management behavior exhibited on the videotapes using a 4-point scale that ranged from unsatisfactory (1) through FAA minimal requirements (2), company standard (3), and above company standard (4). Each level of this scale had a unique well-anchored qualitative meaning for the raters. The segments of behavior portrayed on the videotapes were selected to represent the range of possible resource management behavior, with a focus on behaviors rated in the central portion of the scale (Levels 2 and 3). SMEs established the exact level of performance for each segment. We indexed sensitivity by analyzing the differences in each rater's evaluations for performance segments at different levels.

Reliability

Reliability was assessed on a regular basis (approximately every 6 months) using the multidimensional IRR procedures developed for this project (Williams et al., 1997). This process relies on a group of raters (instructor/evaluator pilots) using normative information for standardizing IRR. All raters individually evaluated a videotape of typical crew performance on the LOE, and these evaluations were statistically compared. The relative amount of congruency of judgment distributions, systematic harsh or lenient judgments, interrater consistency, and agreement were assessed at each session. Each rater received same-day feedback about his or her evaluation performance relative to the other raters. Finally, each single item with agreement below a corporate standard was discussed intensively by the group to isolate and solve causes of rater variability. The focus for this part of the training was the reduction of random variability on each item. Information from these discussions was also used to modify the content of the evaluation scenario, rewrite evaluation items for improved clarity, formally codify explicit grading standards for certain items, and modify or clarify carrier policies and procedures (SOP). After this training, the performance evaluations conducted by these raters were accumulated into a database. After sufficient data were collected, the items that were designed to measure the same aspect of performance were assessed by an internal-consistency reliability metric (coefficient alpha). These estimates served as a final check on the reliability of the performance data.

Validity

The major focus for assessing validity was the internal structural validity of the assessment process. We analyzed the evaluation data with path analysis to verify that the process of evaluation was in fact performed in the correct manner. We used the process of evaluation (from detailed behavioral observations, to judgments of performance components, to overall evaluations of performance) to construct an anticipated structure of relationships among the performance measures. The expected path structure of relationships was found, which supported measurement validity.

Analysis and Interpretation of Evaluation Results

The LOE, line check, and auxiliary measures were all analyzed for the hypothesized fleet performance differences. Evidence from the LOE and the direct cockpit observations were crew-level assessments, and these results were examined for mean differences in crew performance. The evaluation of individual pilots was emphasized in the line check evaluations, the instructor/evaluator survey, and in a survey of individual pilots. Across these measures, both individual and crew levels of performance, which were the targeted levels of change for this study, could be assessed.

Crew Performance. On the LOEs, several specific items concerning CRM behavior were graded with exactly the same grading standards for both fleets. For most of these items the trained and untrained fleets were significantly different in the expected direction. We concluded that the resource management training had the desired effects for the work-sample evaluation.

The second crew-level evaluation was direct observations of cockpit interaction on regular line flights. These observations were carried out by a separate cadre of pilots who rode in the cockpit and watched the crew under voluntary, nonjeopardy conditions. Specific briefing content and other aspects of performance relevant to the training were evaluated by these observers. These direct observations of cockpit interaction showed that crews from the trained fleet were significantly superior on the majority of these items. On the remaining items the trained fleet still had a higher mean, but the observed difference was not statistically significant.

Individual Pilot Performance. The first measure of performance at the pilot level was the instructor/evaluator survey, which involved a comparison of pilots from trained and untrained fleets who were transitioning aircraft or upgrading from First Officer to Captain. Instructor/evaluators who had experience with both sets of pilots gave comparative ratings for average individual pilot performance. These ratings indicated that, in comparison to the untrained fleet, pilots from the trained fleet were significantly better in communication, workload management, and planning and decision making.

The second measure of the effects of training on individual pilots was the pilot survey, which included pilots trained in specific resource management procedures and pilots without this training. Compared to appropriate baselines, trained pilots had acquired a significant amount of knowledge about the resource management procedures, had positive attitudes toward CRM and resource management procedures, frequently performed the trained procedures on routine flights, and strongly indicated that the procedures increased their effectiveness.

Convergent results for performance measurement and confirmatory results for attitudes and knowledge give more confidence in the final evaluation of the effectiveness of this type of resource management training. Multiple evaluations at both the individual pilot level as well as the crew level help rule out various confounds or alternative explanations for the results. For example, positive effects of training were reported by instructor/evaluators, by an independent cadre of observer pilots, and by the pilots themselves. Each of these groups has different potential sources of bias, and the convergence of results reassures us that the positive effects are not simply the result of biased evaluators.

GUIDELINES FOR DEVELOPING EVALUATIONS OF RESOURCE MANAGEMENT TRAINING PROGRAMS

In developing a plan to evaluate a resource management training program, we recommend following the steps outlined in this chapter. These include: selecting a level at which to measure resource management, selecting a research design through which to evaluate the selected level of resource management behavior, and developing measurement instruments that can accurately assess resource management behaviors.

Specifically, Table 9.1 provides an overview of the steps needed to establish and implement an evaluation of a resource management training program. Each step has a set of critical issues that should be resolved for the best possible outcome.

TABLE 9.1
Steps for Developing an Evaluation of a Resource Management Training Program

---

1. Select the level at which resource management will be evaluated

2. Develop the evaluation plan
   - Select an evaluation design
   - Determine appropriate time interval for measuring change

3. Develop measures of resource management performance
   - Operationally define resource management
   - Develop appropriate measures
     - Measure knowledge, attitude, performance, etc.
     - Develop converging measures where possible
     - Decide to focus on mean changes or variability
   - Ensure the quality of measurement
     - Assess sensitivity, reliability, and validity

4. Analyze and interpret the evaluation results

5. Use information to modify training system, personnel selection, etc.

We learned a number of important lessons at each step of the resource management evaluation process:

Guideline 1: An overall framework or theory about the type of performance measured must guide evaluation. A well-developed theory is critical for specifying the levels for the expected effects, operationally defining resource management, and for specifying the other measures necessary to establish construct validity.

Guideline 2: The level of evaluation of resource management is often determined by the context. In commercial aviation the most important levels of evaluation are the individual pilots and the flight crews.

Guideline 3: Multifaceted evaluations of performance are preferred. The effects of resource management training should be examined for a broad range of possible changes. At a minimum, changes in knowledge, attitude, and behavior should be assessed.

Guideline 4: For measuring key effects such as performance, multiple converging lines of evaluation evidence provide stronger support for the effects of resource management training. Creatively consider the various ways that the expected effects could be exhibited by individuals, teams, or organizational units, and then measure them accordingly.

Guideline 5: A long-term, multimeasure evaluation plan is necessary to detect delayed effects of training that may not be immediately apparent. Depending on the type of training, the multiple observations may be collected over weeks, months, or years (as in this study).

Guideline 6: Highly controlled evaluation is desirable. Nevertheless, the selected evaluation design should be a workable compromise between the desire for experimental control and the reality of the training and evaluation setting.

Guideline 7: Control groups are necessary. Having a control (untrained) group helps avoid many confounds that would otherwise hamper single-group evaluations.

Guideline 8: Repeated training in the evaluation process is necessary to maintain calibration of the raters for complex behavior domains such as resource management. Furthermore, calibration must be continually rechecked for statistical levels of sensitivity, reliability, and validity.

Guideline 9: Evaluation of resource management is an iterative process. Ongoing evaluation may cycle back from Step 5 in Table 9.1 to an earlier step in the process. In our research, results from the LOE evaluations in Years 1 and 2 helped change the LOE format to provide more precise, comparative evaluations in Year 3.

Guideline 10: Careful evaluation of resource management will result in a bonus of new knowledge about performance appraisal, the training program, and relevant individual and team processes. More specifically, our research uncovered new information about pilots, crews, and the organization.

Guideline 11: Careful choices must be made at each step of the evaluation process. Each choice involves trade-offs between the desire for the best possible evaluation of resource management and the constraints of time, personnel, and other critical resources.

ACKNOWLEDGMENTS

This research was supported by the Office of the Chief Scientific and Technical Advisor for Human Factors (AAR-100) at the Federal Aviation Administration through Grant 94-G-034 to George Mason University. We thank Dr. Eleana Edens in Aviation Research (AAR-100) and Dr. Thomas Longridge in Aviation Flight Standards Service (AFS-230) for their continuing support of this work.

REFERENCES

Alliger, G. M., & Janak, E. A. (1989). Kirkpatrick's levels of training criteria: Thirty years later. *Personnel Psychology, 42,* 331–342.

Alliger, G. M., & Katzman, S. (1997). When training affects variability: Beyond the assessment of mean differences in training evaluation. In J. K. Ford & Associates (Eds.), *Improving training effectiveness in work organizations* (pp. 223–246). Mahwah, NJ: Lawrence Erlbaum Associates.

Alliger, G. M., Tannenbaum, S. I., Bennett, W., Traver, H., & Shotland, A. (1997). A meta-analysis of the relations among training criteria. *Personnel Psychology, 50*, 341–358.

Anderson, J. R. (1985). *Cognitive psychology and its implications.* New York: Freeman.

Austin, J. T., Klimoski, R. J., & Hunt, S. T. (1996). Dilematics in public sector assessment: A framework for developing and evaluating selection systems. *Human Performance, 93*, 177–198.

Boehm-Davis, D. A., Holt, R. W., & Hansberger, J. (1997). Pilot abilities and performance. In *Proceedings of the Ninth International Symposium on Aviation Psychology* (pp. 462–467). Columbus: Ohio State University Press.

Borman, W. C., & Motowidlo, S. J. (1993). Expanding the criterion domain to include elements of contextual performance. In N. Schmitt & W. C. Borman (Eds.), *Personnel selection in organizations* (pp. 71–98). San Francisco: Jossey-Bass.

Campbell, D. T., & Fiske, D. W. (1959). Convergent and discriminant validation by the multitrait-multimethod matrix. *Psychological Bulletin, 56*, 81–105.

Campbell, D. T., & Stanley, J. C. (1963). *Experimental and quasi-experimental designs for research.* Boston: Houghton Mifflin.

Campbell, J. P. (1990). Modeling the performance prediction problem in industrial and organizational psychology. In M. Dunnette & L. Hough (Eds.), *Handbook of industrial and organizational psychology* (2nd ed., pp. 687–732). Palo Alto, CA: Consulting Psychologists Press.

Chan, D. (1998). Functional relations among constructs in the same content domain at different levels of analysis: A typology of composition models. *Journal of Applied Psychology, 83*, 234–246.

Cook, T. D., & Campbell, D. T. (1979). *Quasi-experimentation: Design and analysis issues for field settings.* Boston: Houghton Mifflin.

Cortina, J. M. (1993). What is coefficient alpha? An examination of theory and applications. *Journal of Applied Psychology, 78*, 98–104.

Crocker, L. M., & Algina, J. (1986). *Introduction to classical and modern test theory.* New York: Holt, Rinehart & Winston.

Cronbach, L. J. (1951). Coefficient alpha and the internal structure of tests. *Psychometrika, 16*, 297–334.

DeNisi, A. S., Cafferty, T. P., & Meglino, B. M. (1984). A cognitive view of the performance appraisal process: A model and research propositions. *Organizational Behavior and Human Decision Processes, 33*, 360–396.

Dubois, C. L. Z., Sackett, P. R., Zedeck, S., & Fogli, L. (1993). Further exploration of typical and maximal performance criteria: Definitional issues, prediction, and white–black differences. *Journal of Applied Psychology, 78*, 205–211.

Goldstein, I. L. (1993). *Training in organizations: Needs assessment, development, and evaluation* (3rd ed.). Pacific Grove, CA: Brooks/Cole.

Goodman, P. S., Lerch, F. J., & Mukhopadhyay, T. (1994). Individual and organizational productivity: Linkages and processes. In D. H. Harris (Ed.), *Organizational linkages: Understanding the productivity paradox* (pp. 54–80). Washington, DC: National Academy Press.

Guttentag, M., & Struening, E. L. (1975). *Handbook of evaluation research* (Vol. 2). Beverly Hills, CA: Sage.

Hays, W. L. (1988). *Statistics* (4th ed.). Chicago: Holt, Rinehart and Winston.

Helmreich, R. L., & Foushee, H. C. (1993). Why crew resource management? Empirical and theoretical bases of human factors training in aviation. In E. L. Wiener, B. G. Kanki, & R. L. Helmreich (Eds.), *Cockpit resource management* (pp. 3–45). San Francisco: Academic Press.

Holt, R. W., Johnson, P. J., & Goldsmith, T. E. (1997). Application of psychometrics to the calibration of air carrier evaluators. *Proceedings of the Human Factors and Ergonomics Society 41st annual meeting.* Human Factors and Ergonomics Society. 916–920.

Howell, D. C. (1997). *Statistical methods for psychology* (4th ed.). Boston: Duxbury Press.

James, L. R., Demaree, R. G., & Wolf, G. (1993). r-sub(wg): An assessment of within-group interrater agreement. *Journal of Applied Psychology, 78*, 306–309.

Joint Committee on Standards for Education Evaluation. (1994). *The program evaluation standards* (2nd ed.). Thousand Oaks, CA: Sage.

Kirkpatrick, D. L. (1976). Evaluation of training. In R. L. Craig (Ed.), *Training and deelopment handbook: A guide to human resource development* (2nd ed.). New York: McGraw-Hill.

Kraiger, K., Ford, J. K., & Salas, E. (1993). Application of cognitive, skill-based and affective theories of learning to new methods of training evaluation. *Journal of Applied Psychology, 78,* 311–328.

Landy, F. J. (1986). Stamp collecting versus science: Validation as hypothesis testing. *American Psychologist, 41,* 1183–1192.

Lauber, J. K. (1984). Resource management in the cockpit. *Air Line Pilot, 53,* 20–23.

Lawshe, C. H. (1975). A quantitative approach to content validity. *Personnel Psychology, 28,* 563–575.

Leedom, D. K., & Simon, R. (1995). Improving team coordination: A case for behavior-based training. *Military Psychology, 7,* 109–122.

Likert, R. (1932). A technique for the measurement of attitudes. *Archives of Psychology, 140,* 44–53.

National Transportation Safety Board. (1994). *A review of flightcrew-involved, major accidents of U.S. air carriers, 1978 through 1990.* (Safety Study NTSB/SS-94/01, Notation 6241). Washington, DC: Author.

Nunnally, J. C. (1967). *Psychometric theory.* New York: McGraw-Hill.

Pedhazur, E. J., & Pedhazur Schmelkin, L. (1991). *Measurement, design, and analysis: An integrated approach.* Hillsdale, NJ: Lawrence Erlbaum Associates.

Prince, C., Oser, R., Salas, E., & Woodruff, W. (1993). Increasing hits and reducing misses in CRM/LOS scenarios: Guidelines for simulator scenario development. *International Journal of Aviation Psychology, 3,* 69–82.

Sackett, P. R., Zedeck, S., & Fogli, L. (1988). Relations between measures of typical and maximal performance. *Journal of Applied Psychology, 73,* 482–486.

Salas, E., Fowlkes, J. E., Stout, R. J., Milanovich, D. M., & Prince, C. (1999). Does CRM training improve teamwork skills in the cockpit?: Two evaluation studies. *Human Factors, 41*(2), 326–343.

Schvaneveldt, R. W. (Ed.). (1990). *Pathfinder associative networks: Studies in knowledge organizations.* Norwood, NJ: Ablex.

Steiner, I. D. (1972). *Group process and productivity.* New York: Academic Press.

Stevens, M. J., & Campion, M. A. (1994). The knowledge, skill, and ability requirements for teamwork: Implications for human resource management. *Journal of Management, 20,* 503–530.

Stout, R. J., Salas, E., & Fowlkes, J. E. (1997). Enhancing teamwork in complex environments through team training. *Group Dynamics: Theory, Research, and Practice, 1,* 169–182.

Stout, R. J., Salas, E., & Kraiger, K. (1997). The role of trainee knowledge structures in aviation team environments. *International Journal of Aviation Psychology, 7,* 235–250.

Wiener, E. L., Kanki, B. G., & Helmreich, R. L. (1993). *Cockpit resource management.* San Francisco: Academic Press.

Williams, D. M., Holt, R. W., & Boehm-Davis, D. A. (1997). Training for inter-rater reliability: baselines and benchmarks. In *Proceedings of the Ninth International Symposium on Aviation Psychology* (pp. 326–343). Columbus: Ohio State University Press.

APPLICATIONS OF RESOURCE MANAGEMENT IN ORGANIZATIONS: LESSONS LEARNED AND GUIDELINES

10

Airline Resource Management Programs

Deborah A. Boehm-Davis
Robert W. Holt
George Mason University

Thomas L. Seamster
Cognitive & Human Factors

This chapter reports on the experience of a research team in developing, implementing, and evaluating resource management procedures at two airlines. This research arose from the synthesis of three factors: (a) the Federal Aviation Administration (FAA) wanted to evaluate a proceduralized form of crew resource management (CRM), (b) a regional airline wanted to upgrade its CRM training by customizing it to its operational needs, and (c) our research team was involved in applied aviation research and was interested in developing and testing an innovative approach to CRM training. In this chapter we describe the development, implementation, and evaluation of a resource management program at a regional airline and the development, implementation, and planned evaluation of one aspect of this work at a major airline. We conclude the chapter with the lessons learned from these efforts by explaining a set of guidelines that can be applied to the development of resource management programs for teams that require a high level of coordination.

RESOURCE MANAGEMENT AT A REGIONAL AIRLINE

Development of the Resource Management Training Program

The goals of the regional airline project were to develop and implement a comprehensive, procedurally based CRM training program and evaluate the effects of this specific CRM training on the performance of flight crews. The effort was unique in that it took the philosophy and principles of CRM that are typically taught in the classroom and implemented them into standard operating procedures. The implementation included changes to normal briefings used on a regular basis as well as changes to checklists and procedures in the Quick Reference Handbook (QRH) that pilots use in abnormal or emergency situations.

Need for the Training Program

The regional airline identified a need for a program that would supplement the basic CRM training in philosophy and principles of CRM. The goal for this supplemental training was to produce objectively observable changes in crew performance for operations in normal and abnormal/emergency conditions. The airline felt that general CRM did not fully address the requirements of their operational conditions. The airline specialized in short-haul operations in a four-season environment with both congested airports and difficult terrain at specific airports and wanted to develop new CRM procedures that would help crews cope with these flight conditions. The airline's recent application under the Advanced Qualification Program (AQP) provided an excellent opportunity to introduce these procedures and the improved CRM training.

The initial sponsor of this training was the CRM coordinator at the airline. This person acted as an opinion leader who convinced the training department and corporate management of the necessity and desirability of additional CRM training. Initially, an existing training program (Mudge, 1993) was assessed for use at this airline, but in the end a training program had to be designed to meet the specific operational needs of this airline.

Background Assessment

Organizational Climate Survey. This project began with an organizational climate survey, which included items about job satisfaction, job description, organizational and safety climate, and the functioning of the training center and the effectiveness of the training provided. This survey was distributed to randomly selected subsamples of pilots, flight attendants, and dispatch and ground personnel.

Overall, pilots were fairly satisfied with their jobs and were generally positive toward the organization, the safety climate, and the training center. These general results indicated that there were no broad-scale organizational problems that would interfere with developing or implementing additional CRM training. A more specific analysis was required to pinpoint specific problem areas for development of CRM procedures.

Identifying Problem Areas. The second step in the development of this program was to identify the areas of operation that most needed improvements to enhance safety. Key issues were determined using the results from the National Transportation Safety Board (NTSB) Safety Study (NTSB, 1994), data from Aviation Safety Reporting System reports, and information from a survey of the airline's instructor/evaluators (I/Es). The data from these sources were used to identify the types of procedural changes that could substantially improve CRM performance in both normal and abnormal/ emergency aircraft operations.

The NTSB Safety Study, which was based on accidents and incidents in U.S. airlines from 1978 through 1990, suggested that there was a pattern to problems in the cockpit:

* in 81% of the accidents, the captain was the pilot flying (PF);
* in 73% of the accidents, it was the crew's first day of flying together; and
* crews who had been awake for more than 11 to 12 hours prior to the accidents made more procedural and tactical decision errors than crews with less time awake.

An analysis of incidents contained in the Aviation Safety Reporting System (ASRS) database also suggested that crew distractions and problems in information transfer played a role in a large number of reports. Specific incident reports in those areas were reviewed to obtain a more detailed problem definition.

Information about specific airline problem areas was generated from a written survey distributed to I/E pilots. This survey tapped observations made by the evaluators during the previous year. Specifically, evaluators were asked to: (a) write down typical CRM problems with suggested reasons for these problems, (b) rank order CRM items where crews had the most problems, and (c) specify CRM items for which crews would benefit from additional training.

This survey led to the identification of three problem areas for this airline: assertiveness, briefings, and decision making. In the area of assertiveness, the I/E pilots felt that crew members were not speaking up with appropriate persistence. Briefings were not done consistently, they were often too general, and there were no guidelines for what constituted an appropriate briefing. Finally, although crews discussed options when making decisions, they did not develop clearly stated plans of action.

Taken together, the data from these three sources lead to the development of three goals for the CRM training program:

1. Reduce distractions to the PF in both normal and abnormal situations
2. Increase structure in briefings to enhance the crew's performance on the first day together and improve information transfer
3. Design checklists, the QRH, and briefings to reduce workload and enhance decision-making skills, especially when crews would be fatigued, running late, or under high workload.

These goals were translated into actual procedures by a design team consisting of the airline CRM coordinator (a pilot); a pilot from a major airline acting as a design consultant; an instructional designer; and researchers specializing in aviation, cognitive human factors, and team research. The diversity of background on the design team resulted in more time required for team interaction but was important to prevent narrowness of viewpoint in the development process.

The initial design of the procedures required only 2 working days. Procedure development was iterative and cycled between the CRM philosophy and principles at the abstract level and operational goals and procedures at the detail level. Procedures could be proposed by any team member, but a proposed procedure had to address one of the major operational goals and derive directly from one or more CRM principles. All team members were consulted for each suggested procedure, and modifications were made until a consensus was achieved.

Further cycles of development involved the fleet management, flight standards, and training personnel of the airline. These cycles were generally critiques of written versions of the proposed procedures, and this part of the process required several months. In retrospect, having a fleet representative as a member of the initial design team might have shortened the time required for fleet revisions and acceptance.

Implementing the Resource Management Program

Discussions with fleet representatives clarified that the implementation of the CRM procedures would require changing the full range of operational documentation to be consistent with the proposed procedures. That is, to avoid confusing pilots, it was essential that the Flight Standards Manual (FSM) and QRH for operating this aircraft be consistent with the developed CRM procedures. To this end, sections of the FSM and QRH were rewritten by two fleet representatives working with the design team. The QRH was completely revised to include the procedures wherever appropriate and to apply human factors principles to QRH indexing, formatting, font, and layout. The abnormal and emergency sections of the FSM were revised to be identical to the QRH, and other appropriate changes were made in the FSM as necessary. These extensive revisions, and the steps required to obtain FAA approval for these revisions, required several more months. The necessity of these collateral changes and the time required for changing official documentation were critical lessons that we learned during implementation.

The next step in the implementation process was training the I/Es at the airline to understand and teach the CRM procedures. Examples of the procedures and methods for the effective instruction of the procedures were introduced to the I/Es. The pilot training course was then developed, along with feedback from those I/Es. While pilots were being trained in small groups over the course of a year, appropriate evaluation materials were developed and implemented as the first step in the systematic evaluation of this resource management program. Intermediate steps in the implementation process were instituting a line operation evaluation (LOE) assessment process and training evaluators for reliable and valid evaluations. The last step in the implementation process was the formal implementation of the CRM procedures as SOP for that fleet.

Formal implementation of the CRM procedures required that the supplemental training for all fleet pilots be completed successfully and that the FAA-approved versions of the revised FSM and QRH be distributed to the fleet. In addition, formal notice was sent to all pilots about the implementation date for the CRM procedures together with an explanation in the form of a refresher videotape that illustrated these procedures. The refresher videotape was developed because the first pilot groups had received CRM procedures training up to a year prior to the formal implementation date. In retrospect, concentrating the pilot training closer to the formal implementation date would have helped to prevent memory loss and degradation of acquired CRM knowledge.

Evaluation of the Resource Management Program

Evaluation Design

The evaluation design selected for this study was a quasi-experimental design (Campbell & Stanley, 1963) in which the fleet that had normal CRM training plus proceduralized CRM was compared with a control fleet that had only the normal CRM training. This design was quasi-experimental because the pilots were assigned to fleets by normal airline rules and seniority rather than by random assignment and because some pilots transitioned between the two fleets during the evaluation, resulting in some knowledge of the CRM procedures in both fleets.

The focal performance measures were crew performance in each fleet on the yearly recurrent LOEs and on random line checks. Because the changes in observed performance were expected to lag behind the formal implementation of these procedures, the design included 3 years of performance evaluations. Performance evaluations covered the year prior to formal implementation of the CRM procedures as a baseline, the implementation year, and the year following implementation to check for slowly emerging effects of the training.

The research team recognized that a structured evaluation process was essential in order to achieve systematic and reliable observations and ratings of performance. Focal performance evaluation measures were selected to provide as complete a set of performance data as could be collected within the constraints of the regional operational environment.

LOE Evaluations

The LOE evaluation, conducted during the pilots' annual evaluation for flight certification, was a work sample performance evaluation with crews performing under a set of scripted conditions. Such evaluations have been used both in the military and civilian sectors (Ford, Kozlowski, Kraiger, Salas, & Teachout, 1997; Leedom & Simon, 1995; Seamster, Hamman, & Edens, 1995; Stout, Salas, & Fowlkes, 1997; Stout, Salas, & Kraiger, 1995). The evaluator adhered to an LOE script to consistently introduce specific problems and distracting conditions. In this way, crew responses to abnormal and emergency situations could be assessed. The evaluation forms emphasized specific crew behaviors for these events, including both technical and CRM performance items. Because the LOE evaluations were conducted on preannounced months, as mandated by FAA requirements, the crew could prepare for the evaluation ahead of time and potentially be on their best behavior for the test situation.

LOE Event Sets. The administration of the actual LOE session was guided by an LOE script that specified all actions and conditions for the evaluator to set for the simulated flight. The script for the entire LOE session was divided into segments called *event sets.* Each event set specified the onset of a focal problem for the crew plus other auxiliary events.

The event set was developed as a refinement of the AQP concept of "event" (FAA, 1991) and was an integral part of training and evaluation. An event set is made up of an event trigger, supporting conditions, and distracters. The event trigger is the condition under which the event was fully activated and presents the crew with some problem or issue in a natural context. *Supporting conditions* are other events taking place within the event set designed to further CRM and technical training objectives and to increase event set realism. Finally, *distracters* are conditions inserted within the event-set time frame that are designed to divert the crew's attention from other events or to increase workload. Most commonly, the LOE event sets corresponded to the usual segmentation of phases of flight. Because of evidence (Seamster, Edens, & Holt, 1995) that assessments made at the event-set level result in more reliable assessments compared to the overall session assessments that are traditionally used at regional airlines, crew evaluations in this project were based on event sets.

LOE Worksheets. Crew performance during the LOEs was assessed using structured worksheets. The evaluation of each LOE event set was guided by a corresponding LOE worksheet page. The LOE worksheet simplified what could be a relatively complex evaluation process, provided instructors with a tool for more reliable ratings, and helped the instructor deliver a more balanced debrief that covered the CRM as well as the technical elements of each event set.

The worksheets helped evaluators standardize the assessment of LOE sessions (American Transport Association [ATA], 1994; Hamman, Seamster, Smith, & Lofaro, 1993). The LOE worksheet had three critical elements that provided a clear structure to the assessment of both CRM and technical crew performance: specific observable behaviors or skills, a standard rating scale, and specific examples of performance for each level of rating. Using worksheets based on each event set allowed the I/E to concentrate on a limited range of observable behaviors and on specific CRM and technical training objectives for each flight segment. Regular I/E calibration training (discussed subsequently) ensured the correct use of these worksheets.

Observable Behaviors. Properly specified observable behaviors were the basis for the evaluation worksheets that guided the I/Es' assessments and provided information for the debriefing process. There are a number of different approaches that can be used in the identification of observable behaviors. In this project we used I/Es' rankings to indicate where crews showed the most CRM problems in the recurrent training environment. We also identified observable behaviors and skills from the fleet-specific task analysis conducted under the AQP.

The specification of observable behaviors was an important step because it affected the implementation of the LOE worksheet and the ultimate outcome of CRM assessment (Seamster, Hamman, & Edens, 1995). These observable behaviors or skills were carefully identified and validated as being central to successful performance on each specific event set. These behaviors or skills provided a point of focus for the I/Es during the observation of the LOE and during the debriefing. These focused observations were a basis for evaluating higher level components of CRM and technical performance. Each evaluator synthesized the observed behavior and rated performance components on a standard rating scale.

Standard Rating Scale. The second critical element of the LOE worksheet was the use of a standard rating scale. In this project standardization was addressed in two different and equally important ways. First, we wanted the rating scale to be standardized in that the same scale would be used throughout the crew assessment process. That is, we wanted a scale that could be used across the full range of evaluation environments, from simulator sessions to line checks, and across a full range of types of items. Currently, many airlines use binary pass–fail scales to assess individual or crew performance but they use rating scales with five or more values to rate CRM performance. The research team felt that a standard scale can and should be used for as many of the different assessment instruments as possible. Using a standard rating scale can reduce the training time required to familiarize I/Es with different assessment instruments. The use of a standardized scale can also increase the amount of practice that I/Es have with that scale. Increased practice with a single scale should also lead to better assessment skills and ultimately better rater reliability. At this airline, a 4-point scale (see Table 10.1) was initially developed for the LOE and later extended to maneuver-validation and line check evaluations.

The second way a rating scale may be standardized is through each scale point being based on a set of specified standards. On the basis of a job analysis, airlines should be able to specify the criteria for "standard performance" for a range of technical and CRM areas (also called *qualification standards*). For this airline, both technical and CRM performance ratings were based on a 4-point scale covering the full range of possible crew performance: *unsatisfactory, satisfactory, standard*, and *above standard*. The labels and precise meanings of each scale point (see Table 10.1) were the result of several cycles of discussion between the I/Es and the researchers. These discussions illuminated the need for both good labels and definitions for each scale point. For example, the label *average* was discussed but discarded as being inherently ambiguous and not clearly tied to the qualification standards for a given item. At the end of this development process, each of the four scale points had a qualitatively unique definition.

Specific Examples. During the first 2 years of evaluation, the I/Es discovered problems in their calibration during I/E recurrent training and requested more guidance in the evaluation of *unsatisfactory, satisfactory,*

TABLE 10.1
Four-Point Rating Scale With Labels and Their Meanings

Rated Value	Label	Precise Meaning
1	Unsatisfactory	Observed crew behavior did not meet minimal requirements.
2	Satisfactory	Observed crew performance met FAA standards but not airline standards.
3	Standard	Observed crew performance met airline standards.
4	Above standard	Observed crew behavior was markedly better than the *standard* performance in some important way.

standard, and *above-standard* performance. Ultimately, a variation of a behaviorally anchored rating scale was developed in which examples of each level of performance for each skill item were defined by training personnel (see Smith & Kendall, 1963). These performance levels included short descriptions of the qualification standard for *standard* performance as well as brief examples of *unsatisfactory, satisfactory*, and *above-standard* performance. For ease of reference, these performance levels were printed on the back of the preceding page of the LOE worksheets so that they would be immediately available for reference during the rating process.

Line Check Evaluations

The initial data collection plan included evaluation of line check data to see if the performance differences between the fleets would be observable in normal operating conditions. In contrast to the LOE evaluations, the line check evaluations were conducted on randomly selected legs of routine commercial flights. There was no prior notification for line checks, and the evaluation situation was natural. However, the limitation in this type of evaluation is that routine commercial flights almost always contain only normal operational conditions. The evaluation forms were therefore structured to emphasize normal conditions occurring on routine flights. Similar to the structuring of LOE items by event sets, the evaluation items on the line checks were structured by phase of flight to reduce evaluator workload.

Initially, an extremely detailed line check form was developed with many items for both the PF and the pilot not flying (PNF) for each phase of flight. However, analyses showed a systematic pattern of missing data for this form that indicated high evaluator workload. The form was subsequently revised with fewer, more general questions, and missing data decreased to a more acceptable level.

The more general line check items could not precisely target the execution of specific CRM procedures. This illustrates an inherent tension in applied research that can occur between operational and research goals. The goal of the research team was to have precise evaluation of the training intervention, whereas the goal of the airline was to have a global evaluation of the safety of line flights. These two goals dictated different sets of items that could not all be accommodated on the line check form without overloading the evaluator. Therefore, an auxiliary nonjeopardy evaluation (jumpseat observations) was designed to be conducted by an independent cadre of evaluators targeting the CRM procedures and expected effects of those procedures.

Jumpseat Observations

The jumpseat observation form was designed to target the occurrence and effects of proceduralized CRM in normal operations. These forms were used in nonjeopardy flight observations carried out by three grant team members and two airline training center staff, all of whom were pilots. All evaluators clarified the nonjeopardy nature of the event and requested voluntary cooperation from the crew before making observations.

The jumpseat observation form was organized by three global phases of flight: departure, cruise, and arrival. The *departure* phase of flight included all performance from preparations at the gate to arriving at cruise altitude. The *arrival* phase of flight included all performance from the beginning of descent to engine shutdown at the end of the flight. For each phase of flight, components of normal briefs required with proceduralized CRM were checked off, and the quality of crew performance on specific CRM procedures and related items was evaluated on a 5-point Likert scale. Finally, each evaluator made a summary evaluation of crew effectiveness for each phase of flight.

I/E Survey

During an evaluator training session, the I/Es maintained that the differences in pilots from the proceduralized CRM fleet compared to the normal CRM fleet were most noticeable when training first officers for an upgrade to captain or when training pilots who were transitioning between fleets. This source of information about the effects of proceduralized CRM had not previously been considered. Consequently, an evaluation form that directly compared crews from these two fleets was developed and administered to all I/Es for both fleets. Results from the 19 I/Es who had trained pilots from both fleets indicated that the effects of proceduralized CRM centered on the three basic performance dimensions of communication, workload management, and planning and decision making. In retrospect, the potential domain of information for comparatively evaluating pilots with and without the procedural CRM training should have been explored more thoroughly.

Pilot Survey

When we expanded our quest for more information on the effects of the training, the evaluations of the pilots themselves became an obvious source. A questionnaire was subsequently designed for pilots that targeted multiple possible effects of training as discussed by Kraiger, Ford, and Salas (1993). All pilots reported their attitudes toward CRM in general and proceduralized CRM in particular. In addition, the ACRM-trained pilots reported on their knowledge of the CRM procedures, practice of the CRM procedures in normal operations, and perceived effectiveness of the CRM procedures. The questionnaire was designed in a voluntary, mail-back format and was distributed to all airline pilots in the last year of the study.

I/E Training

I/E training was found to be the key to combining the LOE worksheets and other elements of the evaluation process into a working, structured assessment system. For this airline we focused on both I/E introductory training and continued standardization training.

Initial I/E Training

The initial training for the I/Es included detailed training on the specific LOE to be used in the evaluation process. This training included the observable behaviors assigned in the LOE, the forms that were used to identify the specific event sets, the assessment criteria for the event sets, and video examples of line crews (not scripted) flying the LOE. This practice supported the identification of the assigned behaviors and began the calibration process for the I/E. We found it critical that the I/E training provide the opportunity to watch video segments of event sets and discuss their ratings of the crew. We used a technique of assigning I/Es a deidentified pilot identification number. After the I/Es watched a video segment of the LOE and made their assessments, their ratings were processed by a data collection program and the results immediately returned so that they could compare their performance against those of the other I/Es. This technique provided actual values for comparison and took some of the subjectivity out of the I/E calibration process.

In addition to the specific training for the LOE assessment, the I/Es' training also included the skills required to brief, administer, and debrief the LOE. Each I/E was required to (a) fly the LOE, (b) observe the administration of the LOE by a previously trained I/E, and (c) practice giving the LOE as an evaluator. The I/Es switched PF and instructor roles during the session to allow them the opportunity to experience the various roles in this type of training. The I/E and the flying instructor pilot performed a facilitated debrief after each segment was completed. The non-flying I/E pilot initially began these debriefings with observations of the CRM skills that were exhibited or that should have been used during each event set. This experience was designed to enhance observation and identification skills, give practice using the LOE worksheets for evaluation, enhance the I/Es' facilitation techniques by the sharing of concepts and techniques by the three pilots, and give practice in the type of crew-centered briefing–debriefing they would be conducting with normal line crews.

Continued Standardization Training

After the initial training, refresher courses and calibration sessions were established. The frequency of this training depended on calibration results, I/E turnover in the training center, perceived issues or problems with the evaluation process, and so forth. Standardization training took place at least twice a year, more frequently when the reliability data suggested it was needed. This training day included feedback on the forms the I/Es were using, specific problems with consistently administering or evaluating the LOE, video samples with assessment exercises of event sets, and immediate instructor feedback on the I/Es' ratings in comparison with other instructors. The regional airline working with the researchers placed great emphasis on these improved training techniques to ensure a systematic and standardized evaluation process.

Summary of Experiences With the Regional Airline

The AQP that the airline had adopted for this fleet required the assessment of CRM with technical skills in the operational environment. The assessment process was initially focused on the LOE, which contained observable behaviors and technical skills, organized by event sets. This organization allowed for the development of tools (LOE worksheets) to support the I/E in making fair and accurate assessments of crew performance. Assessment tools were accompanied by a comprehensive I/E training program that included:

- training with video examples of line crews flying the LOE scenarios,
- clear definitions of rating scales and standardization,
- computerized methods for I/Es to receive objective (data-supported) feedback on their evaluation performance,
- I/E simulator training programs for practice in flying and administering the scenarios, and
- continuing training during the AQP cycle to monitor and maintain I/E calibration.

Although the initial focus was on the LOE, several auxiliary evaluation methods were developed to assess the effects of proceduralized CRM during the course of this project. Because each type of evaluator (I/Es, pilots, jumpseat observers) had a different viewpoint, and each form of evaluation had a different set of strengths and weaknesses, it was important to assess the effects of any training program with a variety of evaluators and methods. In our project this approach led to a more complete evaluation of the resource management procedures that were developed, implemented, and evaluated at the regional airline. Overall, the data (which are described in Holt et al., 1999) suggest that the ACRM program lead to increased effectiveness in the cockpit.

RESOURCE MANAGEMENT AT A MAJOR AIRLINE

At the time we developed the resource management program for the regional airline, a major airline expressed interest in evaluating some kind of procedural CRM as a part of their resource management program. However, this airline had a number of large fleets, and it did not seem feasible to completely rewrite the abnormal procedures, as we had done with the regional airline. Instead, we took some specific aspects of the CRM procedures developed in working with the regional and adapted them for a specific application at the major airline.

Program Development

Need for the Training Program

Although no formal needs assessment had been conducted, officials at the major airline were generally dissatisfied with their current stand-alone CRM training and perceived a need for more effective tools to help pilots, especially in abnormal/emergency situations. This effort was lead by the quality assurance administrator for the training center. At this airline the quality assurance administrator can take an active role in working with fleets to improve the effectiveness of training. This person acted as a liaison to the fleet, which was most interested in improving pilot performance using a proceduralized CRM approach.

Background Assessment

Given the limited scope of this part of the project and the limited resources available, no organizational climate survey was performed. For a major airline, fleets can be relatively independent, and the climate within each individual fleet may differ from the overall organizational climate.

Identifying Problem Areas. On the basis of detailed discussion of critical incidents (NTSB, 1994) and ASRS reports, the airline had previously identified crew performance during abnormal or emergency situations as potential problem areas. The airline's initial goal was to reduce the reliance of the crew on memory for the correct execution of actions during the first 2 minutes of an abnormal or emergency by developing a Quick Reference Card (QRC), which listed the actions to be taken in a checklist fashion. The research team subsequently proposed augmenting the QRC with some aspects of proceduralized CRM to study its effectiveness with the pilots in the cooperating fleet.

One important lesson we learned at this airline was the critical nature of cooperation with the local FAA inspector in making any training changes. Our experience at the regional airline was that the local FAA inspector had been briefed on the project goals and was supportive of the training changes required to achieve these goals. In contrast, the local inspector for the major airline required objective evidence of the effectiveness of the QRC before considering its approval. This necessitated a controlled study in which 120 crews were randomly assigned to QRC or normal (non-QRC) conditions and evaluated on a nonjeopardy LOE. Although the results ultimately supported the use of the QRC in reducing crew errors during these situations, and the use of the QRC was finally approved, there was a delay of several months.

Development of Procedures
for Abnormal/Emergency Situations

In developing the program for the major airline the goal was to assess the effectiveness of developing a *general framework* that captured the resource management practices that had been incorporated into the regional airline's QRH procedures for abnormal or emergency situations. The general framework that was developed after an iterative cycle of development with the airline and the research team emphasized five basic steps or phases of crew activity for each abnormal or emergency situation. These steps captured the general approach to decision making that had been embedded in the FSM and QRH at the regional airline. The first step was clearly assigning PF/PNF duties. The second step was a careful and complete diagnosis of the problem. The third step was a complete situation assessment. The fourth step was planning and decision making. The fifth step was plan execution and monitoring.

At a detailed level, each general step had specific subelements that should be evaluated or performed. For example, in the planning-and-decision-making step the crew was advised to develop "bottom lines" and "backup plans." A *bottom line* specifies the lower limits for some aspect of the situation, at which point the current plan is modified or discarded. A *backup plan* is a prearranged alternate version of the current plan that is enacted when a bottom line is reached. For example, a bottom line may specify 30 minutes of fuel remaining before diverting to an alternate airport. The exact alternate airport and necessary approach procedures constitute the backup plan.

To ensure cross-briefing, situation awareness, and full use of the viewpoints of each crew member, we developed a "Brief, Advocate, and Resolve" (BAR) cycle. In the BAR cycle, the PNF was assigned to the main duties in the problem-diagnosis, situation-assessment, and planning steps. At the end of each major step the PNF was required to brief the PF. The PF was then required to critically evaluate the briefing contents and advocate any different ideas or views about the problem, situation, or plan. Finally, these different ideas or views were to be resolved so that all crewmembers were in agreement before proceeding to the next step.

To ensure consistency with airline operations and pilot expectations, the general framework from the regional airline was revised into a CRM checklist and integrated with the major airline's training philosophy, CRM principles, and nomenclature. Each step in the cycle became a checklist item. Detailed points for each step were designed to give maximum guidance in performing each step and to provide best practices. For example, in the situation-assessment step crews were advised to consult all relevant sources of information, such as air traffic control, dispatch, maintenance, flight attendants, and so on. Clearly, consulting all possible sources of information may take an unrealistically long time. Therefore, the PNF was instructed to complete the five basic steps with as many of the detailed subelements as he or she could complete in the available time.

Evaluation of Procedures for Abnormal/Emergency Situations

Evaluation Design

This project is designed to evaluate abnormal/emergency CRM procedures in a completely randomized field experiment. Crews in training volunteered for participation and were randomly assigned to one of three groups. The control group was trained in the current airline materials and procedures for emergency and abnormal situations. The second group was trained in generic CRM procedures for these situations and used a corresponding standalone CRM checklist. The third group was trained in generic CRM procedures for these situations and used a set of CRM procedures integrated into the abnormal/emergency sections of the FSM.

Development of Evaluation Materials

Crews were evaluated on one of three different LOEs in this field experiment. The LOEs represent three completely different abnormal/emergency problems during the cruise phase of the scenario, each requiring use of a different section of the FSM. The advantage of using LOEs with different content for different crews is that they cannot anticipate the content of the LOE with any certainty. This should result in more natural crew reactions to the problem situations. However, there is a disadvantage associated with using

LOEs with different content: It is possible that the content of the LOE may interact with the training manipulation in unique ways and make it more difficult to determine if there is a general effect of the training and procedures.

Crews were evaluated on their LOEs with the normal event set worksheets used by this airline. These sheets were supplemented by an additional evaluation sheet focusing on whether the crew performed specific CRM procedural elements when the critical trigger event occurred. The additional evaluation page focused on (a) division of PF–PNF duties, (b) problem diagnosis, (c) situation assessment, (d) planning and decision making, and (e) plan execution and monitoring. Appropriate subsidiary items were included in each area. For example, in the planning-and-decision-making section detail items evaluated whether pilots established bottom lines and developed explicit backup plans.

Subjective reactions to the training program also were collected. A brief mail-back questionnaire was given to pilots to evaluate the effectiveness of their training materials. The pilots' evaluations of the materials in the control and procedural CRM groups can be compared to assess which type of training is perceived as more effective.

Summary of Experiences With the Major Airline

The cooperating fleet at this airline is also an AQP fleet and has implemented LOE evaluation, LOE worksheets, and rater calibration training in a fashion similar to that of the regional airline. The scope of the proceduralized CRM intervention was much narrower at the major airline, targeting only crew teamwork performance after an abnormal or emergency situation has occurred. Subsequent airline implementation of any form of proceduralized CRM will depend on the results of this evaluation as well as the local FAA and other major stakeholders.

LESSONS LEARNED IN DEVELOPING RESOURCE MANAGEMENT PROGRAMS IN AVIATION

From our experiences with these two airlines, we have learned a number of lessons that are critical to the success of a resource management program within the airline environment.

Lesson 1: Assess Organizational Commitment

On the basis of our experiences with these airlines it is evident that a resource management program should be implemented as an ongoing process that requires the involvement of the entire organization. This means that a commitment should be made throughout different levels and different sections within the organization.

Guideline 1: Commitment Should Start With Senior Management

Senior management should support resource management improvements and demonstrate that support through actual behavior. If senior management says that it values good resource management, then it should honor that commitment. As an example, single-engine taxiing saves money but results in multiple tasks and added distractions during the taxi phase. These add to the crew's workload and make safety-critical CRM procedures such as scanning the area for traffic and thoroughly briefing the conditions for the departure more difficult. If the organization says it values good CRM but requires single-engine taxiing to save money, it is not honoring its commitment.

Guideline 2: Management Policies Should Be Implemented Using Consistent Incentives

In practice, management policies are not always implemented with consistent incentives; however, this is critical to the success of resource management training. Without appropriate rewards for participating in the resource management program, it will be difficult to convince the "rank and file" to adopt the procedures and practices trained in the program.

Guideline 3: The Training Department Should Demonstrate Its Commitment

The training department often identifies the need for resource management training, so it should not be difficult to get a commitment from that department. However, at a large airline the training department can have diverse sections, with different goals and agendas. In such a case, all sections of the training department should cooperate to make the training work. Furthermore, links should be established and maintained between the training development team and groups representing the customer fleets, the flight standards or flight safety department, and the quality assurance or evaluation department.

Guideline 4: Identify an Internal Champion

Because of the financial considerations involved in establishing and maintaining a resource management training program, it is important that the program have an internal champion who will fight for necessary resources for the program. It is helpful if this individual is well positioned in the company, both in terms of having financial resources available and in terms of commanding respect from personnel within the company.

Guideline 5: Resources Should Be Committed to the Project

Relevant resources always include money and time of relevant personnel and may also include scarce resources, such as simulator time. The procedure-development process itself can be accomplished with a relatively small budget, but associated elements, such as changes to operations documents and training, can result in substantial expenses. The management structure of the organization should be considered in obtaining resources. Obtaining resources in a very hierarchical or balkanized organization will require more time and effort than in a flat, consolidated organization.

More resources may be available at larger airlines, but the costs of implementation may also be correspondingly greater. For the major airlines, one change in a procedure can result in the need to reprint one or more pages for 40,000 manuals. This can become a very large cost, and people who do not fully support or understand the need for CRM may use that as an argument to limit or terminate the program. A strong case can be made that procedural changes can be integrated into a scheduled manual update for little or no additional cost. Thus, with careful planning, CRM procedures can be developed and established efficiently by reducing or eliminating some of the larger expenses.

Additional considerations for effective use of personnel resources include integrating some of the team members' development time with their normal responsibilities. However, new activities should not be assigned without providing the appropriate hours to do them. The best solution is to look for ways to replace some of the existing activities with the CRM procedures development work. It is important to keep the development process on schedule and anticipate and reduce delays that can increase costs in time and resources.

Lesson 2: Background Assessments

Guideline 6: Conduct an Organizational Climate Survey

Conducting an organizational survey during the early stages of training development can help identify organizational elements that need to be involved in the process. For this project, for example, early inclusion of fleet representatives may have facilitated later development steps. The survey can show which departments or fleets understand and want to be involved with the training program and which will require more communication and information. In addition, the organizational survey can be used to increase the entire organization's awareness of the new approach to resource management. The survey development and administration process can be used as a way to get essential parts of the organization introduced to the new program. Asking key individuals from different departments to provide material for the survey or having them review specific survey items will improve the quality of the survey and make those departments aware of the new program. The survey should address issues and concerns of the flight attendants, dispatchers, maintenance, and other key parts of the operation.

Some of the items that should be considered for inclusion in an organization survey are airline safety climate, communication and cooperation, departmental management and structure, job responsibilities and standards, organizational management, professionalism and job performance, and quality and frequency of training. Any severe problems indicated by the organizational survey may have to be addressed before training innovations can be attempted with a reasonable expectation of acceptance and cooperation. For example, if one fleet or section of the training department is dead set against the principles and approaches of the proposed training, that issue must be addressed before proceeding with development.

Lesson 3: Program Development

Guideline 7: Conduct a Needs Assessment

A good resource management program will be based on a needs analysis that is based on data from the other airlines and the airline itself. The needs analysis is used to identify the most important crew performance problems and then determine what form of intervention will best correct those problems. For airlines that do not have data pointing to specific problems, it is useful to review industrywide reports to learn where other airlines have crew performance problems. Finally, informal or qualitative data from the training or evaluation departments within the airline can also be used to identify airline-specific problems, as was the case at the major airline in the project discussed earlier. The preceding activities will provide sufficient data to pinpoint the CRM problem areas and provide possible causes.

Guideline 8: Identify and Prioritize a Limited Number of Training Goals

On the basis of the needs analysis, the development team should choose those problems that point to the need for one or more CRM procedures. The primary determination to be made is whether the performance problem would be resolved best by training, equipment redesign, new or modified procedures, or some combination of these. For example, at the major airline the CRM checklist was implemented in part by a new or modified checklist and in part by appropriate training in best practices for each major CRM step. A airline's philosophy and policy must be accurately understood to determine whether a procedural solution is the right one for that specific operation and organizational climate. The development team should be sensitive to the fact that different departments may have different local versions of the airline's philosophy and that these differences may have to be reconciled before development work can continue.

Lesson 4: Evaluating the Training

Guideline 9: Evaluate the Effectiveness of Your New Training Program

As the training is being developed, the appropriate evaluation materials should also be developed and tested. At each airline evaluated, the development of the best format for the LOE worksheet required several cycles. Similarly, the LOE content and script may need to be refined so that the evaluation situation requires specific resource management skills and behavior. One of the most difficult parts of completing the LOE worksheet is correctly specifying observable behaviors or skills.

Guideline 10: Identify and Clearly Define Specific Observable Behaviors

Specifications that have more than one behavior per observation (such as global observations previously used in research) can make that observable behavior more difficult to check than a single behavior. Concise and simple wording is recommended, and the verb should refer to a clearly observable behavior. Ultimately, the observable behavior has to be worded so that the items can be understood and used by I/Es, who often experience high workload in the assessment process. For example, the statement "Captain coordinated flightdeck activities to establish proper balance between command authority and crew member participation, and acts decisively when the situation requires" which was used on one assessment form, can confuse the evaluator because there are so many elements to consider. If the airline wants to concentrate on the assessment of the captain's ability to assign cockpit activities, a single observable behavior might be stated as "captain assigns tasks effectively." Therefore, it is important to use I/Es to review and validate the observable behaviors. Guidelines for specifying observable behaviors (Seamster, Hamman, & Edens, 1995) include the following.

Guideline 10a. Carefully examine the existing task analyses and related flight and training manuals to ensure that they include all the key performance elements, especially CRM elements.

Guideline 10b. Use I/Es to identify training issues, to determine possible relevant behaviors, and to validate the observable behaviors.

Guideline 10c. Specify observable behaviors as simple statements of actions that can be clearly detected by I/Es who may be under a relatively heavy workload.

Guideline 10d. Specify the relevant interpersonal or technical activities or actions so that I/Es are not asked to observe mental activities.

Guideline 10e. Organize observable behaviors on worksheets for each event set to produce more reliable assessments and more systematic debriefings.

Guideline 10f. Validate the assigned observable behaviors by having the instructors fly the event sets and assess the CRM behaviors chosen.

Guideline 11: Develop Additional Documentation to Support the Training Program

In addition to the procedures developed, there may be a need to develop documentation that helps the pilots learn and implement the new procedures. For example, in this project a refresher videotape was designed to help pilots recall the correct implementation of the new procedures. Similarly, official documentation such as the QRH and FSM may have to be revised for consistency with the new procedures. Modifying official documentation requires FAA approval that may, in turn, require additional time or revisions.

Lesson 5: Report the Program Assessment and Performance Data

Guideline 12: Report the Data You Collect

The reporting of data can either be focused on the effects of a specific training intervention or more broadly focused on an overall picture of crew performance. For the project involving the regional airline the research team's goals of evaluating the proceduralized CRM training required a focused data analysis and report. In contrast, the airline goals for analyzing the data included the main goal of a broad picture of crew performance on different assessments as well as the more minor goal of evaluation of training effectiveness. A more accurate picture of crew performance can be an extremely valuable result of improving the quality of evaluation.

Timely, accurate, and relevant data reporting is one of the most important by-products of this kind of program. Using these improved assessment methods makes it possible to collect a large amount of accurate, reliable, and valid performance data concerning both the technical and teamwork aspects of crews. In addition, tools (see chap. 9) allow airlines to analyze different dimensions of I/E reliability and ensure calibrated evaluations. However, the large amount of data combined with different possible forms of analysis can result in an overwhelming amount of information, so the challenge is to report the essential data in a format that is usable in regard to the appropriate organizational elements.

There are four main organizational elements to consider for data reporting; each of these stakeholders may have different information needs:

1. Crews and union representatives
2. I/Es
3. Training department management and quality assurance
4. Fleet managers and other relevant fleet personnel.

For example, a report designed for fleet managers might emphasize some of the following within-fleet data: distribution of overall performance by position (captain, first officer, second officer), distribution of technical performance by position, distribution of CRM performance by position, maneuver-validation ratings by type of maneuver, distribution of event set performance where there are problems, and overall crew performance trends (quarterly and annual).

The training department management, on the other hand, may also be interested in comparisons across fleets; interesting data might include distribution of overall performance by fleet, distribution of technical performance by fleet, distribution of CRM performance by fleet, maneuver validation ratings by type of maneuver and by fleet, event set performance by fleet where there are identical event sets, and overall crew performance trends by fleet.

APPLICATION TO OTHER DOMAINS

Although our guidelines and conclusions are based on research in aviation, the approach and guidelines apply equally to the development of training programs in other domains. Specifically, the guidelines for developing and evaluating a resource management program will be applicable to the extent that an organization contains teams, groups, or crews that are a key factor in overall organizational performance. The exact applicability of the lessons and guidelines will depend on the nature of the organization and the nature of the domain.

Organizations differ in many respects, some of which can influence the applicability of the lessons and guidelines. For example, the more hierarchical the organization, the more important it is to obtain clear-cut organizational commitment from the top levels. In contrast, in a decentralized organization the division or section leader may have sufficient autonomy, control, and resources to develop, implement, and evaluate a resource management program. In our work, the regional carrier was a small, flat organization with close links among the divisions. In contrast, the major carrier was more hierarchically structured, with traditional lines of authority, personnel assignment, and budgeting. The necessary level of organizational commitment can be influenced by these factors. The background assessments, such as an organizational climate survey, also will differ depending on the nature of the organization for which the training is developed. The climate survey must, for example, represent the relevant functional divisions or departments of the organization.

The nature of the domain also may moderate the precise applicability of the lessons and guidelines. In general, any domain where communication, coordination, or interaction among team members is required to achieve acceptable levels of performance may benefit from the development, implementation, and evaluation of resource management training. Examples of domains having critical performance vested at the team, group, or crew level are nuclear power plants, emergency rooms, military command and control centers, oil refineries or chemical plants, and organizations specializing in the development or maintenance of large-scale computer programs. In many other traditional domains, such as service industries, production lines, and so forth, the critical influence of team or group performance may be equally strong but not as immediately obvious. If the interdependency or interaction among groups of workers is subtle and informal, the contributions of group-level processes to outcomes may be particularly missed. Therefore, an objective look at the nature of the artifacts, processes, and personnel that contribute to key organizational outcomes may reveal unrecognized interdependencies. Wherever interdependencies of some sort exist among the group, team, or crew, the resource management assessment, training, and evaluation processes outlined in this chapter may be beneficial.

Of course, the specific artifacts, processes, and personnel in each domain may dictate differences in the identification of operational problems, program development, and planning for performance assessment. However, in most applied domains there is relevant human factors research that can provide a framework for these steps. That is, the research literature may help

Identify operational needs, effective methods for a training program, and relevant measures or methods for evaluating performance. When available, these off-the-shelf measures and methods greatly reduce the time and expense of needs assessment and program development. Even if the measures and methods in the research literature do not directly apply to a given domain, they can often be adapted for appropriate use in a related domain. For example, the surveys and approaches developed to tap organizational climate, knowledge, and attitudes in the aviation domain could be tailored to be appropriate measures in alternate domains, such as command and control centers.

CONCLUSION

Our experiences suggest that an organization can direct improvements in resource management through the development of and training in management procedures customized to its own operation. Over the last 20 years, resource management has evolved from an emphasis on attitudes and personalities to a focus on specific resource management skills. The procedural approach outlined in this chapter complements that skill-based approach, allowing organizations to move beyond the elaboration of general principles and policies to specific procedures designed to improve worker performance. This procedural approach to resource management facilitates a more focused training and assessment program that allows organizations to analyze the effect of the new procedures on individual and team performance. When wisely used, these tools can produce a large quantity of high-quality performance information, which can be used by the organization in many different ways. Once relevant information is obtained, the challenge remains to isolate the essential information and present it effectively to all relevant stakeholders. This approach is potentially relevant for a wide variety of organizations that use groups, teams, or crews to perform critical jobs or tasks.

ACKNOWLEDGMENTS

This research was supported by the Office of the Chief Scientific and Technical Advisor for Human Factors (AAR-100) at the Federal Aviation Administration through Grant 94-G-034 to George Mason University. We thank Dr. Eleana Edens in Aviation Research (AAR-100) and Dr. Thomas Longridge in Aviation Flight Standards Service (AFS-230) for their continuing support of this work. This chapter is based in part on the article: Boehm-Davis, D. A., Holt, R. W., Hansberger J. T., & Seamster, T. L. (1999). Overview of lessons learned developing ACRM for a regional carrier. In *Proceedings of the Tenth*

International Symposium on Aviation Psychology (pp. 966–972). Columbus: Ohio State University Press.

REFERENCES

Air Transport Association. (1994). *Line operational simulations: LOFT scenario design, conduct and validation* (LOFT Design Focus Group, AQP Subcommittee Report). Washington, DC: Author.

Campbell, D. T., & Stanley, J. C. (1963). *Experimental and quasi-experimental designs for research.* Chicago: Rand McNally.

Federal Aviation Administration. (1991). *Advisory circular 120-54: Advanced Qualification Program.* Washington, DC: Author.

Ford, J. K., Kozlowski, S. W. J., Kraiger, K., Salas, E., & Teachout, M. S. (Eds.). (1997). *Improving training effectiveness in work organizations.* Hillsdale, NJ: Lawrence Erlbaum Associates.

Hamman, W. R., Seamster, T. L., Smith, K. M., & Lofaro, R. J. (1993). The future of LOFT scenario design and validation. In *Proceedings of the Seventh International Symposium on Aviation Psychology* (pp. 589–593). Columbus: Ohio State University Press.

Holt, R. W., Boehm-Davis, D. A., Hansberger, J. T., Beaubien, J. M., Incalcaterra, K., & Seamster, T. L. (1999). *Evaluation of proceduralized CRM training at a regional airline* (Tech. Rep.). Fairfax, VA: George Mason University.

Kraiger, K., Ford, J. K., & Salas, E. (1993). Application of cognitive, skill-based and affective theories of learning outcomes to new methods of training evaluation. *Journal of Applied Psychology, 78,* 311–328.

Leedom, D. K., & Simon, R. (1995). Improving team coordination: A case for behavior-based training. *Military Psychology, 7,* 109–122.

Mudge, R. W. (1993). Pilot judgment—And the management system. In *Proceedings of the Seventh International Symposium on Aviation Psychology* (pp. 216–220). Columbus: Ohio State University Press.

National Transportation Safety Board. (1994). *Safety study: A review of flightcrew-involved, major accidents of U.S. air carriers, 1978 through 1990* (PB94-917001 NTSB.SS-94/01). Washington, DC: Author.

Seamster, T. L., Edens, E. S., & Holt, R. W. (1995). Scenario event sets and the reliability of CRM assessment. In R. S. Jensen (Ed.), *Proceedings of the Eighth International Symposium on Aviation Psychology* (pp. 613–618). Columbus: Ohio State University Press.

Seamster, T. L., Hamman, W. R., & Edens, E. S. (1995). Specification of observable behaviors within LOE/LOFT event sets. In R. S. Jensen (Ed.), *Proceedings of the Eighth International Symposium on Aviation Psychology* (pp. 663–668). Columbus: Ohio State University Press.

Smith, P. C., & Kendall, L. M. (1963). Retranslation of expectations: An approach to the construction of unambiguous anchors for rating scales. *Journal of Applied Psychology, 47,* 149–155.

Stout, R. J., Salas, E., & Fowlkes, J. E. (1997). Enhancing teamwork in complex environments through team training. *Group Dynamics: Theory, Research, and Practice, 1,* 169–182.

Stout, R. J., Salas, E., & Kraiger, K. (1995). The role of trainee knowledge structures in aviation team environments. *International Journal of Aviation Psychology, 7,* 235–250.

11

Applying Crew Resource Management on Offshore Oil Platforms

Rhona Flin
Paul O'Connor
Aberdeen University, Scotland

The aviation industry recognized the significance of human error in accidents almost 30 years ago and has been instrumental in the development of effective training programs, designed to reduce error and increase the effectiveness of flight crews, known as *crew resource management* (CRM; Wiener, Kanki, & Helmreich, 1993). Because of the success of CRM in the aviation industry, it has been adopted by a number of other professions, including anesthesiologists (Howard, Gaba, Fish, Yang, & Sarnquist, 1992), air traffic control (ATC; introduced in the United Kingdom as *team resource management*); the merchant navy (Byrdorf, 1998), the nuclear power industry (Harrington & Kello, 1992); aviation maintenance (Marx & Graeber, 1994) and, the subject of this chapter, teams on offshore oil and gas installations (Flin, 1995).

In this chapter we outline how the basic principles of CRM training and assessment, as used in the aviation industry, have been used as a basis for the design of offshore CRM courses. We briefly describe the U.K. offshore oil and gas industry and explain why CRM training is deemed appropriate for this work environment. Examples of CRM training for offshore emergency response teams and control room teams are presented, followed by guidelines for adapting CRM training for more general purposes in the energy industry.

THE U.K. OFFSHORE OIL AND GAS INDUSTRY

Natural gas was discovered in the North Sea in 1965 and arrived ashore 15 months later in 1967. Oil was discovered in 1970 and brought ashore in 1975 (United Kingdom Offshore Operators Association, 1998). There are approximately 136 offshore fields feeding more than 200 installations in the British waters of the Atlantic Ocean, the North Sea, and the Irish Sea. These installations range from small gas platforms to mobile and jackup drilling rigs, production ships, and large production platforms. The offshore workforce must carry out complex and potentially hazardous operations in a constrained, isolated, and remote environment characterized by a combination of factors that are not typically found in other industries. These include helicopter transportation to and from the work site, living near high-inventory dangers, a 24 hour society, a mixture of cultural backgrounds, limited leisure opportunities, 12-hour workdays, rotational 2- to 3-week periods at work and at home, and rough weather conditions (Flin & Slaven, 1996; International Labour Office, 1993).

As the U.K. Continental Shelf has matured as an oil and gas province, there have been changes in the organization and management of offshore operations. In the early days of North Sea exploration and production, a typical platform would have 100 to 250 personnel on board, with the majority of the workforce employed by the operating company. By the 1990s, technological innovations, increases in operational efficiency, and cost reductions had resulted in downsizing and multiskilling, which contributed to a minimum-staffing approach. Now smaller crews (80–120), mainly contractor staff are required to run the installations, and effective team working is the essence of safe and productive operations (Boyd, 1996). Although current fatality and injury rates tend to be comparable to or better than those of other hazardous occupations (Health and Safety Executive, 1995) there is significant concern for safety, with a current industry initiative to reduce accidents by 50% by the year 2000 (see www.oil-gas-safety.org.uk).

In the United Kingdom, the upstream oil industry has been at the forefront of technological development; however, in terms of safety management, psychosocial or human factors issues have tended to play a secondary role to engineering solutions (Flin & Slaven, 1996). Around 1990, this technical myopia began to change, because of a major offshore accident that highlighted the risks of failing to understand the human and organizational dimensions of accident causation and prevention.

The Piper Alpha Disaster

The Piper Alpha disaster is a chilling illustration of how a chain of operators' and managers' errors, in an organizational culture of "precipice management," can have catastrophic consequences. On July 6, 1988, *Occidental's* Piper Alpha platform, situated in the North Sea 110 miles northeast of Aberdeen, Scotland, suffered a series of explosions that resulted in the deaths of 167 men (67 survived). Night shift operators had inadvertently switched over a pump that had been shut down for repair by members of the day shift, who had removed a pressure valve from the pump's relief line. Most of the men survived the initial gas explosion, apart from the one or two who were probably killed instantly (Cullen, 1990). However, there was no blast wall around the area where the explosion occurred, only fire walls, and so an oil fire quickly took hold. Furthermore, two other platforms (*Texaco's* Tartan and *Occidental's* Claymore) that were feeding into the same oil export line did not shut down until 1 hour after the initial mayday call, resulting in oil from these other platforms flowing back toward Piper Alpha and fuelling the fire. Riser pipelines eventually ruptured in the heat, and the explosion engulfed the platform in thousands of tons of burning gas. As with most industrial accidents (Reason, 1997; Turner & Pidgeon, 1997), a chain of human and organizational errors led to the disaster, both in terms of safety management and emergency response. Several examples of behavior that could be described as poor CRM include the following:

• There was a lack of communication between the day crew and the night crew on Piper Alpha, resulting in the "*failures in the transmission of information under the permit to work system and at shift handover*" (Cullen, 1990, para. 6.188). Thus, the night crew were unaware that the pressure valve in Pump A had been temporarily replaced with a blind flange that was not leaktight. However, as there was no inspection of the work, an error in the fitting, if it had happened, could not have been detected and fixed.

• On the Claymore, although the operating superintendent requested that the OIM (offshore installation manager; the most senior individual on an offshore installation) cease production on six separate occasions, "*he clearly deferred to him for his decision*" (para. 7.49). Lord Cullen concluded from the evidence that the OIM "*was reluctant to take responsibility for shutting down oil production*" (para. 7.49). The failure to shut down the Claymore earlier was due to a lack of assertiveness on the part of the operating superintendent and poor leadership, situation awareness and decision making by the OIM.

• On the Tartan, once the OIM had finally reached the decision to depressurize the pipeline, the process took over 45 minutes. The OIM commented that "*on this particular night I think personnel were suffering from shock, so they would be additionally cautious in what they did, so maybe it took longer than expected*" (para. 7.42). Furthermore, although it was discussed, the OIM also did not consider using a fast rate of blowdown to depressurize the pipeline. Thus, there were errors on the part of the OIM in terms of decision making and poor situation awareness. Furthermore, because of the failure of the crew to deal with the stress of the incident—and, possibly, poor leadership—depressurizing the pipeline took longer than expected.

• The OIM on Piper Alpha did not exercise good leadership in response to the emergency. *"The OIM. . . . appeared to be in a state of panic"* (para. 8.9). Furthermore, the radio operator said that *"he himself was also panicking and the message* [that the platform was to be abandoned] *was haphazard"* (para 8.9). *"It is unfortunately clear that the OIM took no initiative in an attempt to save life"* (para. 8.35). Thus, the OIM demonstrated a failure in leadership and decision-making skills, and both he and the radio operator were unable to cope with the extreme stress of the situation.

The Piper Alpha disaster highlighted not only the human factors aspects of accident prevention but also underlined the need for training to cope with the stress of decision making and team management in an emergency situation. The regulator (Health and Safety Executive [HSE], Offshore Safety Division) commissioned a survey of how organisations select and train their managers to take command in a crisis (Flin & Slaven, 1994), and the oil companies began to turn to other high-reliability industries to learn from their emergency management training techniques. Of particular interest was CRM, which was already being used in major airlines, such as *British Airways*. Several operating companies began to liaise with the airlines to learn these techniques and to introduce them into their emergency response training programs.

CRM FOR PLATFORM EMERGENCY RESPONSE TEAMS

CRM training has been used by the international oil company *Shell Expro* as part of their control room operators' emergency response training (Flin, 1995) and for OIMs and their emergency response teams (Flin, 1996). A description of these courses is given below. Similar courses have also been adopted in the Scandinavian oil industry: *Elf Petroleum Norge* has developed a CRM course called "Emergency Resource Management" (Grinde, 1994).

Offshore Control Room Operators

A four-module CRM course was used for human factors training during a program of offshore control room operator competence assessments and emergency response training (Flin, 1995). Control room operators on an offshore oil installation carry out a role very similar to that of control room operators in a power station or petrochemical plant. The assessment was carried out on an onshore simulator, and the company, having examined aviation CRM packages, decided to incorporate four modules that they had decided were particularly relevant: decision making, communication, stress, and assertiveness. These were originally based on the type of CRM modules used by commercial airlines; however, by drawing from industrial psychology research (e.g., Flin & Slaven, 1994) and the expertise of the trainers, a customized CRM package was designed. A brief description of the content of each module (which lasted 2.5 hr) is as follows:

• *Decision making.* This module aimed to emphasize to participants the difference between making a decision in a normal situation and making one in an emergency. Instruction in the identification of cognitive and social factors that aid or hinder decision making under stressful conditions was presented, and methods for making good decisions when under pressure identified. The architecture of the human memory system was presented to demonstrate the limitations and sensitivity of working memory against the strengths of long-term memory (Baddeley, 1992). One of the naturalistic decision-making models, recognition-primed decision making (Klein, 1993), which relies on rapid retrieval from long-term memory, was used as a framework for discussing different styles of decision making.

• *Communication.* Through the discussion of an actual offshore incident involving a communication problem, factors that promote and hinder communication were considered. Exercises were used to illustrate the importance of listening skills, feedback, the role of nonverbal communication, and effective communication, techniques.

• *Assertiveness.* Assertiveness was defined, and its relevance to control room operators was examined. Also, the effect of different behavioral styles on oneself and others was considered. Exercises were used that involved role playing of different styles of behavior (passive, assertive, aggressive) in control room situations that would merit an assertive response, such as: *"You have a major process upset and you receive the second call from the Toolpusher saying that he wants fresh water immediately. What do you say to the Toolpusher?"*

• *Stress.* This module was designed to advance understanding of the causes and effects of stress (Cox, 1993), particularly its effects on decision making. Personal experiences of stress were discussed, and a variety of coping strategies that could be used in the control room were considered.

The teaching methods included lectures, exercises, and role play in addition to discussion of personal experiences related to the topics. The materials were developed initially by the psychologists and then refined in conjunction with the control room trainers. The trainers were then instructed on how to deliver the materials. The first course was given by the psychologists and observed by the trainers, the trainers taught the second course and were observed by the psychologists, and the trainers taught the courses thereafter.

OIMs and Their Emergency Response Teams

CRM training has also been used with OIMs and their teams undergoing emergency response team training, normally based in an emergency control center simulator facility. The nontechnical skills deemed critical for effective team performance in the emergency command center of an offshore platform were identified by experienced staff within the oil company in conjunction with the psychologists, with reference to research material from a study of OIMs' crisis management (Flin & Slaven, 1994) and the industry's standards of competence for OIMs' emergency management (see Flin, Slaven, & Stewart, 1996, for an example). In addition, material from command training in other organizations (Flin & Slaven, 1995) and other CRM research groups, such as United States Naval Air Warfare Center (USNAWC) and the National Aeronautics and Space Administration (NASA Ames), also were used (e.g., Orasanu & Salas, 1993). The topics included understanding team roles, communications, group decision making and problem solving, assertiveness, team attitudes, stress management, and shared mental models. Stress management and communication have been outlined above, so only the other three elements will be described. Each module lasted 2 hr and was set between simulator exercises.

• *Roles and responsibilities.* This is of particular importance to offshore crews because crew members may have roles to play during an emergency (e.g., muster checker, on-scene commander) that differ from their normal work roles. Furthermore, for platforms that have small crews, an individual's roles may change as an incident escalates. An exercise was developed that involved groups of three team members outlining their own roles and their understanding of other members' roles.

• *Assertiveness.* As on a flightdeck of an aircraft, junior members of a team may need training in assertiveness to challenge a decision of a more senior team member if they believe that it may be incorrect. The training package for this module involved a lecture and a video demonstrating the need for assertiveness in operational conditions, a role play exercise, and a scenario in the simulator where skills could be practiced.

• *Team decision making.* Along with lecture material, a group decision-making exercise was conducted to emphasize the need for team members to share and review incoming information in order to build a mental model of the situation. The concept of team situation awareness, presented as "sharing the big picture," was illustrated and examined.

Team members' attitudes were assessed with an adaptation of the questionnaire used by Harrington and Kello (1993) with control room operators in a nuclear power plant. This is an adaptation of the Cockpit Management

Attitudes Questionnaire (Gregorich, Helmreich, and Wilhelm, 1993). The results of the questionnaire were used to generate discussion and to facilitate the introduction of the human factors training package during the course (see Flin, 1996, for sample items).

Emergency Resource Management in Norway

There are also reports of CRM training being used with Norwegian offshore crews; for instance, emergency resource management training is used by *Elf Petroleum Norge*. According to Grinde (1994), this is an initial 3-day course with a 2-day refresher every 2 years. The objectives of the course are to give the platform emergency organization a comprehensive understanding of the resources available during an emergency and to provide training in information transfer and logging, task allocation, decision making, communication, and situation assessment. Furthermore, the management was also given an understanding of crisis psychiatry, human stress, terrorism, and drugs. The course consists of lectures and four scenarios run in an onshore simulator that are *"designed to cover as many aspects of emergency response as possible, with every member of the team being involved at some stage"* (Grinde, 1994, p. 415). After completion of the course there are extensive debriefings about the team's performance in the scenarios.

CRM FOR OFFSHORE OPERATIONS

The examples of CRM training for emergency response teams outlined above demonstrate that CRM training has been successfully adapted for the offshore industry. However, properly designed CRM courses can have wider application than emergency response situations and may be used to enhance safety and improve productivity in routine operations in which effective teamwork is important. As part of a safety research project sponsored by the Offshore Safety Division of the HSE and 13 oil companies, a new prototype CRM course has been designed for offshore platform crews. The course is delivered onshore over 2 days. The method of training includes lectures, practical exercises, case studies, and video clips. It consists of six main components based on identified nontechnical skills (also see Table 11.1):

- Introduction to CRM
- Situation awareness
- Decision making
- Communication
- Team coordination
- Personal resources

TABLE 11.1
Offshore Operations: Nontechnical Skills Framework

Categories	Skills
Situation awareness	Plant status awareness Environmental awareness Anticipation Concentration/avoiding distraction Shared mental models
Decision making	Problem definition/diagnosis Risk and time assessment Recognition primed decision making/procedures analytical Option generation/choice outcome review
Communication	Assertiveness/speaking up Asking questions Listening Giving appropriate feedback Attending to nonverbal signals
Teamwork	Maintaining team focus Considering others Supporting others Team decision making Conflict solving
Personal resources	Identifying and managing stress Reducing/coping with fatigue Physical and mental fitness
Supervision/Leadership	Use of authority/assertiveness Maintaining standards Planning and coordination Workload management

The course objectives are:

- To raise crew awareness and enhance knowledge of human factors that can cause or exacerbate incidents related to safety or production.
- To develop nontechnical skills and attitudes which, when applied, can prevent or mitigate the effects of an accident or incident whether instigated by human or technical failings.
- To integrate CRM knowledge, skills, and attitudes into current work practices.

Although error management is a key focus, our own research into offshore safety emphasizes that company culture and work climate create conditions for human error (Flin, 1996; Mearns, Flin, Fleming, & Gordon, 1997), and this broader approach is accentuated.

The course was designed on the basis of a number of principles:

• The main human factors causes of accidents and incidents in the off-shore environment were examined in order to identify the nontechnical skills necessary to prevent or mitigate the consequences of human error.
• These skills should be well defined, objective, and measurable.
• The knowledge and attitudes required to develop these skills should be equally well specified, so that they can be integrated into the participants' work behavior.

On the basis of our experience developing and delivering CRM modules on emergency response training and our involvement with CRM research in the aviation industry (e.g., Flin & Martin, in press; O'Connor et al., 2000), we suggest a number of guidelines for designing and delivering CRM programs in an industrial setting.

Guideline 1: Determine the Key CRM Skills

The first stage of designing any new CRM program is to conduct a training-needs analysis to identify the training requirements (Griffiths & Lees, 1995). In aviation, an international survey of major airlines used CRM training revealed that these courses have a common set of components, as identified by the regulators (Flin & Martin, in press). Although assorted labels are used by different airlines, the core elements are leadership, team work, situation awareness, decision making, and personal limitations. Communication skills underpin the first four elements, but this is sometimes included as a separate topic. (These had also been used on the emergency management courses described earlier in which they were deemed critical elements by the company trainers.) The core CRM topics in aviation were identified through analysis of flight data recorders and cockpit voice recorders, sophisticated accident analysis, pilot interviews, and flight simulator studies (David, 1996). Therefore, we also attempted to use a multifaceted approach to examine the type of human factors problems that occur in the offshore industry.

Data collected and classified by the Mineral Management Service in the United States indicate that human error accounted for approximately one third of industrial accidents and incidents in the United States (outer Continental Shelf) between 1995 and 1996 (Mineral Management Service, 1997). This is similar to the proportion of human factors errors found in an examination of a representative selection of the 1997 accident and incident reports ($n = 276$) for a major U.K. offshore operating company (Bryden, O'Connor, & Flin, 1998).

Members of the UK offshore workforce also recognize the consequence of human factors in accident causation. A survey of six platform crews ($N = 622$) carried out by Flin, Mearns, et al. (1996), 70% of the workers agreed that "most accidents are due to human failure" (negligence, inattention). In addition, more than one third of the respondents cited "lack of care and attention" as the most common cause of accidents (p. 75). A similar sentiment was also established from a survey of 200 OIMs in the U.K. Continental Shelf, carried out by O'Dea and Flin (1998). The OIMs rated factors relating to the person, such as not thinking the job through, carelessness, failing to follow the rules, and lack of communication as the main causes of accidents. Data collected by Mearns et al. (1997) allow a closer examination of the proportion of accidents and near-misses offshore that are attributed to problems with the core CRM topics taught in aviation. Mearns et al. studied accident reports from seven companies that were extracted over a 2-year period from 1994 to 1996. A total of 1,268 incidents were recorded (lost-time injuries, minor incidents, or near misses). These incidents were then assigned to 55 human factors categories according to their underlying causes on the basis of the International Safety Rating System of coding (International Loss Control Institute, 1990). The incidents produced 1,123 codes, with some incidents containing no human factors codes and others having multiple codes. A total of 46% of the codes fell within one of the broad CRM topics (i.e., teamwork, leadership, situation awareness, decision making, communication, and personal limitations). The remaining percentage of codes were concerned with causes such as lack of knowledge, lack of skill, or poor engineering or design. Of the six CRM categories, 6% of the codes were due to errors in teamwork; 2% were attributed to poor leadership; 9% were due to a lack of situation awareness; 11% were due to poor decision making; 5% occurred as a result of poor communication; and 13% were attributed to personal limitations, such as stress and fatigue. This provided some confirmation that the core issues being examined in aviation CRM courses were likely to have relevance for offshore operations, and it indicated that two key areas were personal limitations and decision making.

Thus, our analysis of human factors causes of accidents and incidents and analysis of workforce surveys, as well as 12 years' experience of working with the offshore industry (see Flin & Slaven, 1996) allowed us to develop a proposed taxonomy of key nontechnical skills for CRM training (see Table 11.1).

Guideline 2: Draw on Material From Existing CRM Courses and Advice From Experts

Because the principal causes of human errors offshore did not differ from those in aviation, we were able to adapt the CRM training from aviation. Visits were made to aviation psychologists working in CRM, and members of our research team were permitted to take part in CRM courses being run by the major helicopter companies servicing the UK sector of the North Sea, *Bristows, Bond,* and *British International.* We also used information from the Internet; for example: the Industrial CRM Developers Group (www.crm-devel.org/), which also runs an electronic mail discussion group in which questions and best practice can be shared with CRM experts from many different industries around the world, the Federal Aviation Administration's (FAA's) human factors in aviation maintenance and inspection (www.hfskyway.com/hfami/document.htm), and the Aerospace Research Group at University of Texas/NASA Ames (www.psy.utexas.edu/psy/helmreich/nasaut.htm).

Guideline 3: Ensure That the CRM Materials Are Customized for the Audience

If aviation CRM is to be adapted for an industrial audience, the training materials must be customized for the particular industry. In our experience in developing the emergency response and control room CRM materials, we have found it essential to work with experienced offshore staff in order to select case studies and examples that will be relevant. Moreover, the staff can assist in the translation of psychological jargon (called *psychobabble* by pilots) into the language of offshore operations and procedures. The prototype course materials need to be reviewed and, if possible, tested on a number of experienced personnel from the relevant domain. If CRM is adopted by an industry, then the "roll out" program should be ultimately delivered by practitioners rather than the psychologists, who should move to a monitoring and advisory role.

In addition, work was undertaken to develop specific training materials; for example, exercises and case studies based on examples of offshore incidents and typical production and maintenance work situations. Videotaped re-enactments of accidents are very powerful tools when based on real voice recordings and are frequently used in aviation CRM based on cockpit voice recorder data. However, we found that although aviation videos and examples were of interest to participants in the offshore CRM course, they were not as powerful as examples from their own industry.

Guideline 4: Establish Appropriate Participants

Once it had been established that standard CRM topics were likely to be applicable to offshore teams for general operations, it was necessary to establish which group of individuals would benefit the most from the training. After experimenting with a variety of course participants (e.g., participants from two different platforms, supervisory staff, members of a single crew) we found that, for the offshore industry, a cross-section of participants from the most senior manager (OIM) to the most junior offshore workers from a single shift for one platform was optimal. An ideal number for a given session was found to be 10–12 participants. This allowed the course to be used as a forum for individuals to comment on the behaviors of other individuals within the shift and discuss issues particular to their own platform. This is consistent with Swezey and Salas's (1992) statement that teams should be trained as entire units. To illustrate with an analogy from aviation, there is little point training copilots to be assertive if captains are not trained to listen.

Guideline 5: Ensure Management Support, and Identify "Workforce Champions"

There must be a committed corporate-level champion, as well as local champions at each site (Drury, 1998). It is important that senior management at a corporate level show their commitment to any training program. If the management do not demonstrate that they believe the training is important, then it is unlikely that the workforce will show much enthusiasm. Moreover, it is important that the management allow suitable access and that expertise is made available for the preparation of material for the course. Management commitment can be demonstrated to the course participants by getting a senior manager to give a short opening speech prior to the commencement of the course. In our experience, the support demonstrated by the onshore platform manager or asset manager has proved to be very valuable.

A champion from the workforce at a local level who has experience of working "in the field" is required. These individuals are necessary to aid in the construction of the course and can also help to present the courses to give credibility and help answer any very technical questions. A suitable champion is an individual who has many years of experience and commands the respect of his or her peers. In addition, he or she must also have an understanding of and enthusiasm for human factors (Drury, 1998).

Guideline 6: Assess the Impact of Training

It is important that measures are put in place to evaluate the impact of CRM training on attitudes and behaviors as well as on accident rates and efficiency. However, the fundamental question of whether CRM training can fulfill its purposes of increasing safety and efficiency does not have a simple answer (Helmreich, Merritt, & Wilhelm, 1999). In commercial aviation, because the overall accident rate is so low and the training programs are so variable, it is not possible to form strong conclusions about the impact of training on the basis of accident data (Helmreich, Chidester, Foushee, Gregorich, & Wilhelm, 1990). Therefore, it is necessary to use other methods of assessment to draw inferences about the effects of CRM training more indirectly.

Salas and Cannon-Bowers (1997) outlined a number of principles for evaluating team training that have emerged over time. The best approach used by CRM research teams in aviation is one that is multifaceted and considers several separate methods of assessment. It is necessary to measure performance and attitudes at pretraining to establish a baseline and to use corresponding posttraining measures to determine the impact of training on team members' knowledge, skills, performance, and attitudes.

In aviation, the two criteria used to assess CRM are behavior on the flightdeck and attitudes showing acceptance or rejection of CRM concepts (Helmreich et al., 1999). Flight crew are assessed in the simulator as well as in normal flight operations using a behavioral marker system such as the aviation CRM behavioral markers University of Texas (UT)/FAA Line/Los checklist (Helmreich, Butler, Taggart, & Wilhelm, 1997) or the European nontechnical skills (NOTECHS) marker system (Avermaete & Kruijsen (1998). Data from this type of audit have demonstrated that CRM training that includes line-oriented flight training and recurrent training does lead to desired changes in behavior (Helmreich & Foushee, 1993). Another method of assessing the effects of CRM training in aviation is to measure the participants' attitudes that have been identified as playing a role in air accidents and incidents (Helmreich & Foushee, 1993; Helmreich, Merritt, Sherman, Gregorich, & Wiener, 1993). Data from a number of airlines show that attitudes about flightdeck management also change in the desired direction as a result of training (Helmreich & Wilhelm, 1991).

In the offshore CRM course evaluation, we also used a multifaceted method of assessing the impact of training: attitude measurement, course evaluation, and analysis of accident and incident data.

Attitude Measurement

The participants' attitudes were measured with the 30-item Offshore Crew Resource Management Questionnaire (Flin, O'Connor, & Mearns, in press). This was developed with reference to the Cockpit Management Assessment Questionnaire, which was developed for aviation by Gregorich, Helmreich, and Wilhelm (1990); the Maintenance Resource Management/Technical Operations Questionnaire, designed by Taylor (1998) for aviation maintenance, the aviation CRM behavioral markers in the UT/FAA Line/Los checklist (Helmreich et al., 1997); and the European NOTECHS marker system (Avermaete & Kruijsen, 1998). The questionnaire consists of statements designed to measure attitudes to each of the topics covered in the training (see Table 11.1). It is administered before, and then immediately after, the training course. The results suggest that there is a shift in attitudes in the desired direction; however, we need to increase the number of participants who have completed the training before any firm conclusions can be drawn.

In addition, we also intend to follow up the participants 6 months after completing the training to assess any long-term changes in behavior.

Course Evaluation Form

Training evaluation tools must provide feedback on how to improve training (Salas & Cannon-Bowers, 1997). After each block in the offshore CRM course the participants rate the trainers, exercises, interest, and videos in each module on a 5-point scale that ranges from 1 (*very poor*) to 5 (*excellent*). Participants are also encouraged to write comments and engage in a group discussion at the end of the course. The feedback obtained thus far is encouraging, with participants reporting particular interest in the situation awareness and decision-making modules.

Analysis of Accident and Incident Data

We are tracking the accident and productivity data to assess the impact of CRM training. In regard to military aviation, Diehl (1991) stated that CRM training decreased the accident rate by 81% for U.S. Navy A-6 Intruder crew members. This method of assessment has been used in other industries; for example, after 4 years of human factors and CRM training, the incidents and accidents in the Danish shipping company Maersk decreased by one third, and all insurance premiums have been lowered by 15% (Brydorf, 1998). Taylor (1998) also found some suggestive evidence for an improvement in both occupational injury and aircraft damage in the aviation maintenance industry. However, there are distinct limitations with using accident and productivity data to assess the effectiveness of CRM offshore:

• The accident rate offshore is so low, as in aviation, that it does not provide a robust test for the effectiveness of CRM programs. Similarly, the number of unplanned trips and amount of resulting down time (a *trip* is when the plant ceases to produce oil and gas, and *down time* is the time taken to bring production back on-line) is also fairly low.

• It is difficult to draw useful comparisons between different oil installations (e.g., equipment, age of platform, conditions, type of operations, etc.); thus, this complicates any evaluation of the effect on accident rates or production.

Guideline 7: Make Ongoing Revisions of CRM Training on the Basis of Safety Data, Psychological Research, and Feedback From Participants

The characteristic best observed in the aviation industry is that the CRM courses are well grounded in ongoing psychological research, safety reporting systems, and incident investigations into the work conditions and the human factors causes of accidents. Our research group has worked on safety and emergency response projects with the offshore oil industry for more than 10 years (see Flin & Slaven, 1996), and we are feeding the findings from ongoing research projects (e.g. Gordon, 1998; O'Dea & Flin, 1998) into the longer term development of CRM for this industry. As the nature of offshore oil and gas operations continues to change, these data can and should be used to tailor CRM courses to deal with emerging CRM issues.

Equally as important is the quantitative and qualitative feedback gained from the course participants' course evaluation forms, discussions after the course, and comments during course. This has proven invaluable, particularly for the early courses, in making improvements to the training.

In conclusion, our experience to date suggests that aviation CRM can successfully be adapted to other high-reliability industrial environments, such as offshore oil and gas platforms. One of the great strengths of this field is the willingness of training providers and companies to share experiences of developing and delivering CRM. The common goal of improving safety transcends organizational competitiveness and industrial parochialism, because the core philosophy of CRM provides a basic drive for the step change in work culture required to reduce accidents towards the desired target zero.

REFERENCES

Avermaete, van, J. A. G., & Kruijsen (1998). *The evaluation of non-technical skills of multi-pilot aircrew in relation to the JAR-FCL requirements* (project report: NLR-CR-98443). Amsterdam: National Aerospace Laboratory.

Baddeley, A. (1992, January). Working memory. *Science, 255,* 556–559.

Boyd, R. (1996). *Preparing a team to operate the Judy platform—Through empowerment and motivation, success?* Unpublished master's thesis, Robert Gordon University, Aberdeen, Scotland.

Bryden, R., O'Connor, P., & Flin, R. (1998, October). *Developing CRM for offshore oil platforms.* Paper presented at the Human Factors and Ergonomics Society conference, Chicago.

Byrdorf, P. (1998, September). *Human factors and crew resource management: An example of successfully applying the experience from CRM programmes in the aviation world to the maritime world.* Paper presented at the 23rd conference of the European Association for Aviation Psychology, Vienna.

Cox, T. (1993). *Stress research and stress management: Putting theory to work* (HSE Contract Rep. 61/1993). Suffolk, England: Health and Safety Executive Books.

Cullen, D. (1990). *The public inquiry into the Piper Alpha disaster* (Vols. 1 & 2). London: Her Majesty's Stationery Office.

David, G. (1996). Lessons from offshore aviation: Towards an integrated human performance system. In R. Flin & G. Slaven (Eds.), *Managing the offshore installation workforce* (pp. 219–238). Tulsa, OK: PennWell.

Diehl, A. E. (1991, November). *Does cockpit management training reduce aircrew error?* Paper presented at the 22nd international seminar of the International Society of Air Safety Investigators, Canberra, Australia.

Drury, C. G. (1998). *The World Wide Web edition of the Human Factors issues in aircraft maintenance and inspection 3.0 CD-ROM 1998, chapter 2* [on-line]. Available: http//:www.hfskyway.com/hfami/document.htm.

Flin, R. (1995). Crew resource management for teams in the offshore oil industry. *Journal of European Industrial Training, 19*(9), 23–27.

Flin, R. (1996). *Sitting in the hot seat: Leaders and teams for critical incident management.* Chichester, England: Wiley.

Flin, R., & Martin, L. (in press). Behavioural markers for CRM: A survey of current practice. *International Journal of Aviation Psychology.*

Flin, R., Mearns, K., Fleming, M., & Gordon, R. (1996). *Risk perception and safety in the offshore oil and gas industry* (Rep. No. OTH 94454). Suffolk, England: Health and Safety Executive Books.

Flin, R., O'Connor, P., & Mearns, K. (in press). Crew resource management training for offshore operations. *Factoring the human into safety: Translating research into practice* (Vol. 3). (Rep. HSE OTO 2000063). London: Health and Safety Executive Books.

Flin, R. & Slaven, G. (1994). *The selection and training of offshore installation managers for crisis management* (Rep. No. OTH 92374). London: Health and Safety Executive Books.

Flin, R., & Slaven, G. (1995). Choosing the right stuff: The selection and training of on-scene commanders. *Journal of Contingencies and Crisis Management, 3,* 113–123.

Flin, R., & Slaven, G. (Eds.). (1996). *Managing the offshore installation workforce.* Tulsa, OK: PennWell.

Flin, R., Slaven, G., & Stewart, K. (1996). Emergency decision making in the offshore oil and gas industry. *Human Factors, 38*(2), 262–277.

Gordon, R. (1998). Human factors coding of accidents in the offshore oil industry. *Journal of Loss Prevention in the Process Industries, 61,* 95–108.

Gregorich, S., Helmreich, R., & Wilhelm, J. (1993). The structure of cockpit management attitudes. *Journal of Applied Psychology, 75,* 682–690.

Griffiths, C. W., & Lees, A. (1995). Training needs analysis—A human factors analysis tool. *Quality and Reliability Engineering International, 11,* 435–438.

Grinde, T. A. (1994). Emergency resource management training. In *Proceedings of the second international conference on health, safety, & the environment in oil and gas exploration and production* (Vol. 2, pp. 413–417). Richardson, TX: Society of Petroleum Engineers.

Harrington, D., & Kello, J. (1992, November). *Systematic evaluation of nuclear operator team skills training.* Paper presented to the American Nuclear Society, San Francisco.

Health and Safety Executive. (1995). *Annual report 1994/95.* London: HSE books.

Helmreich, R., Butler, R., Taggart, W., & Wilhelm, J. (1997). *The NASA/University of Texas/FAA Line/LOS Checklist: A behavioural-based checklist for CRM skills assessment.* (Version 4.4). Austin: NASA/University of Texas/FAA Aerospace Group.

Helmreich, R. L., Chidester, T. R., Foushee, H. C., Gregorich, S. E., & Wilhelm, J. A. (1990). How effective is cockpit resource management training? Issues in evaluating the impact of programs to enhance crew co-ordination. *Flight Safety Digest, 9*(5), 1–17.

Helmreich, R. L., & Foushee, H. C. (1993). Why crew resource management? Empirical and theoretical bases of human factors training in aviation. In E. Wiener, B. Kanki, & R. Helmreich (Eds.), *Cockpit resource management* (pp. 3–45). San Diego, CA: Academic Press.

Helmreich, R. L., Merritt, A. C., Sherman, P. J., Gregorich, S. E., & Wiener, E. L. (1993). *The Flight Management Attitudes Questionnaire (FMAQ*; NASA/UT/FAA Tech. Rep. 93-4). Austin: University of Texas.

Helmreich, R. L., Merritt, A. C., & Wilhelm, J. A. (1999). The evolution of crew resource management training in commercial aviation. *International Journal of Aviation Psychology, 9*(11), 19–32.

Helmreich, R. L., & Wilhelm, J. A. (1991). Outcomes of crew resource management training. *International Journal of Aviation Psychology, 1,* 287–300.

Howard, S., Gaba, D., Fish, K., Yang, G., & Sarnquist, F. (1992). Anaesthesia crisis resource management training: Teaching anaesthesiologists to handle critical incidents. *Aviation, Space, and Environmental Medicine, 63,* 765–770.

International Labour Office. (1993). *Safety and related issues pertaining to work on offshore petroleum installations.* Geneva, Switzerland: Author.

International Loss Control Institute. (1990). *International safety rating system.* Loganville, GA: Author.

Klein, G. (1993) The recognition-primed decision model of rapid decision making. In G. Klein, J. Orasanu, R. Calderwood, & C. Zsambok (Eds.), *Decision making in action* (pp. 138–147). Norwood, NJ: Ablex.

Marx, D. A., & Graeber, R. C. (1994). Human error in aircraft maintenance. In N. Johnston, N. McDonald, & R. Fuller (Eds.), *Aviation psychology in practice* (pp. 87–104). Aldershot, England: Avebury Technical.

Mearns, K., Flin, R., Fleming, M., & Gordon, R. (1997) *Human and Organisational Factors in Offshore Safety.* (Rep. No. OTH 543). Suffolk, England: Health and Safety Executive Books.

Mineral Management Service. (1997). *1995–1996 Accident statistics.* Available: http://mms.gov/eod/safety.htm

O'Connor, P., Hörmann, H.-J., Flin, R., Lodge, M., Goeters, K.-M., & The JARTEL group (2000). *Developing a method for evaluating CRM skills: A European perspective.* Manuscript submitted for publication.

O'Dea, A., & Flin, R. (1998, September). Safety initiatives: Room for improvement. *Petroleum Review,* 26–27.

Orasanu, J., & Salas, E. (1993). Team decision making in complex environments. In G. Klein, J. Orasanu, R. Calderwood, & C. Zsambok (Eds.), *Decision making in action* (pp. 327–345). Norwood, NJ: Ablex.

Reason, J. T. (1997). *Managing the risks of organizational accident.* Aldershot, England: Ashgate.

Salas, E., & Cannon-Bowers, J. A. (1997). Methods, tools and strategies for team training. In M. Quinones & E. Ehrestein (Eds.), *Training for a rapidly changing workplace: Applications in psychological research* (pp. 291–322). Washington, DC: American Psychological Association.

Swezey, R. W., & Salas, E. (1992). Guidelines for use in team-training development. In R. W. Swezey & E. Salas (Eds.), *Teams: Their training and performance* (pp. 219–247). Norwood, NJ: Ablex.

Taylor, J. C. (1998). *Evaluating the effects of maintenance resource management (MRM) interventions in airline safety.* Report presented to the Federal Aviation Administration, Washington, DC.

Turner, B. A., & Pidgeon, N. F. (1997) *Man-made disasters* (2nd ed.). Oxford, England: Butterworth-Heinemann.

United Kingdom Offshore Operators Association. (1998). *The story of oil and gas.* Available: http://www.ukooa.co.uk

Wiener, E., Kanki, B., & Helmreich, R. (Eds.). (1993). *Cockpit resource management.* San Diego, CA: Academic Press.

12

Resource Management for Aviation Maintenance Teams

Michelle M. Robertson
Liberty Mutual Research Center

Aviation maintenance operations are complex, demanding, and dependent on good communication and teamwork for their success. Success in aviation maintenance is measured by the safety and quality of a maintenance operation. Aviation maintenance operations are most successful when crews function as integrated, communicating teams rather than as a collection of individuals engaged in independent actions. Over the last decade, the importance of teamwork has become widely recognized; (Maurino, Reason, Johnston, & Lee, 1995; Robertson, 1998, Rogers, 1991; Taylor & Robertson, 1995). This has resulted in the emergence of maintenance resource management (MRM) training programs and other safety-related programs within the aviation community.

MRM training is a human factors intervention designed to improve communication, effectiveness, and safety in airline maintenance operations. Effectiveness is measured through the reduction in maintenance errors and the increase in individual and unit coordination and performance. MRM training is also used to change the work "safety culture" of the organization by establishing a positive attitude toward safety on the part of maintenance personnel. Attitudes, if positively reinforced, can lead to changed behaviors and performance. Safety is typically measured by occupational injuries, ground damage incidents, reliability, and aircraft airworthiness. MRM improves the reliability of the technical operations processes by increasing the coordination and exchange of information among team members and among teams of airline maintenance crews.

Within their programs, airlines may use MRM principles differently. Based on human factors principles (e.g., engineering, cognitive psychology, work physiology), and research from the social sciences (e.g., industrial psychology, organizational behavior), MRM programs link and integrate traditional individual aviation human factors topics, such as equipment design, human physiology, cognitive workload, and workplace safety and health. MRM principles are best understood through training programs; however, the goal of any MRM training program is to improve work performance and safety and reduce maintenance errors through improved coordination and communication.

One of the early activities when starting an MRM program is to gain the understanding, commitment, and visible support of the senior management in the company. Management must actively support and value MRM. The relevance of the MRM program to business objectives must be clear, or management will question the investment in time and the costs associated with such a program. It may also be necessary to develop some simple return-on-investment models to justify implementing the program. For example, the cost of one ground damage incident, inflight shutdown, or turnback versus the cost of an MRM training course plus the benefits of reducing maintenance errors and increasing safety results is a positive return on investment.

Once support has been established, the first step toward designing an MRM training program is to step back and view the entire maintenance operation as a large system. This system is composed of numerous subsystems, including: aviation maintenance technicians (AMTs), engineering, quality control, planners, document support, inspectors, maintenance control, materials and stores, management, and administrative support. When viewed as a system, it is apparent that the overall success of the maintenance operation is dependent on the quality of information exchanged among the team members making up each function, and among the functions themselves.

Once the functions involved are identified, and their roles understood, an MRM training program can be designed. However, as with senior management, it is important to establish a clear rationale for all employees in maintenance operations about the relevance of the MRM program to the airline business. For instance, if an objective of the airline is to reduce errors and increase safety, then the training program should include examples of how the principles and concepts being taught in the MRM training are directly related to these goals. It is important for employees to understand the relevance of any changes they make in their work, and the effort they must put into that change, to the broader business objectives of the airline.

An MRM training program has many facets, all focused on improving communication, coordination, and safety. A typical MRM training program addresses each of the following components (Dupont, 1997; Robertson, 1998):

- Understanding the maintenance operation as a system
- Identifying and understanding the basics of human factors issues
- Recognizing contributing causes to human errors

- Situation awareness
- Decision making skills, leadership
- Assertiveness (how to effectively speak up during critical times)
- Peer-to-peer work performance feedback techniques
- Stress management and fatigue
- Coordination and planning
- Teamwork skills and conflict resolution
- Communication (written and verbal)
- Norms

This chapter describes the development of MRM training and the performance issues and problems in aviation maintenance that MRM addresses. Discussions on current practices and description of some real-life implementation experiences are given. This chapter presents a systems approach to designing and developing an MRM training program. Generic guidelines and successful elements to be used in developing and implementing an MRM training program are provided.

DEVELOPMENT OF MRM TRAINING

Evolution of MRM

MRM is the result of a series of events that drive its development. The catalyst for the development of MRM grew from a reaction to a tragic event. In 1988, Aloha Airlines Flight 243 suffered a near-catastrophic failure. Eighteen feet of fuselage skin ripped off the aircraft at 24,000 feet, forcing an emergency landing. The post accident analysis revealed there were more than 240 cracks in the skin of this aircraft. The ensuing investigation identified many human-factors-related problems leading to the failed inspections National Transportation Safety Board [NTSB] (1984). These findings focused attention on maintenance as a potential accident-causal factor and led to the development of MRM and human factors training.

In 1991, due in part to this new focus, Continental Airlines expanded and modified its crew resource management (CRM) training to crew coordination concept (CCC) training designed for its maintenance technical operations. CCC is the precursor of what has become known as MRM (Robertson, 1988; Robertson, Taylor, Stelly, & Wagner, 1994). MRM shares certain basic features with CRM, including addressing the issues of communication and team coordination. However, the target audience for MRM includes AMT's staff support personnel, inspectors, engineers, and managers—a much more diverse group than cockpit crews.

Similarly, in response to the 1989 crash of the Air Ontario Flight #1363, Transport Canada developed the Human Performance in Maintenance workshop. Crew coordination was identified as a contributing factor to this accident. This workshop was successful in providing a heightened awareness of human factors problems and solutions in the maintenance environment. The outcome of this workshop was the identification of the "Dirty Dozen"— human factors elements that affect people's ability to perform effectively and safely (Dupont, 1997).

In response to these initial successes, the industry began to develop its own organization-specific MRM training programs. US Airways developed an MRM program that continues to evolve (Driscoll, Kleiser, & Ballough, 1997). This program is the product of a partnership consisting of maintenance management, labor (International Association of Machinists and Aerospace Workers), and the Federal Aviation Administration's [FAA's] Flight Standards District Offices. Their MRM activities include: (a) participatory methods to reduce paperwork errors, (b) a paperwork training course and preshift meetings, and (c) problem-solving meetings. US Airways designed and developed an MRM training course using a participatory design process.

Several other airlines (e.g., United, Northwest, American Eagle) have designed human factors training courses for maintenance operations (Robertson, 1998; Taggart, 1990). These courses are typically based on what is known as human performance improvement methods (Dupont, 1997). Common training elements include human factors training materials developed for the FAA and the airline's *own* human-factors-related experiences and case studies. Some airline companies are currently implementing a second phase of MRM training, incorporating team situation awareness training for maintenance operations (Robertson & Endsley, in press; Endsley & Robertson, 1997).

Why an MRM Training Program?

The ultimate goal of MRM training is to increase aviation safety. To accomplish this goal, MRM addresses a number of factors contributing to maintenance and inspection errors. The most salient issues and problems addressed by MRM include workplace communication, corporate culture, and situation awareness.

There is little argument that air travel, at least in the developed world, and via scheduled commercial flights, is, by any measure, very safe. However, maintenance errors contribute significantly to aviation accidents and, perhaps more important, to events that initiate accidents or unsafe flight conditions (Maurino et al., 1995). It has been shown that 39% of [widebody] aircraft accidents began with a problem in aircraft systems and maintenance, and that pilot error comes later in the sequence of events after something has gone wrong with the airplane itself (Wiegers & Rosman, 1986). One study reported that maintenance and inspection errors accounted for 12% of major aircraft accidents (Marx & Graeber, 1994). These data show that maintenance and inspection errors contribute significantly to many aircraft accidents. Twelve percent of major aircraft disasters involving Boeing aircraft were attributable to maintenance and inspection errors and to 15% of onboard fatalities (Marx & Graeber, 1992).

Ineffective Communication

Conclusions from several studies show that poor communication practices and skills exist throughout the aviation industry (e.g., Drury, 1993; Lock & Strutt, 1981; NTSB, 1989; Robertson, 1998; Taylor, 1991; Taylor & Robertson, 1995). Poor communication among management teams and individual workers can compromise safe and cost-effective maintenance operations. One study confirmed that effective communication is the most important factor for ensuring team coordination and effective work performance (Shepherd, Johnson, Drury, Taylor, & Berninger, 1991; Taylor, 1991). Communication plays an important role in the quality of maintenance and has a significant impact on flight safety. Maintenance workers do not always understand their company's policies and goals or their individual roles in meeting those goals. The effects of assigning a low priority to communication may include delays, high rates of employee turnover, and low morale (Robertson, 1998; Robertson & Taylor, 1996; Taylor, 1991).

Poor Corporate Culture and Management Commitment

The FAA's National Plan for Human Factors (FAA, 1991) reported that the overall culture of commercial aviation still emphasized individual, rather than team aspects of cockpit work. That report and others (Hackman, 1990) emphasized that the same individualistic culture of the flight cockpit resource management techniques used successfully in flight operations could be applied to aviation maintenance with positive results. However, approaching the issue within maintenance operations requires a larger system view, including consideration of the individual, team, technology and environment (i.e., the sociotechnical systems view; Drury, 1993; Robertson & Taylor, 1996; Taylor, 1991). When a poor organizational culture exists, coupled with a low level of management commitment to change the system, the result is ineffective, and sometimes unsafe, performance. To overcome these issues many companies are finding that solving communication and coordination problems requires changes in management, work organization, and corporate culture. These changes involve the following: (a) soliciting top management support by defining a vision for the purpose of the change; (b) providing training resources and support to supervisors; (c) a quality intervention, such as MRM training or participatory redesigning activities; and (d) timely, valid, and appropriate feedback through a range of communication mechanisms and shared measurement processes (e.g., How well is the MRM program doing? What effect does it have on maintenance performance?).

Lack of Situation Awareness

One of the most pervasive problems facing AMTs is the loss of situation awareness. Failures in situation awareness have been linked to conditions that lead to reduced flight safety, flight delays, ground damage, and other problems that directly increase costs. In certain severe cases, the viability of a company has been affected by situation awareness errors (Endsley & Robertson, 1996; Robertson & Endsley, 1997). The results of poor situation awareness can be seen in a variety of common maintenance-related errors. These include loose objects left in an aircraft, missing fuel and oil caps, loose panels and other unsecured parts, and preflight pins not removed prior to operation (Marx & Graeber, 1994). Putting together observed cues to form a proper understanding of malfunctions is a challenging and significant problem in diagnostic activities. For example, in more than 60% of cases, an incorrect avionics component is replaced in an aircraft (Ruffner, 1990).

In team-oriented environments, such as aviation maintenance, the level of situation awareness within and among teams is an issue of concern. The overall goal of providing safe aircraft can be compromised if any team member loses a sufficient level of situation awareness. Aviation maintenance tasks are typically coordinated within teams and among teams on different shifts or in different geographic locations. The Eastern Airlines incident, involving an L-1011 with a missing chip detector O-rings, has been directly linked to the lack of coordination across shifts, along with other contributing factors (NTSB, 1989).

TRAINING CONCEPTS FOR MRM PROGRAMS

In this section the training concepts that are most directly applicable to establishing an MRM training program are presented. Each core concept is described, and in the next section specific methods to use in the development of an MRM training program are provided.

Systems Approach

MRM, as with other human-factors-oriented processes, is based on a systems approach. It incorporates a variety of human factors methods, such as job and work design, and considers the overall sociotechnical maintenance system. For example, as shown in Fig. 12.1, the *SHELL* model (S-Software, H-Hardware, E-Environment, L-Liveware) defines how we view human factors as a system and illustrates the various interactions that occur between subsystems and the human operator (Hawkins, 1993; Robertson, 1998).

The interactions in this model can affect both individual and team performance. MRM training typically focuses on the interaction between the individual AMT and other team/crew members—liveware–liveware interactions, in SHELL terminology. This person-to-person interaction can be considered the micro level of communication and team building, whereas the interactions among teams and departments is at the macro level. There are also external forces that can affect individual and team performance. These include political and regulatory considerations (e.g., FAA, Occupational Safety and Health Administration, NTSB) and economic factors (e.g., global competition). Achieving the goals of MRM requires improving interactions at both the micro and macro levels. These improvements must occur within the context of external factors, and they require an understanding of their effects. To this end, the SHELL model depicts the systems approach to integrating human factors methods and principles to design an MRM program (Robertson, 1998).

Instructional Systems Design

The concept of the systems approach is exemplified by a hierarchical, top-down, and bottom-up structured approach to instructional design and development. Identifying organizational needs and performance gaps focused on

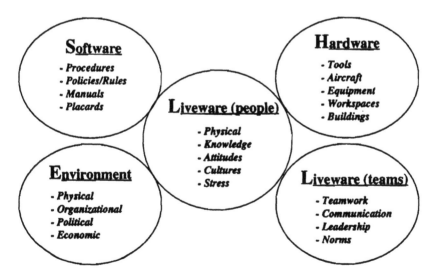

FIG. 12.1. SHELL model: Human factors system.

the macro level provides the program structure. Involving end-users and subject matter experts in designing and developing training programs incorporates the bottom-up systems approach. Using a systems approach to design, implement, and evaluate an MRM training program ensures that it will meet the needs of the learners and the organization.

The systems approach consists of the following five processes: analysis, design, development, implementation, and evaluation (Gagne, Briggs, & Wagner, 1988; Goldstein, 1993; Knirk & Gustafson, 1986). This process includes setting goals and defining objectives; developing and implementing the training program; involving end users, subject matter experts, or both; measuring the effects of the training; and providing feedback to the developers.

Continuous Learning and Improvement

The idea of continuous change and adaptation is fundamental to making any system responsive to the needs of its users. Continuous improvement is not a short-term activity; rather, it requires a long-term management commitment of financial and organizational resources to continuously adapt and improve the program.

Measurement and Evaluation

Once an MRM program is in place, we must determine how well (or whether) it is working. This measurement and evaluation process is typically the weak link in the systems approach. However, it is necessary to measure the effects of the MRM program *over time*. It is also necessary to use multiple measures in order to gauge the effectiveness of the program. It is also important to keep in mind that we are measuring not only the effects of the MRM training program but also the effects of on-the-job MRM practice.

Evaluation tools can include assessment instruments and questionnaires, behavioral observation and analysis, and unit and organizational performance measurements (Cannon-Bowers, Salas & Converse, 1993; Kirkpatrick, 1979; Robertson, 1988). These tools can be used to assess the cost of designing and delivering an MRM training course. In addition, performance measures establish the basis for calculating the company's return on its investment regarding the effects of the training program.

Feedback

Feedback provides information to accomplish two performance improvement goals: identifying necessary corrective actions to improve the program and reinforcing the positive outcomes of using MRM skills. Internal newsletters can be used to describe specific MRM outcomes. For example, one organization reported how an MRM group exercise led to initiation of a positive change in a specific maintenance operation. A more active feedback method is to have AMTs write their own "MRM story," describing their experiences using MRM principles and skills (Driscoll et al., 1997).

Team Situation Awareness

Situation awareness is one of the foundation concepts of MRM. Typically, we think of situation awareness in terms of the individual AMT. However, many of the most common maintenance errors involve the loss of situation awareness among different individuals often across different teams or shifts. The concept of team situation awareness relates to maintaining a collective awareness of important job-related conditions (Cannon-Bowers, et al., 1993; Endsley & Robertson, 1996). Researchers have identified five elements and activities that are necessary to improve team situation awareness in the maintenance environment (Endsley & Robertson, 1996, in press; Robertson & Endsley, 1995, 1997). These are shared mental models, verbalization of decisions, better team meetings, teamwork and feedback, and individual situation awareness.

Participation

Participating in the creation, development, and implementation of an MRM program promotes a feeling of individual ownership and a sense of commitment to supporting the MRM program goals. Being a member of a team that is developing and implementing an MRM program is motivating, rewarding, and beneficial to the individual and organization. Working together on a cross-functional, interdisciplinary team also provides a unique strength in designing and developing an individual MRM program (Robertson, 1998).

Active Learning

Effective instructional methods, sometimes called *inquiry* or *discover* learning, emphasize the involvement of learners (Gordon, 1994). Active learning gets students involved by having them participate in problem-solving activities and group discussions. The strength of this approach for the maintenance environment is that the use of group exercises and maintenance-related case studies promotes an active and motivating learning environment, since students are doing more than just passively receiving information-they are actively applying and using the various concepts and skills. To further strengthen this approach, training courses can be cofacilitated by subject matter experts in maintenance operations (e.g., AMTs, inspectors, quality assurance personnel). They can encourage students to participate by bringing "real workload" experiences into the classroom.

Transfer of Training

For training to be effective, AMTs must be able to apply their newly acquired skills in their real work environment—a positive transfer of training. Transfer of training is enhanced by reinforcement from coworkers and supervisors. A practice period occurs when AMTs return to their workplace after completing a training course. The reactions of others during the practice period either reinforces newly learned MRM skills or discourages their use. Therefore, it is important that managers receive MRM training in advance of workers. Transfer of training is also enhanced when classroom exercises are similar to actual workplace experiences.

GUIDELINES FOR DEVELOPING RESOURCE MANAGEMENT TRAINING PROGRAMS

In this section general guidelines for developing and implementing an MRM training program are provided. A framework of training principles and processes in this section with supporting tools and methods is described. Using the instructional systems design (ISD) approach for developing an MRM training program is effective and beneficial. Each of the major activities in the ISD framework, as shown in Fig. 12.2 is discussed.

Guideline I: Conduct a Front-End Analysis (Phase I)

A front-end analysis is conducted to determine strategic training needs and to assess the company's return its investment in training. The MRM design team identifies the organizational and trainee needs and constraints *before* it begins designing the training program. Analysis is the foundation for all later work related to implementing an MRM training program. A thorough needs assessment helps reduce the risk of funding inappropriate or unnecessary training. MRM training should be developed by a team that includes, among others, training professionals, a human factors practitioner, and various maintenance supervisors and workers.

Needs Assessment

Assessing needs ensures that MRM training addresses real workplace problems. A *need* is the difference between what exists and what is desired. Once a need is identified it must be examined to see whether it has an instructional solution. Some problem statements can then be further refined by using task analysis to isolate and relate specific worker activities. In other situations, a task hierarchy is constructed to determine how various work tasks are related. There are three levels of a needs assessment: (a) Organi-

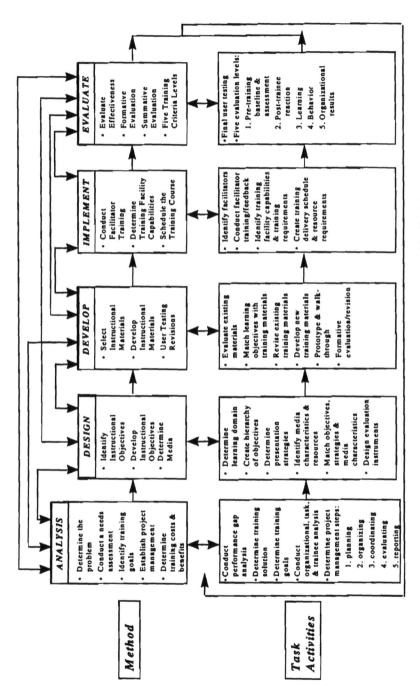

FIG. 12.2. Instructional system design phases, activities and feedback loops.

245

zational Analysis, (b) Task Analysis, and (c) Trainee Analysis (Goldstein, 1993; Knirk & Gustafson, 1986).

Organizational Analysis. An organizational analysis consists of an evaluation of the organization or industry in which the trainee performs the job and an evaluation of the organization expected to provide the training. There are a number of methods one can use to obtain information and collect data for an organizational analysis. Probably the most efficient approach is a combination of document analysis, interviews, and questionnaires. The training designer would start by obtaining and analyzing organizational documents. Once this has been completed, a few key interviews will yield additional relevant information. This information is verified on a wider population sample by means of questionnaires. The MRM training design and development team is selected during this phase of the process. It is critical to the success of the training program to include representatives from all of the essential areas of the maintenance organization. Likewise, senior management must allow team members to devote the necessary time to the project. In determining the reasons why an MRM training program is being undertaken, it is important to answer two basic questions: What is the current performance of the organization and workers? and What is the desired performance of the organization and workers? If a difference between current and desired performance exists, then we say there is a *performance gap.* Organizational analysis determines the probable cause(s) of performance gaps and includes a distinction between needs that can be solved by training and needs that must be addressed by a change in organizational procedures or policies. For example, issues that should be addressed by developing a company policy might include redesigning workcards or reformatting an Engineering Authorization form. A training need might be teaching engineers and technical writers how to write a workcard or Engineering Authorization effectively. An MRM training need would be to teach AMTs the skills that will help them to recognize how the environment impacts human performance.

An evaluation of the resources available for the development and delivery of the MRM training program also is necessary. This consists of identifying various constraints, such as the availability of equipment, time, money, and instructors. This information is transformed into a set of functional design specifications, a specific list of training goals, and system requirements that will provide the boundaries of the training program. Benefits can be measured by the company's typical performance measures related to maintenance tasks, such as dependability departures, remain overnight, safety [ground damage, occupational injuries], and efficiency and quality [component shop statistics].

Training costs are typically compared with benefits (both tangible and other) to calculate a return on investment for the training program. The benefits of an MRM training include reducing errors and increasing safety, dependability, and efficiency, which can be quantified and estimated by unit and overall performance measures.

Tasks Analysis. A task analysis determines the tasks required in a job, the subtasks that compose each task, and the knowledge and skills required to successfully complete these. Task analyses help the instructional designer determine exactly what the learner needs to be able to do. It also allows the designer to develop objectives based on the task elements. A task analysis may be accomplished by: (a) observing a skilled and knowledgeable employee, (b) reviewing documents and manuals that pertain to the tasks, (c) interviewing employees who perform the tasks, and (d) performing the tasks. It is critical to remember that the best sequence for instruction might be quite different from the sequence of tasks in the workplace. An instructional designer may be able to generate the instructional objectives directly from the task analysis results.

Trainee Analysis. A trainee analysis is performed to identify the relevant characteristics of the learners who will be participating in the MRM training program. These characteristics and associated learning issues have been presented by Goldstein (1993) and Gordon (1994).

Guideline 2: Design the Training Program (Phase 2)

This phase of the ISD model involves developing the goals and objectives of the MRM training program, selecting the instructional content, specifying the instructional strategies, designing evaluation instruments, and specifying the training media. The design process consists of four levels: Program, Curriculum, Course, and Lesson (Hannum & Hansen, 1992). Program and curriculum are associated with a *macro* (general) type of organizational analysis. Training is linked with the strategic plans of the organization, and a series of course needs is identified for different groups of trainees. The course and lesson levels are defined as the *micro* type of planning, such as developing instructional objectives, learning task hierarchies, evaluating and testing procedures, and selecting media. At this point decisions are based on instructional theory and research; thus, the designers are concerned with the learners' ability to understand, remember, and transfer the training concepts to the worksite. A training program can include several curricula. Figure 12.3 illustrates the four levels of MRM training program design. Each curriculum consists of a series of courses, and each course will usually include a number of different lessons. The time required to design a curriculum containing several courses varies from a few weeks to a few months, depending on the level of effort expended and the complexity of the curriculum. Designing a course might require several days to a week. A lesson usually can be designed in 1 to 3 hours.

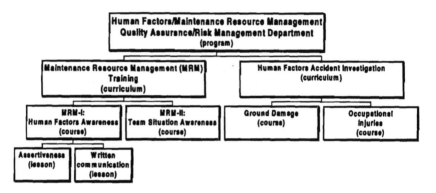

FIG. 12.3. Learning hierarchy: Team situation awareness maintenance resource management training.

A typical MRM training program is managed by a systems training group or a human factors, quality assurance, or risk management department. A specific MRM course could be developed for each curriculum. For example, some companies have developed an MRM curriculum including MRM-I: human factors awareness; MRM-II: team situation awareness, human factors and MRM skill development; and MRM -III: scenarios and simulations (Robertson, 1998). Each MRM course consists of lessons or modules, such as assertiveness, communication, team building, conflict resolution, stress management, decision-making, and human factors performance elements (Robertson, 1998).

It is important to note that successful training design for MRM courses includes a high level of interaction, for example, group exercises, case studies, and practice sessions. This type of design is known as *adult inquiry learning*, because learners manipulate materials and equipment, participate in problem-solving discussion groups, respond to open-ended questions, and collect data from direct observations of instructional events (Gordon, 1994; Knirk & Gustafson, 1986). This type of learning promotes the effective acquisition and processing of information. Concrete experiential activities (actually doing something) are highly motivating and tend to promote better retention.

Instructional Design Versus Media Technology. It is important to note that it is *not* the choice of a particular technology that ultimately determines the effectiveness of a training program; rather, the soundness of the instructional design will set the stage for the efficiency and effectiveness of the training. The design process should not be driven by media technology. One should not decide on a particular delivery system or medium until the analyses are complete. Sometimes we want to deliver a training lesson by means of videotape or lecture without considering the underlying instructional objectives. A medium is neither inherently good nor bad, but it can be appropriate or inappropriate. If sound instructional design principles are used, the designer will choose instructional materials and technologies that meet identified learning objectives and functional requirements.

Micro Design. During micro design we develop the course and instructional objectives and plan the courses and lessons. We also develop the learning task hierarchies and the testing evaluation procedures and, finally, select the training media. Course-level design requires a careful analysis of goals and objectives to ensure they are observable and measurable. Learning objectives—what the trainees are expected to know and be able to do after training—guide the selection of "enabling objectives." Course prerequisite knowledge and skills are established. The differences between pretraining and posttraining knowledge and skills are transformed into learning objectives for individual lessons. The learning objectives are organized into a learning objective hierarchy. Lesson plans are then developed to implement the objectives and determine the sequencing of the course.

Organizing and Sequencing Instructional Content. To organize and sequence MRM training program content, we first assign large units of related content to courses. Then the related content within each course is grouped into individual lessons. Finally, the content of each lesson is analyzed to determine the necessary supporting content or prerequisite knowledge and skills. Once the content of individual lessons is set the instructional events that are components of a lesson are developed and sequenced. The first step in organizing and sequencing courses for a given curriculum is to link the courses with their expected outcomes and instructional objectives. Related instructional goals are organized into course groups. A course typically contains several lessons completed over several days.

Learning Task Analysis. It is important to determine exactly what a person must learn in an MRM training course in order to reach the desired performance goals. The task analyses that were conducted in the needs-assessment phase are used as the starting point for a learning task analysis. To complete a learning task analysis one identifies the types of learning and the learning domain implied by each instructional goal. The different learning domains are:

1. Intellectual skills—including understanding and articulating concepts, rules and procedures,
2. Information—the ability of individuals to verbalize declarative information
3. Cognitive strategies—acquiring strategic knowledge,

4. Attitudes—emotions and values adopted by trainees, and

5. Psychomotor skills—skills involving muscle development and coordination (Gagne, Briggs, & Wagner, 1988).

Because the learning outcomes are different in each domain, analyzing these learning outcomes requires different techniques. Once the particular domain is identified the appropriate analytical technique can be applied. Because human performance is organized by these categories of learning outcomes, understanding and identifying these learning outcomes help determine the appropriate instructional conditions for the trainees.

Sequencing Instructional Content. There are two aspects to sequencing lessons: instructional content and instructional events (Hannum & Hansen, 1992). *Instructional content* includes the facts, ideas, concepts, skills, and so on, that are defined in the instructional objectives. This is the material one expects the learners to master by going through the lesson. *Instructional events* are the features of a lesson that, when present, facilitate learning. These include informing the learner of the objectives of the lesson, providing examples, and giving feedback. The content in different learning domains has a different natural organization. For example, motor skills have procedural structures, intellectual skills have learning prerequisite structures, and attitudes have a behavioral structure. These different structures imply different instructional sequences. There is not one instructional sequence that is effective for all types of instructional outcomes—learning is more complex than that. Different types of instructional content, that is, different domains of learning—require different instructional sequences.

Learning Objectives and Hierarchy. The learning objectives related to each learning outcome are ordered so that an instructional developer can specify the level or depth to which the potential learners must be in order to know the information. The use of learning categories ensures that instruction is properly focused. Identifying and understanding the learning outcomes can ensure that it is implemented in a logical and sequential manner.

An example of a learning task hierarchy specifically constructed for an MRM team situation awareness course is presented in Fig. 12.4 (Robertson & Endsley, 1995). The terminal objective is specified at the top of the hierarchy, and the enabling objectives are in the lower portion. Prior knowledge and entering skills are stated at the bottom, depicted below the line.

Performance Objectives. Developing and writing performance or instructional objectives should follow the ABCD format, as described by Knirk and Gustafson (1986). It is more important to include all four components than to follow the exact sequence implied by the format. Clearly stat-

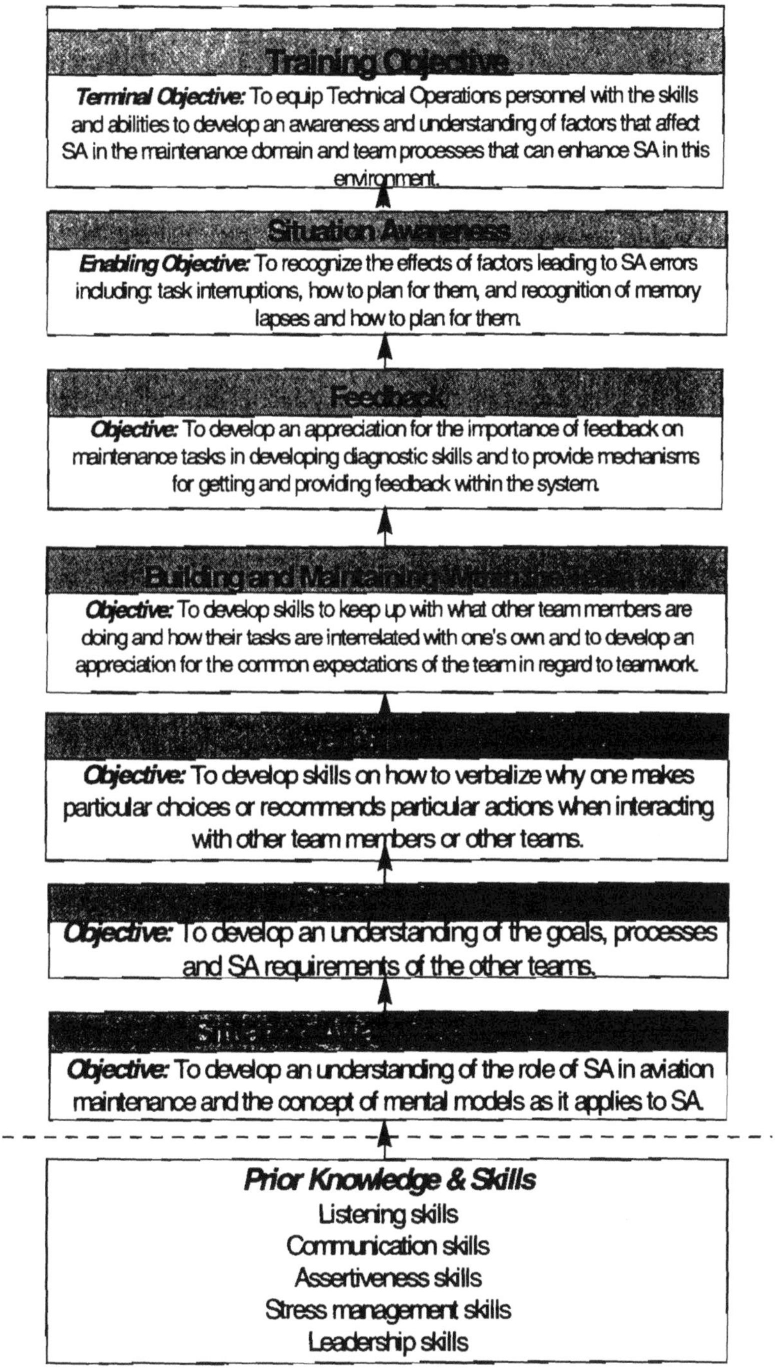

Training Objective

Terminal Objective: To equip Technical Operations personnel with the skills and abilities to develop an awareness and understanding of factors that affect SA in the maintenance domain and team processes that can enhance SA in this environment.

Situation Awareness

Enabling Objective: To recognize the effects of factors leading to SA errors including: task interruptions, how to plan for them, and recognition of memory lapses and how to plan for them.

Feedback

Objective: To develop an appreciation for the importance of feedback on maintenance tasks in developing diagnostic skills and to provide mechanisms for getting and providing feedback within the system.

Building and Maintaining Workload

Objective: To develop skills to keep up with what other team members are doing and how their tasks are interrelated with one's own and to develop an appreciation for the common expectations of the team in regard to teamwork.

Objective: To develop skills on how to verbalize why one makes particular choices or recommends particular actions when interacting with other team members or other teams.

Objective: To develop an understanding of the goals, processes and SA requirements of the other teams.

Objective: To develop an understanding of the role of SA in aviation maintenance and the concept of mental models as it applies to SA.

Prior Knowledge & Skills
Listening skills
Communication skills
Assertiveness skills
Stress management skills
Leadership skills

FIG. 12.4. Example of a learning task hierarchy for a situation awareness (SA) maintenance resource management training program.

251

ed and written performance objectives establish the basis for evaluating the training. Basically, one is trying to determine whether the trainees have successfully accomplished the stated instructional objectives. Learning objectives can serve as an organizer for trainees. They explicitly state what trainees are expected to learn and demonstrate.

Sequencing Instructional Events. A useful design format that supports learning consists of nine steps or instructional events. These instructional events are based on how we process information as we learn. They are appropriate for all learning situations and outcomes. These nine instructional events are:

1. Gain and maintain the trainee's attention by the use of novel, surprising, incongruous, or uncertain events instruction.
2. Describe the objectives of training—state in simple terms what the trainee will have accomplished once he or she has completed the course.
3. Stimulate recall of prerequisite skills—recall concepts previously learned.
4. Present the content to be learned—present a definition of the concept.
5. Provide guidance and support, providing examples and nonexamples.
6. Elicit performance—ask trainees to use the skills and knowledge.
7. Provide feedback, reinforcement of and information about the responses.
8. Assess performance, including application of the skills, knowledge, understanding, and level of mastery.
9. Enhance retention and transfer—present novel and related work situations to which trainees can apply newly acquired skills, knowledge, and abilities.

Specifying Training Media. After instructional objectives have been developed, designers must select instructional strategies and media. These decisions tend to be interrelated and should be made concurrently. Many instructional strategies use a combination of methods and media to deliver training. The instructional medium should present instructional stimuli in an efficient, easily understood manner. Complex media, which tend to be costly and time consuming, are often inefficient and unreliable. Use the least expensive medium that will result in trainees' attainment of the desired objectives within a reasonable amount of time. Choose as the primary medium of instruction one that is appropriate for the majority of objectives—one that can be used throughout the instructional program. Additional media, such as simulations or animated visualization, can be used for emphasis or motivation. It is desirable to have a mix of instructional methods that actively involve the learners. Frequent media changes are often confusing, time consuming, and expensive. Group exercises, role playing, and games or simulations that involve the trainee, promote the sense that the training is relevant and useful.

One method of making media selection decisions is to relate the general domain of each objective to student grouping requirements (Goldstein, 1993; Salas, Dickinson, Converse, & Tannenbaum, 1992). If the course objectives are at the lower end of the cognitive taxonomy (e.g., knowledge of specifics and comprehension), then certain types of teaching activities or media are more appropriate for individualized instruction. Others may be more appropriate for group instruction. Typically, MRM and human factors courses are designed for group learning, and the media selections should be appropriate for these learning and group requirements.

Training Plan: Course Outline and Lessons. A training plan provides a blueprint for training development. This includes providing a description of the training objectives, content, media, training aids, and other elements required for actual instruction and the estimated time required for each training topic.

Design of Evaluation Instruments. Evaluating the training program involves measuring the degree to which the learning objectives were met. Using the information from the task analyses and the learning objectives hierarchy, one can establish performance criteria that can be subsequently measured and evaluated. These training evaluation instruments may include: questionnaires, observations, interviews, verbal protocols, task performance measures, and work unit and organizational performance measures. Certain instruments collect essential training-related performance data before and after the training program take phase. These assessment tools are used at different times during the training evaluation process (Cannon-Bowers et al., 1993; Goldstein, 1993; Gordon, 1994; Kirkpatrick, 1979). Table 12.1 presents a typical evaluation assessment process (Cannon-Bowers et al., 1993; Gordon, 1994; Hannum & Hansen, 1992; Kirkpatrick, 1979). Creating an evaluation plan during the design phase will ensure that the process is more efficient and provides useful information. The evaluation process establishes the link between the goals and objectives of the training program and its results. Information collected from the course evaluation creates an important feedback loop, demonstrating the overall effectiveness of the training. This information is also useful for course revision.

Guideline 3: Develop the Training Program (Phase 3)

Training Plan Development. The primary activities in the development phase are developing training materials and media as well as developing and testing prototypes. Training materials are modified during this phase, on the

TABLE 12.1
Evaluation Process: Six-Level Process and Evaluation Measures

Level	Examples of Evaluation Measures
1. Pretraining baseline assessment	Questions asked of trainees concerning the training objectives. Example: What are four human factors elements that impact performance?
2. Trainee reaction	Questions asked of trainees concerning usefulness, value, and relevancy of the training. Example: How useful was the lesson on norms? (Scale ranges from 1 [waste of time] to 5 [extremely useful])
3. Learning	Questions asked of trainees are same as pretraining questions. Example: What are three types of human error?
4. Performance	Observations, interviews, and behaviors; questions can be asked of the trainee of his or her behavioral intentions: Example: How will you use this training on your job? What changes have you made as a result of this training? Observations: Supervisor and leads observe crews and rate them on a scale (1–5) as to when they use the newly acquired behavior. Example: Crew members speak up when potentially unsafe situations develop.
5. Organizational results	Organizational results: Performance measures gathered at the unit and organizational levels. Example: aircraft safety (ground damage), occupational safety (injuries), dependability (on-time departures, on-time maintenance), efficiency (contained overtime cost).
6. Organizational performance and attitudes	Correlate attitude changes with performance. Example: Individual data correlated with maintenance unit performance.

basis of the results of prototype and user testing. An outline of typical activities for selecting and developing training materials follows (Gordon, 1994; Hannum & Hansen, 1992; Knirk & Gustafson, 1986):

- Create a development plan
- Search for existing content-related training materials
- Evaluate existing instructional materials
- Match objectives with training content and materials
- Make trade-offs of objectives and training materials (economic and effectiveness)
- Examine copyright requirements (obtain copyright permissions)
- Revise/modify existing training materials
- Develop and produce new training materials
- Develop facilitator and participant handbooks

- Develop prototypes and walk through new training materials
- Revise/modify new training materials and handbooks
- Conduct final user testing of training materials and handbooks
- Conduct final development of training materials and handbooks

A training development project should specify the following elements (Gordon, 1994):

- Personnel—training and human factor specialists, content specialists (maintenance)
- Budget—money to develop the training materials and handbooks, personnel cost of developing the course, travel time and expenses, evaluation costs
- Equipment—technical equipment, audio/video facility, video cameras, studio equipment, editing equipment, audio equipment
- Outside services and consultants—scriptwriters, actors, graphic designers, videographers, computer programmers.
- Tasks and activities to be completed and by whom
- Training tasks and activities timeline

Evaluating Instructional Materials. When selecting commercial instructional materials, each candidate material should be evaluated using a prescribed evaluation procedure, as provided by Knirk and Gustafson, (1986); Gordon, (1994), Goldstein, (1993), and Hannum and Hansen (1992).

Developing Training Materials. The media selection model(s) chosen in the design phase will serve as the framework for developing training media. Before beginning production, instructional objectives should be reviewed to confirm their sequential order. Conceptual sketches and outlines of the audio-visual aids are developed and then reviewed by other members of the design team and relevant subject matter experts.

Storyboard Scripting. Developing a script requires the instructional designers and other team members to think visually. Sound and the written word are often not as reliable as a visual presentation for developing the trainees' retention. Visuals can carry the message, and narration can be used to clarify and reinforce the visuals. The development team should be certain that the graphics, written material, and audio support one another. Scripting the training materials requires the instructional designer to: (a) clarify difficult points through visual illustrations; (b) simultaneously present to two different human senses, for example, seeing and hearing; (c) determine the best approach to convey the message quickly and clearly;

and (d) isolate and focus the trainees' attention on the central points specified by the performance objectives (Gagne et al., 1988; Knirk & Gustafon, 1986).

Walkthrough and Formative Evaluation. It is much easier to modify the training materials during the design and development cycle rather than after the training program has been implemented. Conducting a formative evaluation of the training program while the training materials are in a draft form allows essential and meaningful feedback to be collected from the learners. A simple formative evaluation consists of having the trainees read and review a storyboard. It should also measure the usefulness of the materials. Users will give the development team a good idea of whether the general training approach is a sound one. Table 12.1 lists various formative evaluation methods (Cannon-Bowers et al., 1993; Gordon, 1994; Kirkpatrick, 1979, 1990; Knirk & Gustafson, 1986). The advantages of user testing is that it allows one to solicit meaningful feedback from the users. Early user testing can reduce program costs and increase the probability that the product will perform as required.

Final Development and User Testing. After the prototyping and walkthroughs have been completed the training materials are put through the final development and production steps. After the production training materials are available, but before they are actually implemented, they should be subjected to one more stage of user (learner) testing.

Facilitator and Trainee Handbooks. After the final user testing is completed and the training program is ready to move into full development and production, facilitator and trainee handbooks should be developed. The facilitator handbook typically contains the following elements: (a) a detailed outline of the instructional sequence, (b) a description of the course and its lesson goals and objectives, (c) a narration of the visuals to be presented, (d) the training time frame, (e) administration issues and guidelines, (f) a description of the group exercises, (g) a list of the participants and facilitator materials, (h) a list of reference materials, (i) a description of how to set up the training classroom, and (j) a list and description of the evaluation instruments (Dupont, 1997; Robertson, 1998; Robertson & Endsley, 1997).

Guideline 4: Implementation and Delivery of the Training Program (Phase 4)

In this phase the overall training implementation plan is developed and the training is actually conducted. If the training is delivered in stages, it is possible to conduct further formative evaluations and revise the training program and materials before full production and implementation. A spreadsheet can be created that includes the schedule for delivering the training. It is also possible to purchase various computer-based management programs that can provide a framework for structuring the training schedule. An important program element that must be addressed during this phase is to gain (or reconfirm) management's commitment to deliver the training program and provide the necessary resources to successfully implement the course.

Facilitator Training. During the implementation phase the facilitators must be formally trained. It is possible that the facilitators are part of the design and development team and require only a minimal level of training; however, in many instances the facilitators or instructors will not have taken an active part in designing the training program and will need assistance. If facilitators have been identified previously, they can practice delivering the training lessons during the formative evaluation and user testing stages. In this way immediate feedback can be provided to the instructors as part of the overall evaluation process. For MRM courses it is important to have cofacilitators, as they can provide subject matter experts with different perspectives.

Guideline 5: Evaluate the Training Program

Planning for evaluation should take place during the design phase. Evaluation is important in order to: determine if the training meets the objectives, determine if the entire training program meets its goals, provide feedback to the facilitators and the top managers and the organization as a whole; and review and improve the training program. The evaluation process should measure the effects of training on the variables that have identified as being important. Evaluation criteria are variables that represent the specific factors the course designers targeted during the development process. These criteria are based on the training objectives and goals and are established in the needs–assessment and design phases.

Evaluation Process. There are two types of evaluation: formative and summative (Gagne et al., 1988; Goldstein, 1993; Knirk & Gustafson, 1986). Previously discussed was formative evaluation used in the design and development phases. Examples of pre-, post-, and follow-up training questionnaires, which measure changes in learning, attitudes, behaviors, and organizational performance were given by (Robertson & Endsley, 1997; Taylor & Robertson, 1995).

A summative evaluation is conducted after the MRM training course has been developed, implemented, and delivered. A summative evaluation typically determines the extent to which the training program has been successful in meeting its stated training, behavioral, and organizational objectives. It also determines the value of the training program and what modifications need to be made to make it more effective.

A summative evaluation should be conducted following the following general principles: (a) conduct the evaluation in an environment that is as similar to the ultimate job environment as possible, (b) conduct the evaluation after a realistic period of time (preferably, 2, 6, and 12 months after training), and (c) conduct the evaluation based on the targeted job tasks and conditions (Gordon, 1994; Knirk & Gustafson, 1988).

Evaluation Model. Table 12.1 lists various methods for evaluating training courses (Cannon-Bowers et al., 1993; Goldstein, 1993; Gordon, 1994; Hannum & Hansen, 1992; Kirkpatrick, 1979; Knirk & Gustafson, 1986). These levels of evaluation and the types of data that can be collected provide for a solid framework in evaluating an MRM training program.

MRM EVALUATION: ONE COMPANY'S EXPERIENCE

Evaluation results of an MRM training program at a major airline company demonstrate positive and significant effects of the MRM training program (Robertson & Taylor, 1996; Robertson et al., 1994; Taylor & Robertson, 1995). A systematic evaluation of the effects of the MRM team training program on maintenance personnel attitudes and behaviors was used that was based on the six-level evaluation model (Robertson & Taylor, 1996; Taylor & Robertson, 1995). Multiple measures and assessments of the managers' attitudes and self-perceptions of behaviors as well as maintenance performance results were used, spanning a 4-year period. This provided an unique opportunity to longitudinally measure and track the long-term training effects. Analyses of the association between attitudes and organizational performance over time also were conducted. Data were gathered through the use of the Maintenance Resource Management/Technical Operations Questionnaire (MRM/TOQ), on-site interviews and observations, trends of maintenance performance measures, and attitude-performance analysis (Robertson & Taylor, 1996; Taylor & Robertson, 1995).

Evaluation Results

Overall, the results of the evaluation demonstrated a positive and significant effect of the MRM training on attitudes, behavior, and organizational performance. The significant and positive improvements in maintenance personnel attitudes reflected the expected and intended training effects on the participants' attitudes and their stability over time. The results of each of the evaluation steps are presented below.

Step 1: Baseline Assessment. Two baseline measurements were made before the training intervention occurred so that any changes in the trainees' MRM attitudes and knowledge could be assessed. These two measurements help create a stronger quasi-experimental field research design. There were no significant changes found in the AMTs' attitudes and behaviors as measured by the baseline and pre–training MRM/TOQs.

Step 2: Reaction. This level of evaluation involved the participants' written reactions to the value and usefulness of the team training program as measured by the MRM/TOQ. Several questions were developed to assess the trainees' reactions to the training course materials, objectives, organization, training climate, and instructor skills. This level of evaluation also serves as a formative evaluation of training materials and delivery methods in the initial phases of the training program. The Level 2 evaluation showed that the participants' immediate responses to training were positive—more than 90% rated the training as "very useful" or "extremely useful," and more than 96% said they felt that it was one of the best training courses they had attended. Other positive aspects of the course included having a mix of participants in the class. This was beneficial because the managers were able to gain an appreciation of other managers' job functions, what their constraints and problems are, and how the outputs of their jobs affect others in the work system.

Step 3: Learning. The knowledge gained and the immediate changes in the participants' attitudes and the stability of these changes in time were measured with the pre- and posttraining MRM/TOQ questionnaires. Changes in relevant attitudes measured immediately before and after training were significant, with positive changes following training for three of the four attitudes measured ("command responsibility," "communication and coordination," and "recognizing stressor effects"). The attitude measure of assertiveness rose significantly between the posttraining measure and the 2-month follow-up survey. Follow-up results indicated that all four attitude scales remained high and stable over the 2-, 6-, and 12-month surveys following training.

Step 4: Behavior and Performance. The Step 4 evaluation results, which were derived and content coded from the open-ended responses on the follow-up surveys, indicated how the trainees actually used the training on the job. The trainees' self-perceptions of their behavior on the job significantly shifted from passive responses (e.g., "be a better listener," "being more aware of others") to more active responses, (e.g., "having more daily meetings to solve problems," "gathering more opinions," and "getting more feedback from others"). Field interviews and observations were conducted over a 1-year period to validate the contents of the self-reported behaviors.

Step 5: Organizational Results. In Step 5, evaluation trends in maintenance performance before and after the onset of the MRM team training program were examined. One of these performance trends represented occupational safety (lost time injuries—rate of lost time injuries, per 1,000 hours worked, for 55 work units). Overall, the injury rate remains at a low level for the year and a half after training was introduced.

Step 6: Organizational Performance and Attitudes. To correlate attitude changes with performance, the individual maintenance personnel data are combined into averages for the units to which they belong. The organizational–performance measures included were aircraft safety (ground damage); personal safety (occupational injury), dependability, based on departures within 5 minutes, and on-time maintenance. The results from this analysis for the follow-up surveys show a significant number of correlations between maintenance unit performance and attitudes.

Future Directions. Using a systematic training evaluation process provides a framework to demonstrate the effects of an MRM training program. This company's MRM training program is still being conducted with several new courses still being developed and implemented (Endsley & Robertson, in press; Robertson & Endsley, 1997). Evaluations of these MRM courses are being conducted using the same process described earlier, demonstrating significant and positive effects of the MRM training on maintenance personnel attitudes, behaviors, and organizational performance (Taylor, Robertson, & Wong, 1998). Other companies are currently evaluating their MRM training programs and are showing positive and significant effects of the training on maintenance personnel attitudes and behavior (e.g., Driscoll, Kleiser, & Ballough, 1997; Dupont, 1997).

ELEMENTS OF SUCCESSFUL RESOURCE MANAGEMENT PROGRAMS

Even though every company designs and implements a slightly different MRM training program, there are common elements across the industry. On the basis of the experiences of these companies, five common and critical elements of successful resource management (RM) programs were identified.

Guideline 1: Senior Management Active Support Is Needed

The foundation of any successful organizational program is senior management support. Senior managers must have the vision and commitment to reduce maintenance errors and increase safety through the use of RM. When top decision makers clearly support the mission and purpose of RM, an organizational culture change can occur. Without such a commitment a pervasive organizational change is unlikely.

Guideline 2: Train Supervisors and Middle Managers

Linked to the first element is training for supervisors and middle managers. These individuals interact daily with the workers who are ultimately responsible for implementing the new strategies. Mid-level managers also need the support of upper-level management in implementing the new RM skills and approaches in the field. This support can take many forms but certainly includes devoting the time necessary to attend appropriate RM training courses. With this commitment, supervisors will have the opportunity to use their own RM skills in addition to managing a cultural change.

Guideline 3: Foster Continuous Communication and Feedback

It has been said that nature abhors a vacuum. The same can be said for organizations undergoing pervasive changes. To sustain the change process, continuous communication and feedback must occur. Several communication channels exist to distribute the results of RM training programs. These include newsletters, group meetings, public bulletin boards, electronic mail, and so on. The idea is to provide managers and workers with information on the type of actions occurring in the workplace and their effects on the company's overall performance (i.e., quality, safety, dependability).

Guideline 4: Use a Systems Approach to Training

The fourth element of a successful RM program is the use of a systems approach in designing and implementing RM training. Following an instructional system design approach when managing and developing training ensures a comprehensive and effective program. Applying a systems approach to RM leads to a well-planned program and relevant interventions.

Guideline 5: Provide for Employee Participation

The fifth element of successful RM programs is the participation by all people who have a significant stake in its outcome. Active participation increases feelings of personal ownership and motivation to implement new ideas. A typical MRM stakeholder group includes AMTs, inspectors, engineers, quality assurance personnel, managers, and FAA regulators (typically from the Flight Standard District Office). There are several examples in the industry demonstrating the positive effects of full participation in the design of training programs; redesigning logbooks and manuals; and facilitating courses in MRM, human factors, and self-directed team building (Drury, 1993; Robertson, 1988; Taylor & Robertson, 1995).

Guideline 6: Create a Responsive Environment

Facilitation of the positive transfer of RM skills to the workplace necessitates the creation of an organizational responsive environment. When the line supervisors and managers are rewarded by the organization as a result of their actions in reinforcing and using the RM skills, a positive, cultural organization begins to emerge. Using the results of the evaluation process can provide valuable information related to the effects of the RM program and demonstrate *how* the RM skills and behaviors are being used. Reinforce positive employee actions, such as highlighting the results of work unit actions in improving job processes and procedures as a direct result of applying RM skills.

SUMMARY

In this chapter, the development of MRM and the performance issues and problems in aviation maintenance that MRM addresses were discussed. Current practices and presented airline companies' development, implementation, and evaluation experiences were presented. A description of a systems approach to designing and developing an MRM human factors training program was given. Throughout the chapter, representative case examples and experiences from airline companies were provided to illustrate *how* they applied the principles to their MRM training. Generic guidelines and successful elements to be used in developing and implementing an effective and successful MRM training program were presented.

REFERENCES

Becker-Lausen, E., Norman, S., & Pariante, G. (1987). *Human error in aviation: Information sources, research obstacles and potential.* Moffett Field, CA: NASA Ames Research Center.

Cannon-Bowers, J. A., Salas, E., & Converse, S. A. (1993). Shared mental models in expert team decision making. In N. J. Castellan, Jr. (Ed.), *Current issues in individual and group decision making* (pp. 221–246). Hillsdale, NJ: Lawrence Erlbaum Associates.

Driscoll, D., Kleiser, T., & Ballough, J. (1997). *US Airways Maintenance Resource Management, Aviation Safety Action Program*, Pittsburgh, PA: (Quality Assurance).

Drury, C. (1993). Training for visual inspection of aircraft structures. In G. S. Corporation (Ed.), *Human factors in aviation maintenance—Phase Three, Volume 1 progress report* (DOT/FAA/AM-

93/15, pp. 133–154). Washington, DC: Federal Aviation Administration, Office of Aviation Medicine.

Dupont, G. (1997). The dirty dozen errors in maintenance. In *Proceedings of the 11th Federal Aviation Administration Meeting on Human Factors Issues in Aircraft Maintenance and Inspection.* Washington, DC: Federal Aviation Administration, Office of Aviation Medicine.

Endsley, M. R., & Robertson, M. M. (1996). *Team situation awareness in aircraft maintenance.* Lubbock: Texas Tech University.

Endsley, M. R., & Robertson, M. M. (in press). Situation awareness in aircraft maintenance teams. *International Journal of Industrial Ergonomics.*

Federal Aviation Administration, (1991). *The aviation human factors: National plan.* Unpublished manuscript, Washington, DC.

Gagne, R., Briggs, L., & Wagner, R. (1988). Principles of instructional design (3rd ed.) New York: Holt, Rinehart and Winston.

Goldstein, I. L. (1993). *Training in organizations* (3rd ed.). Belmont, CA: Wadsworth.

Gordon, S. (1994). *Systematic training program design: Maximizing and minimizing liability.* Englewood Cliffs, NJ: Prentice-Hall.

Hackman, J. R. (1990). *Groups that work.* San Francisco: Jossey-Bass.

Hannum, W., & Hansen, C. (1992). *Instructional systems development in large organizations.* Englewood Cliffs, NJ: Educational Technology.

Hawkins, F. (1993). *Human factors in flight* (2nd ed.). Aldershot, England: Gower Technical Press.

Kirkpatrick, D. (1979). Techniques for evaluating training programs. *Training and Development Journal, 31*(11), 9–12.

Knirk, F. G., & Gustafson, K. L. (1986). *Instructional technology: A systematic approach to education.* New York: Holt, Rinehart and Winston.

Lock & Strutt, (1981). *Reliability of in-service inspection of transport aircraft structures* (CAA Paper 5013). London: Civil Aviation Authority.

Marx, D. A., & Graeber, R. C. (1994). Human error in aircraft maintenance. In N. Johnston, N. McDonald, & R. Fuller (Eds.), *Aviation psychology in practice* (pp. 87–104). Aldershot, England: Avebury.

Maurino, D., Reason, J., Johnston, N., & Lee, R. (1995). *Beyond aviation human factors.* Brookfield, VT: Ashgate.

National Transportation Safety Board. (1989). *Aircraft accidents report: Eastern Air Lines, Inc., L-1011, Miami, Florida, May 5, 1983.* Washington, DC: Author.

Robertson, M. M., (1998). Maintenance resource management. In M. Maddox (Ed.), *Human factors guide for aviation maintenance,* Washington, DC: Federal Aviation Administration, Office of Aviation Medicine.

Robertson, M. M. & Endsley, M. R. (1997). Creation of team situation awareness training for maintenance technicians. In human factors in aviation maintenance—Phase Seven, Volume 1 progress report (pp. 173–197). Washington, DC: Federal Aviation Administration, Office of Aviation Medicine.

Robertson, M. M., & Endsley, M. R. (1995). The role of crew resource management (CRM) in achieving situation awareness in aviation settings. In R. Fuller, N. Johnston, & N. McDonald (Eds.), *Human factors in aviation operations* (pp. 281–286). Aldershot, England: Avebury.

Robertson, M. M., & Taylor, J. C. (1996). Team training in an aviation maintenance setting: A systematic evaluation. In B. Hayward & A. Lowe (Eds.), *Applied aviation psychology: Achievement, change and challenge* (pp. 373–383). Sydney, Australia: Avebury.

Robertson, M. M., Taylor, J. C., Stelly, J. W., & Wagner, R. H. (1994). Maintenance CRM training: Assertiveness attitudes and maintenance performance in a matched sample. In N. Johnston, N. McDonald, & R. Fuller (Eds.), *Aviation psychology in practice.* Aldershot, England: Avebury.

Rogers, A .G., (1991). Organizational factors in the enhancement of aviation maintenance. In *Proceedings of the fourth conference on human factors issues in aircraft maintenance and inspection* (pp. 45–59). Washington, DC: Federal Aviation Administration, Office of Aviation Medicine.

Ruffner, J. W. (1990). *A survey of human factors methodologies and models for improving the maintainability of emerging Army aviation systems.* Alexandria, VA: U.S. Army Research Institute for the Behavioral and Social Sciences.

Salas, E., Dickinson, T. L., Converse, S., & Tannenbaum, S. I. (1992). Toward an understanding of team performance and training. In R. W. Swezey & E. Salas (Eds.), *Teams: Their training and performance* (pp. 3–29). Norwood, NJ: Ablex.

Shepherd, W. T., Johnson, W. B., Drury, C. G., Taylor, J. C., & Berninger, D. (1991). *Human factors in aviation maintenance—phase 1, progress report* (Chap. 6). Washington, DC: Federal Aviation Administration, Office of Aviation Medicine.

Taggart, W. (1990). Introducing CRM into maintenance training. In *Proceedings of the Third International Symposium on Human Factors in Aircraft Maintenance and Inspection.* Washington, DC: Federal Aviation Administration, Office of Aviation Medicine.

Taylor, J. C. (1991). Maintenance organization. In W. Shepherd, W. Johnson, C. Drury, J. Taylor, & D. Berninger (Eds.), *Human factors in aviation maintenance—phase 1, progress report.*). Washington, DC: Federal Aviation Administration, Office of Aviation Medicine.

Taylor, J. C., Robertson, M. M., & Wong, S. (1998). Effects of MRM training on aviation maintenance technicians attitudes and behaviors. In *Proceedings of the Ninth International Symposium of Aviation Psychology.* Columbus: Ohio State University Press.

Taylor, J. C., & Robertson, M. M. (1995). *The effects of crew resource management (crm) training in airline maintenance: results following three year's experience* (Contractor's report) Moffett Field, CA: NASA Ames Research Center, Office of Life and Microgravity Sciences and Applications.

Wiegers, T., & Rosman, L. (1986). The McDonnell-Douglas Safety Information System (SIS). In *Proceedings of the international sir safety seminar.* Flight Safety Foundation.

13

Medical Applications
of Crew Resource Management

Jan M. Davies
University of Calgary

With the introduction of highly reliable aircraft and advanced weapons systems in World War II came the recognition that failures in aviation were more related to the weaknesses of humans than to the machines themselves. This recognition triggered more extensive research into the nature of human error, with the general findings that problems in communication, teamwork, leadership, and decision making played a larger role among professionals than did problems with equipment or lack of technical competence (Cooper, White, & Lauber, 1980). These research data, including the appreciation of the role of human error in more than two thirds of accidents, triggered suggestions that the (then) new field of human factors might contribute even more to safety than the traditional fields of ergonomics and automation.

Twenty years ago, the understanding of the importance of interpersonal behaviour as a cause of error in aviation led to training first known as *cockpit resource management* (Cooper, White, & Lauber, 1979) and what is now generally known as *crew resource management* (CRM; Helmreich & Foushee, 1993). Increased knowledge of human behavior has contributed further to the evolution of understanding about human error. Because humans design, manufacture, maintain, and operate systems, human error has a role in 100% of accidents. Cockpit resource management has undergone similar evolution, with the development of CRM. Recently, Helmreich and his colleagues described a fifth generation of CRM, specifically designed to decrease the probability of errors occurring, to correct errors before they have an impact, and to contain or decrease the severity of errors that have consequences.

This fifth generation is seen by Helmreich and colleagues as a refocusing of basic CRM concepts and behavioral strategies, which are now identified as error countermeasures (Helmreich, Merritt, & Wilhelm, 1999; see also chap. 15, this volume, for a detailed discussion).

Medicine has undergone a similar industrial revolution, with the development of sophisticated anesthetic equipment, heart–lung machines, kidney dialysis machines, and radiology equipment, to name but a few examples. In addition, there are more sophisticated control systems, such as "intelligent monitoring, the use of neural networks for diagnostic purposes, of fuzzy logic systems for drug delivery and expert systems for therapeutic purposes" (de Leval, 1996, p. 350). However, medicine has not had a systemwide development of a program of CRM, let alone the evolution of CRM. One of the reasons may be that medicine has not had the safety-impetus equivalent of the crash of two jumbo jets. Although legal claims against doctors have increased since the 1980s, as have payments to patients and malpractice insurance fees by doctors (Davies & Robson, 1994), these increases have not equaled the extent of the payments seen for victims of aviation accidents.

The aim of this chapter is to describe the application of CRM to a system quite different from and yet similar to aviation—that of medicine. A brief history of the evolution of CRM will be given, and lack of a similar program in medicine will be identified. Three questions will be posed: (a) why should CRM be applied to medicine?, (b) how should CRM be applied to medicine?, and (c) what should be the outcome of the application of CRM to medicine? Answers to the first question will reflect the current understanding of human error in medicine, the limitations of human performances and team interface problems. The process by which CRM should be established in medicine will include a review of the efforts to date, the differences between current CRM programs in medicine and those in aviation, and four elements that future efforts should consider. Last, the results of a medical CRM program will be discussed with respect to those who are likely to be affected—patients, personnel, the organization, and the regulator.

WHY SHOULD CRM BE APPLIED TO MEDICINE?

The Role of Human Error

For most health care workers and administrators, there is still little recognition of the universal role of human error in both day-to-day problems and in catastrophes. From their earliest days, medical students are "exhorted to learn from their mistakes" (McIntyre & Popper, 1983).

> No species of fallibility is more important or less understood than fallibility in medical practice. The physician's propensity for damaging error is widely denied, perhaps because it is so intensely feared. . . . Physicians and surgeons

often flinch from even identifying error in clinical practice, let alone recording it, presumably because they themselves hold . . . that error arises either from their own or their colleagues' ignorance or ineptitude. (Gorowitz & MacIntyre, 1976, quoted in McIntyre & Popper, 1983)

Traditionally, doctors, nurses, and paramedical personnel are taught that they are to function without error, despite the knowledge that this goes against all understanding of human error. In fact, the reason why people so readily accept human error is an error itself - the *fundamental attribution error.* This error is embedded deep in the psyche and relates poor performance to some aspect of the individual's personality or even a character defect (Reason, 1997). In addition, the threat of peer and organizational sanction, as well as the specter of malpractice litigation, strongly encourage health care providers not to report mistakes. This need for flawless performance has created a "strong pressure to intellectual dishonesty, to cover-up mistakes rather than admit them, and to overlook opportunities for improvement" (de Leval, 1997, p. 725). When errors are discovered, often the people who have committed them are blamed, and "ineffective countermeasures," such as "exhortations to be more careful," are then invoked (Reason, 1994).

There have been some attempts to take a wider, systems approach to medical catastrophes, to improve knowledge "by the recognition of error" (McIntyre & Popper, 1983). For example, Eagle, Davies, and Reason (1992) reported their investigation of the events surrounding the death of a patient (who suffered severe, and ultimately fatal, breathing difficulties immediately after the start of a general anesthetic). This was actually the first application of the Reason model (Reason, 1990) to a medical catastrophe (Eagle et al., 1992). Five "latent failures" were identified as existing in the system before the "active failure" occurred that resulted in the patient inhaling large quantities of stomach contents into his lungs. Another example is that of the hospitals in the Calgary Regional Health Authority, which trigger a systematic critical review policy when either the "process of care was not as planned or anticipated, even though the outcome of care was acceptable" or when the care resulted in "death or serious bodily harm" (Davies, 1996). Other authors have advocated development of a "new ethos" (McIntyre & Popper, 1983) or a "culture of error" (de Leval, 1997) wherein there would be rejection of the attitude that "an authority is not expected to err; if he does, his errors tend to be covered up to uphold the idea of authority" (McIntyre & Popper, 1983, p. 1920). Instead, there would be adoption of the importance of exposing errors so that "after discussion and analysis, change in practice may prevent their repetition" (McIntyre & Popper, 1983, p. 1920). With this change in attitude would come recognition that human error is only a description of behavior. In addition, doctors, nurses, and administrators should learn to say "sorry" to patients and their families. They would also share with them the results of the recommendations resulting from the investigation of errors that result in medical complications (Davies,

1996). However, these new concepts are, for too many, still just waiting to be widely accepted and set into practice.

The Role of Human Performance Limitations

All individuals are susceptible to what is known as *human performance limitations*. Many health care workers fail to recognize that their abilities may be limited, particularly when they are fatigued, unwell, or otherwise stressed. This was shown by Helmreich and Schaefer (1993), who adapted the Flight Management Attitudes Questionnaire to develop the Operating Room Management Attitudes Questionnaire (ORMAQ). This latter questionnaire contains a core set of items that measure the same concepts as the parent questionnaire, that is, communication, negotiation, team member responsibilities, leadership, and recognition of stressors. The ORMAQ also contains additional items targeted to each subgroup of the operating room (OR) team (i.e., anesthesiologists, surgeons, anesthetic nurses, and surgical nurses). A final section contains open-ended questions designed to elicit information about major problems and ways of improving the function of the OR team.[1]

Data were collected using the ORMAQ in teaching hospitals in three countries. Responses to items about attitudes were used to develop five human factors scales, dealing with leadership, followership, team roles, information sharing, and stress recognition. The Stress Recognition scale contains items that reflect "awareness of the deleterious effects of stressors" (e.g., fatigue, personal problems, crises) on performance. What was most striking about the results was the "low absolute recognition of the effects of stress" (Helmreich & Davies, 1996, pp. 283–284). This strong denial of personal vulnerability paralleled results from aviation, where pilots considered themselves "bullet-proof" (Helmreich & Foushee, 1993; Helmreich & Wilhelm, 1991).

The nonaviation results are quite striking, as one might assume that health care workers "would have more sensitivity to psychological effects, given some training in behavioral sciences, but this does not appear to be the case" (Helmreich & Davies, 1996, p. 285). Indeed, many doctors (and nurses) have an "appalling insensitivity to factors which they would both proscribe (tobacco, recreational drugs) and prescribe (moderate alcohol intake, good nutrition, regular exercise, a balance between work and home, seeking medical care when unwell)" (pp. 285–286). Part of the explanation for these responses could lie in underlying personality factors. However, the results (for the doctors, at least) are probably related to "learning to be a

---

[1]A revised version of the ORMAQ—the Operating Room Teamwork, Safety, & Management Questionnaire—was developed by Sexton and Helmreich (1998). This new questionnaire targets an expanded group of individuals, including perfusionists, nursing assistants, and department leadership.

doctor" (Helmreich & Davies, 1996, p. 286), that is, "learning not to feel and react like an ordinary person—i.e., like a patient" (Anonymous, 1993, p. 1250). As doctors progress through medical school and residency, they "cope with the anxieties of the potential catastrophic illness, passing from the 'medical student hypochondria' (i.e., suffering from each condition as it is learned) to advanced denial even of mortality" (Helmreich & Davies, 1996, p. 286). However, awareness of the human condition is vital. At the individual level all health care workers need to learn and use measures to counteract the stresses of the workplace. At the team level, workers need to recognize that working with other team members will help in the management of errors, although ideally this requires the admission of the effects of stress and illness on performance.

The last hundred years have produced major changes in the practice of medicine, including the shift from "doctor-centered" care to "team-centered" care. For example, "many of the tasks formerly regarded as solely those of physicians are now commonly shared by nurses," for example, monitoring irregular heart rhythms, blood chemistry, and blood gases (Mechanic & Aiken, 1982, p. 748). With the increased emphasis on care by teams has come a change in the nature of risks associated with medical practice. Although always a high-risk activity, the origin of the risk has shifted. Many of the problems now seen in medicine occur with difficulties in teamwork and communication. These interface issues have been shown to occur in the OR (Buck, Devlin, & Lunn, 1987; Helmreich & Schaefer, 1994) as well as in aviation (Helmreich, 1992).

HOW SHOULD CRM BE APPLIED TO MEDICINE?

Review of Current Programs

Current programs started with the first application of CRM concepts in medicine by Steven K. Howard and David M. Gaba and their colleagues in the Department of Anesthesia at the Stanford University School of Medicine and the Anesthesiology Service of the Palo Alto Veterans Affairs Medical Center. In the article describing this application, Howard et al. started by stating that anesthesiology was a "complex, dynamic world" that was "in some respects . . . similar to other domains, particularly aviation." They continued by describing the requirements of "safe patient care in anesthesiology" (p. 763), which were "medical and procedural knowledge and technical skills" as well as the application of these skills involving the "effective coordination of multiple resources in an unusually complex environment" (Howard, Gaba, Fish, Yang, & Sarnquist, 1992). However, they went on to relate how safe anesthesia was, quoting Cooper and coworkers from the Massachusetts General

Hospital: "Perhaps the most insidious hazard of anesthesiology is its relative safety" (Cooper, Newbower, & Kitz, 1984, p. 41). Because of this degree of safety, "significant negative outcomes related to anesthetic mishaps" were "uncommon" and anesthesiologists were "only occasionally called upon to manage a crisis situation" (Howard et al., 1992, p. 763). Cooper and colleagues (1984) stated that the handling of such crises depended "almost solely on the anesthetist's ability to react instinctively and flawlessly every time a problem arises" (p. 41). Yet Howard et al. acknowledged that "even expert clinicians do not always manage real or simulated crises optimally, both because of a lack of previously learned responses to specific situations, and because of the poor utilization and coordination of available resources" (Howard et al., 1992, p. 763).

Howard et al. (1992) therefore decided to develop a model for "anesthesia crisis management training" by using "training techniques of the aviation industry" (p. 764). They cited the importance of the concept of the "over-learning" of skills and emergency responses. However, they also indicated that many airline problems were "not due to lack of flying skills, but to the crew members' inability to use resources" (p. 764) that were readily available. Quoting Lauber (1987), they listed the principles taught in CRM at that time: "delegation of tasks and assignment of responsibilities; priority assessment; monitoring/cross-checking; use of information; communication; leadership; problem assessment; and avoidance of preoccupation" (Howard et al., 1992, p. 764).

The "analogous course for anesthesia trainees and practitioners" was called "Anesthesia Crisis resource management," or ACRM. (This is described in detail as it serves as the model for most CRM programs in medicine.) The two goals for this course were (a) "to provide participants with precompiled responses to critical incidents which (could) be called upon when needed" and (b) "to instruct participants in the coordinated integration of all available resources to maximize safe patient outcomes (crisis resource management)" (p. 764). The course was developed around the use of a realistic anesthesia simulator. The Stanford University group had been one of two in the United States in 1988 that had developed highly realistic anesthesia simulators for the training of anesthesiologists (Gaba & DeAnda, 1988), the other being a group at the University of Florida (Good, Lampotang, Gibby, & Gravenstein, 1988).

The ACRM course started with an introduction to the "concepts of anesthesia crisis resource management," given in a half-day of lectures and discussions. These concepts included: (a) the course and its goals; (b) "human performance, decision-making, and human error in the dynamic world of anesthesia"; (c) a videotape screening of the simulator re-enactment of the crash of Eastern Airlines Flight 401; (d) a videotape screening of an "actual anesthetic mishap recorded in an operating room routinely equipped with video cameras" (p. 765), for which permission had been received from the not-identifiable anesthesiologists involved. (During the operation, the patient received a relative overdose of the one of the anesthetic agents, developed severe slowing of the heart and a decrease in blood pressure, and then required cardiopulmonary resuscitation from which there was full recovery.) Last, there was (e) discussion of both the successful and unsuccessful aspects of crisis management as shown in the two videotapes as well as a "set of crisis management principles developed by the authors" (Howard et al., 1992).

Participants then spent time in the OR, becoming familiar with the simulator, the anesthetic machine, monitors, and drug supplies. During the second day of the course anesthesiologists participated in crisis management scenarios. Each individual was the "primary" anesthesiologist for 30 minutes and then the backup for another 30 minutes. A surgical resident and two OR nurses "were hired to interact with the subject according to the scenario script." Six scenarios were "created to present a variety of anesthesia crisis situations involving equipment failures, underlying patient disease and complications of surgery." Each session was followed by a 2-hr debriefing period with one of the senior authors. Participants were "encouraged to critique themselves, while the debriefing faculty attempted to link their actions to the concepts covered in the framework of anesthesia decision making and the principles of crisis management." The overall goal of the debriefing session was the "interactive, constructive, nonjudgmental critique and analysis of options and alternatives" (Howard et al., 1992, p. 766).

Following the lead of Gaba and colleagues at Stanford University, similar ACRM programs have been established in a number of centers. These include Harvard University and the Universities of Copenhagen (Christensen, Laub, & The Sophus Group, 1995; Lindekaer, Jacobsen, Andersen, Laub, & Jensen, 1997) and Toronto (Devitt et al., 1997; Kurrek & Fish, 1996).

A slightly different approach was undertaken at the University of Basel in Switzerland, by the late Hans-Gerhard Schaefer and R. L. Helmreich; the program was termed *Team Oriented Medical Simulation*. The first difference was in the simulator. Collaboration with the surgeon involved in the program, Thomas Kocher, led to the development of a laparoscopic surgery simulator that was positioned in the abdomen of the "patient" (nicknamed "Wilhelm Tell"). This laparoscopic insert allowed the surgeons to actually perform laparoscopic surgery during the course of the simulation. In addition, if the surgeon accidentally cut a blood vessel, then Wilhelm Tell would bleed. The second difference was in the participants. The entire OR team, including the orderlies, the anesthetic and surgical nurses, and the staff anesthesiologists and surgeons, would arrive in the simulator center as though that were their regular OR list. The third difference was in the program. The session would start with a 1-hour briefing and review of human factors concepts (as in other programs). However, at Basel the team also performed a preoperative review of Wilhelm Tell's condition (something which, unfortunately, rarely occurs with real OR teams). After the briefing the team would then participate in the simulator session, during which one or more problems (such as excessive bleeding or a decrease in the patient's oxygenation) would develop. This would be followed by a self-directed debriefing of the team, during which participants would view and comment on a videotape of the simulation. The fourth difference was in the specialized human factors training for the staff anesthesiologists and surgeons and the senior nurses who were responsible for the training and evaluation of the residents and anesthesia nurse trainees. This training was designed to ensure that the organizational commitment to effective teamwork was "consistent, strong and visible" (Helmreich & Davies, 1996, p. 291) and that the trainers received "formal instruction in techniques of effective instruction, briefing and performance feedback" (p. 291). In addition, the training was considered essential so that these senior faculty members, who were the institutional role models whom the rest of the staff followed, would be "leaders in the interpersonal as well as technical aspects of their professions" (Helmreich & Davies, 1996, p. 291).

Differences Between CRM and ACRM

The specialty of anesthesiology has led the way in introducing the concepts of aviation CRM to medicine. However, there are two major differences between CRM and ACRM, that is, between the aviation model of CRM and those developed in medicine.

First, most of the latter programs are based on the Stanford model of ACRM. This is in effect a "program to assess the ability of an individual 'in the hot seat' to handle specific crises in anesthesiology, although 'resource' management and communication are an integral part" of the program (Helmreich & Davies, 1997, p. 909). This type of CRM training is addressed to only a subgroup of the OR team—the anesthesiology staff—rather than to the full team, and only to the anesthetic aspects of the operation. Even the debriefing is focused on the options and actions of the anesthesiologist. As a result, any improvements in team performance will be less than optimal. Understanding of the importance of the concepts will be limited to the group in focus. Also, when a team forms, a "team mind" forms also, through the sharing of information and pooling of knowledge (Stokes & Kite, 1997). Thus, when the focus is on only one subgroup of the team the team will not truly form, and there will be no "team experience," with all members participating in problem solving, decision making, and conflict resolution.

Second, most CRM programs in anesthesiology (and even the program at the University of Basel, to some extent) are centered around the simulation of crises in the OR, with CRM as just one aspect of the program. This is actually the reverse of CRM programs in aviation, in which programs are designed to address team-interface issues that require an understanding of how teams manage errors, and simulation is but one tool in this overall program. Simulation should represent only one of the components necessary to achieve the maximum impact of safety in the OR and in other areas of medical practice. This is not to discount the enormous benefit in allowing doctors, nurses, and other health care workers to acquire basic skills (Good et al., 1992), to establish individual competence, and to practice coping with rare and challenging crises (Chopra et al., 1994; Davies & Helmreich, 1996; Howard et al., 1992). There is, however, also a need for CRM or team programs. The two are complementary but should be clearly differentiated. Health care workers need to learn how to work together, not only during crises but also, more important, under normal circumstances—during the routine, day-to-day activities when "nothing goes wrong." Teams that normally work together well will probably encounter fewer problems. Then, if problems are encountered, the team will have already dealt with its interpersonal and organizational difficulties before the crisis, during which both active failures occur and latent failures or conditions are exposed (Davies & Helmreich, 1996; Eagle et al., 1992; Reason, 1990, 1997). As stated earlier, this is more like the approach taken by the program at the University of Basel, although there is no medical CRM program that is truly analogous to those in aviation.

Elements of a CRM Program

Davies and Helmreich (1996) outlined four descriptive elements for a comprehensive human factors program. These elements were represented by the terms *initiated, integrated, recurrent,* and *data driven.* These same elements can equally represent those needed for the application of a comprehensive program of CRM in medicine.

Initiated.   Probably the most important element concerns the initiation of CRM training. Although medical (and nursing) education has continued to evolve to include new scientific concepts, the addition of such aspects as interpersonal dynamics is almost universally lacking. Nor is there joint training (of medical and nursing students together) in the knowledge, skills, and attitudes for teamwork. From the first introduction to these professions, concepts such as interpersonal dynamics should be incorporated into the curricula by the faculty and acquired and practiced by the students (Davies & Helmreich, 1996). Without this, there will no change in the culture of the profession of medicine to support and enhance the "culture of error."

Integrated.   Next, the CRM program should be integrated into an organization whose culture is such that effective communication and teamwork are recognized and supported. The organization also must develop a culture that accepts that errors are inevitable and that violations must not be tolerated. Such a culture requires active promotion by the senior staff who are responsible for training and evaluation. These staff will therefore need additional training themselves in CRM. In particular, they will require training as trainers and evaluators, for both technical and interpersonal aspects. This training should encompass knowledge, skill, and attitudes (Davies & Helmreich, 1996).

The details of the program should be established by both the health care workers and the administrators. The inclusion of the health care workers in the design and evaluation of the program is paramount to acceptance. This is, in effect, the essence of peer review, a well-established and strongly held tenet of medicine. Without this, the individuals who are the recipients of the program may be less likely to accept either the program or the results of any evaluation.

In addition, as stated above, the real need for the application of CRM to medicine lies in the area of interface problems. These are the difficulties between the different specialties that make up teams. Thus, effective CRM programs in the OR will encompass representatives involving surgeons, anesthesiologists, OR nurses, nurse anesthetists, anesthetic assistants, perfusionists, anesthetic (respiratory) therapists, technicians, orderlies, and clerical staff. In the emergency room the representatives would include the doctors, nurses, orderlies, clerical staff, respiratory therapists, paramedics, and other ambulance personnel. In the labor and delivery suite the representatives would include the obstetricians, general practitioners, anesthesiologists, neonatologists, midwives, nurses, respiratory therapists, and clerical staff.

Recurrent. As has been clearly shown in aviation, CRM training cannot be a single lesson that is taken once, assumed to fix the problem, and never repeated. Repetition and reinforcement are vital if the desired outcomes are to be achieved (Helmreich & Foushee, 1993).

How often should medical CRM courses be conducted? In aviation, there are no formal rules for intervals, but many companies offer recurrent training in CRM annually. In addition, CRM concepts are now incorporated routinely into every simulator line-oriented flight training session without the session being labeled specifically as *CRM.* In the field of anesthesia, Kurrek and Fish (1996) surveyed participants at an ACRM simulator workshop in Toronto. Half of the respondents to the survey thought that they should take such a course every year. However, the frequency of courses should probably depend on the course content. A greater frequency may be required in the initial stages of a CRM program, as there will be a high "moment of inertia" to effect changes in both the culture of the organization and that of the professions.

Data Driven. Finally, all education needs to be driven by data. For example, evidence-based medicine is a concept that suggests that medical decision making be based ideally on the best available evidence. The highest weighting of such evidence is given to data obtained from double-blind, randomized control trials (Naylor, 1995). In 1975, when I graduated from medical school, there was no mention of the term *evidence-based medicine* in the medical literature, either as a "textword" (i.e., mentioned in the text of an article), or as a subject heading. This did not change until after 1990, when a Medline search for 1991–1995 shows 12 subject headings and 75 textword citations. After this period there was then a sudden explosion of interest and development in this topic, as Medline shows 1,582 subject heading citations and 408 textword citations for 1996–May 1999. Medical education now reflects this change.

Similarly, CRM programs should be guided by information about the performance of the team, the organization, and medicine in general. Data should include results about the actions of the team members and their attitudes, the operation and climate of the organization, and new trends in the provision of health care. The CRM program will therefore require a quantitative database to be effective. Five specific uses for such a database have been described (Helmreich & Davies, 1996).

- Areas of concern in training are based on data from the organization itself to achieve credibility; for example, actual cases, not the "zebras" of medicine (Hunter, 1996).
- The strengths and weaknesses of the organization's practices and training program must be measured to achieve maximum safety and training effectiveness; for example, comparison against published results from other institutions (Davies, 1994).
- Individuals and teams require performance feedback to work toward improvement; for example, both verbal debriefing and written reports (Shysh, 1997).
- Trends in performance must be measured over time to detect areas that require more or less emphasis; for example, monthly, quarterly, and yearly mortality reports (Davies, 1994).
- Both short- and long-term performance data must be collected to serve as criteria for selection of new trainees; for example, strategies for the evaluation of candidates (Gough, Bradley, & McDonald, 1991).

In aviation, use of accident data has been made difficult, in effect, because of the safety of the system. A similar problem exists in medicine, particularly in such specialties as anesthesiology, where the mortality rate for anesthesia alone may be less than 1 in 200,000 (Davies, 1994). One solution is to include evaluation of what aviation calls the "near miss" (but what I prefer to call the "near hit"). Evaluation of the case should include the mechanisms by which the event was both initiated and salvaged and a consideration of the circumstances of the event, including the qualifications and numbers of personnel and the environment at the time of the incident. Scenarios can then be used for development of the CRM curriculum.

However, studying accidents, and even the near-hits, is somewhat flawed because that is the study of abnormality. Optimally, knowledge, skills, and attitudes are imparted by the demonstration of the ideal forms, not the less than ideal. If one is to adopt the "culture of error," then one must denounce the concept that errors are committed by an individual "bad apple" (Berwick, 1989) and instead focus on how the entire organization (or even system) functions every day. In effect, this requires linking the CRM program to the quality-assurance (QA) activities of the organization. In its most basic form, QA is simply a method to determine if a system is organized, working, and producing to its optimum level. An important point of understanding about QA is that the "A" represents *assurance* rather than simple *assessment*. The assessment of quality implies that a system will be examined but that no efforts will be made to improve the quality of the system. In contrast, the assurance of quality implies that recommendations will be formulated to improve the system and that further observations will be made to ensure that these recommendations not only were successful but also did not produce any new problems.[2] This important characteristic is known as "closing the loop, from problem back to problem" (Davies, 1991).

More than 30 years ago, Donabedian (1966) suggested a classification of the components of the health care system, structure, process, and outcome, which provides an outline for data collection and therefore evaluation of the quality of the system. Table 13.1 shows the classic triad of QA and its components.

Three types of measurements of the system should be carried out. Ongoing measurements run daily, monthly, and yearly. These measurements provide the system with the essentials of the database: how many patients, how many procedures, how long the procedures lasted, what the outcome was, and so on. Episodic measurements are represented by incident or case reports, both of which are ways of alerting the system to problems. Focused measurements may be stimulated reactively, by results from ongoing or episodic measurements, or proactively, from a desire to carry out prospective evaluation of the system. All of these will provide the data needed to drive a comprehensive CRM program.

---

[2]Some authors have used the terms *continuous quality improvement* (CQI) or *total quality management* (TQM) as opposed to *QA*. However, I believe that a comprehensive QA program will encompass all that TQM or CQI has to offer and use the term *QA* throughout this chapter.

TABLE 13.1
The Components of Quality Assurance

	Elements	Examples
Structure	Where With what By whom	Environment Equipment Administration, Personnel
Process	What is done How it is done	Tasks Methods
Outcome	How one knows what is done How one knows how well a task is done	Audit Evaluation

WHAT SHOULD BE THE OUTCOME
OF THE APPLICATION OF CRM TO MEDICINE?

Ideally, the outcome of a CRM program should be determined even before the program is developed. That is, the first question that should be asked is "What do we want to have happen?" or "What is the desired outcome of the CRM program?" The health care system has four main groups for which there is an outcome: patients, the health care workers, the institution, and the regulatory agency. The four corresponding outcomes are shown in Table 13.2.

For patients (and their families), the outcome should be an improvement in safety of the care delivered, as measured through review of mortality (lethal events) and morbidity (nonlethal events). For health care workers, the outcome should be an improvement in morale, in part because of improved patient outcome and in part because of an improvement in the process of the delivery of care. Teams that communicate and work together more effectively will also work more efficiently (Schaefer, Helmreich, & Scheidegger, 1995). The organization should therefore see an improvement in efficiency. This can be measured against the use of both physical and fiscal resources.

TABLE 13.2
Outcomes for Crew Resource Management Programs in Medicine

Group	Outcome	Example
Patients	Improved safety	Decreased rate of complications
Health care workers	Improved morale	Decreased number of sick days
Organization	Improved efficiency	Increased number of procedures performed
Regulatory agency	Improved appropriateness	Consolidation of low-volume duplicated services

There could also be a change in appropriateness at the level of the regulatory agency. Critics might argue that this is not possible and that the very idea represents a corruption of the true intent of CRM. However, CRM programs in aviation are now expanding to encompass not only flight, cabin, and maintenance crew but also administrative personnel (R. L. Helmreich, personal communication, June, 1998). There is, therefore, no reason why representatives of the regulatory agency should not be involved. For example, Laffel et al. (1992) reviewed the relation between experience and outcomes in heart transplantation. They observed an "institutional learning curve," with patients having a higher mortality rate if they received one of the center's first five transplants (20%) than patients who received a subsequent transplant (12%). A similar exercise could be applied to the question of whether a province with a population base of 1 million could or should support a pediatric open heart surgery program. If the answer is that the population is too small to provide sufficient numbers for team and institutional competence, then the application of CRM principles to error avoidance would suggest that such a program should not be established.

SUMMARY

A worthwhile start has been made in the application of aviation CRM to medicine, although to date the evidence that such programs have more than short-term effectiveness is similar to that from the early days of aviation CRM. More can, and needs to, be done, particularly in the area of team performance. In addition, by following the lead and examples set by aviation, it may be possible for medicine to accelerate its progress in this area, with the hope of reaching the same level set by aviation, but in fewer years. Thus, the continued application of CRM to medicine must incorporate three basic principles:

• *Principle 1*: The field of medicine must develop an understanding of what CRM is and how it has evolved in aviation. Without such an understanding of the history of CRM, medicine will continue to experience difficulties with teamwork and communication.

• *Principle 2*: The field of medicine must learn to accept human error as a ubiquitous characteristic of human performance and to develop ways of both managing error and apologizing to those affected by catastrophic complications of error. Without such an understanding, medicine will continue to foster the culture of perfection, with its attendant hazards.

• *Principle 3*: The field of medicine must establish programs of CRM with input from all involved. Without global input, any CRM program in medicine will be viewed as unnecessary at best and as a threat to professional autonomy at worst.

In addition, four guidelines are considered:

• *Guideline 1: The CRM program should be initiated into the earliest training of doctors, nurses, and other health care workers.* The education of these professionals should therefore include not only the basic knowledge, skill, and attitudes that have traditionally defined medical (and nursing) competence but also the concepts that are essential to true team care: communication, decision making, problem solving, and conflict resolution. Changes should be made to the curriculum to include CRM.

• *Guideline 2: The CRM program should be integrated into an organizational culture that recognizes and supports effective teamwork.* The organization should be prepared to provide the resources to do this and should be shown to be promoting the requirement for teamwork. These resources should reflect both the financial and temporal demands of participating in CRM activities, including teaching, research, and QA.

• *Guideline 3: The CRM program should be recurrent rather than a one-time event.* The individuals who run the programs, as well as those who participate in them, should recognize the need and provide the opportunities for continuing education. This continuing education should be made easily available and emphasized as being crucial to ongoing professional development.

• *Guideline 4: The CRM program should be data driven rather than imported without regard to the culture and climate of the organization.* The basis of the curriculum should be derived from the structure, process, and outcome of the organization and ideally should be linked to the organization's QA activities. These activities should provide ongoing information for future CRM efforts, thus providing positive feedback to all personnel.

The combination of these three principles and four guidelines will provide the basis for a program that should help in the continuing efforts to improve the safety, satisfaction, efficiency, and appropriateness of medical care.

REFERENCES

Anonymous. (1993). The doctor is unwell. *The Lancet, 342,* 1249–1250.

Berwick, D. M. (1989). Continuous improvement as an ideal in health care. *New England Journal of Medicine, 320,* 53–56.

Buck, N., Devlin, H. B., & Lunn, J. N. (1987). *The Report of a Confidential Enquiry Into Perioperative Deaths.* London: Nuffield Provincial Hospitals Trust, King's Fund Publishing House.

Chopra, V., Gesink, B. J., de Jong, J., Bovill, J. G., Spierdijk, J., & Brand, R. (1994). Does training on an anaesthesia simulator lead to improvement in performance? *British Journal of Anaesthesia, 73,* 293–297.

Christensen, U. J., Laub, M., & The Sophus Group. (1995). The Sophus anaesthesia simulator. *British Journal of Anaesthesia, 5*(Suppl. 1), A72.

Cooper, G. E., White, M. D., & Lauber, J. K. (1979). *Proceedings of the NASA workshop on resource management training for airline flight crews* (CP-2120). Moffett Field, CA: National Aeronautics and Space Administration, Ames Research Center.

Cooper, G. E., White, M. D., & Lauber, J. K. (1980). *Resource management on the flight deck: Proceedings of a NASA industry workshop* (CP-2455). Moffett Field, CA: National Aeronautics and Space Administration, Ames Research Center.

Cooper, J. B., Newbower, R. S., & Kitz, F. J. (1984). An analysis of major errors and equipment failures in anesthesia management: Considerations for prevention and detection. *Anesthesiology, 60*, 34–42.

Davies, J. M. (1991). Quality assurance: Learning from the American experience. *Hospimedica: The International Journal of Hospital Medicine*, January–February, 14–16.

Davies, J. M. (1994). Mortality/morbidity and audit. In W. Nimmo, D. J. Rowbotham, & G. Smith (Eds.), *Anaesthesia* (pp. 641–676). Oxford, England: Blackwell Scientific.

Davies, J. M. (1996). Risk assessment and risk management in anaesthesia. In A. R. Aitkenhead (Ed.), *Quality assurance and risk management in anaesthesia: Bailliere's clinical anaesthesiology. International Practice and Research, 10*, 357–372.

Davies, J. M., & Helmreich, R. L. (1996). Simulation: It's a start. *Canadian Journal of Anaesthesia, 43*, 425–429.

Davies, J. M., & Robson, R. (1994). The view from North America and some comments on "Down Under." *British Journal of Anaesthesia, 73*, 105–117.

de Leval, M. R. (1996). Human factors and outcomes of cardiac surgery. *Paediatric Anaesthesia, 6*, 349–351.

de Leval, M. (1997). Human factors and surgical outcomes: A Cartesian dream. *The Lancet, 349*, 723–725.

Devitt, J. H., Kurrek, M. M., Cohen, M. M., Fish, K., Fish, P., Murphy, P. M., & Szalai, J.-P. (1997). Testing the rates: Inter-rater reliability of standardized anaesthesia simulation performance. *Canadian Journal of Anaesthesia, 44*, 924–928.

Donabedian, A. (1966). Evaluating the quality of medical care, part 2. *Millbank Memorial Fund Quarterly, 11*, 166–206.

Eagle, C. J., Davies, J. M., & Reason, J. (1992). Accident analysis of large-scale technological disasters applied to an anaesthetic complication. *Canadian Journal of Anaesthesia, 39*, 118–122.

Gaba, D. M., & DeAnda, A. (1988). A comprehensive anesthesia simulation environment re-creating the operating room for research and training. *Anesthesiology, 69*, 387–394.

Good, M. L., Gravenstein, J. S., Mahla, M. E., White, S. E., Banner, M. J., Carovano, R. G., & Lampotang, S. (1992). Anesthesia simulator for learning basic skills. *Journal of Clinical Monitoring, 8*, 187–8.

Good, M. L., Lampotang, S., Gibby, G. L., & Gravenstein, J. S. (1988). Critical events simulation for training in anesthesiology. *Journal of Clinical Monitoring, 4*, 140.

Gorowitz, S., & MacIntyre, A. (1976). Toward a theory of medical fallibility. *Journal of Medical Philosophy, 1*, 51–71.

Gough, H. G., Bradley, P., & McDonald, J. S. (1991). Performance of residents in anesthesiology as related to measures of personality and interest. *Psychological Reports, 68*, 979–994.

Helmreich, R. L. (1992). Human factors aspects of the Air Ontario crash at Dryden, Ontario: Analysis and recommendations. In V. P. Moshansky (Commissioner), *Commission of Inquiry into the Air Ontario Accident at Dryden, Ontario* (final report). Ottawa, Ontario, Canada: Minister of Supply and Services.

Helmreich R. L., & Davies, J. M. (1996). Human factors in the operating room: Interpersonal determinants of safety, efficiency and morale. In A. R. Aitkenhead (Ed.), *Quality assurance and risk management in anaesthesia: Bailliere's clinical anaesthesiology. International Practice and Research, 10*, 277–295.

Helmreich, R. L., & Davies, J. M. (1997). Anaesthetic simulation and lessons to be learned from aviation. *Canadian Journal of Anaesthesia, 44*, 907–912.

Helmreich, R. L., & Foushee, H. C. (1993). Why crew resource management? Empirical and theoretical bases of human factors training in aviation. In E. Wiener, B. Kanki, & R. L. Helmreich (Eds.), *Cockpit resource management* (pp. 3–45). San Diego, CA: Academic Press.

Helmreich, R. L., Merritt, A. C., & Wilhelm, J. A. (1999). The evolution of crew resource management training in commercial aviation. *International Journal of Aviation Psychology, 9*, 19–32.

Helmreich, R. L., & Schaefer, H.-G. (1993). *The Operating Room Management Attitudes Questionnaire* (Tech. Rep. 93-8). Austin: National Aeronautics and Space Administration/University of Texas/Federal Aviation Administration.

Helmreich, R. L., & Schaefer, H.-G. (1994). In M. S. Bogner (Ed.), *Human error in medicine* (pp. 225–253). Hillsdale, NJ: Lawrence Erlbaum Associates.

Helmreich, R. L., & Wilhelm, J. A. (1991). Outcomes of crew resource management training. *International Journal of Aviation Psychology, 1*, 287–300.

Howard, S. K., Gaba, D. M., Fish, K. J., Yang, G., & Sarnquist, F. H. (1992). Anesthesia crisis resource management training: Teaching anesthesiologists to handle critical incidents. *Aviation, Space & Environmental Medicine, 63*, 763–770.

Hunter, K. (1996). "Don't think zebras": Uncertainty, interpretation, and the place of paradox in clinical education. *Theoretical Medicine, 17*, 225–241.

Kurrek, M. M., & Fish, K. J. (1996). Anaesthesia crisis resource management training: An intimidating concept, a rewarding experience. *Canadian Journal of Anaesthesia, 43*, 430–434.

Laffel, G. L., Barnett, A. I., Finkelstein, S., & Kaye, M. P. (1992). The relation between experience and outcome in heart transplantation. *New England Journal of Medicine, 327*, 1220–1225.

Lauber, J. K. (1987). Cockpit resource management: Background studies and rationale. In H. W. Orlady & H. C. Foushee (Eds.), *Cockpit resource management training: Proceedings of the NASA/MAC workshop* (pp. 5–14). Moffett Field, CA: National Aeronautics and Space Administration, Ames Research Center.

Lindekaer, A. L., Jacobsen, J., Andersen, G., Laub, M., & Jensen, P. F. (1997). Treatment of ventricular fibrillation during anaesthesia in an anaesthesia simulator. *Acta Anaesthesiologica Scandinavica, 41*, 1280–1284.

McIntyre, N., & Popper, K. (1983). The critical attitude in medicine: The need for a new ethics. *British Medical Journal, 287*, 1919–1923.

Mechanic, D., & Aiken, L. H. (1982). A cooperative agenda for medicine and nursing. *New England Journal of Medicine, 307*, 747–750.

Naylor, C. D. (1995). Grey zones of clinical practice: Some limits to evidence-based medicine. *The Lancet, 345*, 840–841.

Reason, J. (1990). *Human error.* New York: Cambridge University Press.

Reason, J. (1994). Foreword. In M. S. Bogner (Ed.), *Human error in medicine* (pp. vii–xv). Hillsdale, NJ: Lawrence Erlbaum Associates, Inc.

Reason, J. (1997). *Managing the risks of organizational accidents.* Aldershot, England: Ashgate.

Schaefer, H.-G., Helmreich, R. L., & Scheidegger, D. (1995). Safety in the operating theatre—Part 1: Interpersonal relationships and team performance. *Current Anaesthesia and Critical Care, 6*, 48–53.

Sexton, B., & Helmreich, R. L. (1998). *The Operating Room Teamwork, Safety & Management Questionnaire.* Austin: National Aeronautics and Space Administration/University of Texas/Federal Aviation Administration.

Shysh, S. (1997). The characteristics of excellent clinical teachers. *Canadian Journal of Anaesthesia, 44*, 577–581.

Stokes, A., & Kite, K. (1997). *Flight stress: Stress, fatigue and performance in aviation.* Aldershot, England: Ashgate.

14

Applying Resource Management Training in Naval Aviation: A Methodology and Lessons Learned

Randall L. Oser
Naval Air Warfare Center Training Systems Division

Eduardo Salas
University of Central Florida

Danielle C. Merket
Naval Air Warfare Center Training Systems Division

Clint A. Bowers
University of Central Florida

In naval aviation environments effective team coordination within the cockpit is crucial for successful mission performance (Prince & Salas, 1993, 1999; Salas, Prince, et al., 1999). As research has demonstrated, teams that effectively coordinate resources perform better, whereas teams lacking such coordination are more susceptible to team failures (Cannon-Bowers & Salas, 1998; Salas, Prince et al., 1999). Because coordination is important for conducting missions in an effective and safe manner, military aviation communities have designed and implemented training interventions that foster effective crew coordination.

This current chapter focuses on a program of aircrew coordination training[1] research efforts performed within U.S. Naval Aviation (Navy and Marine

---

[1] It should be noted that the term *aircrew coordination* within naval aviation is synonymous with other terms, such as *crew resource management* and *team resource management*, used in other settings.

Corps). The major thrusts of this applied program of research were to: (a) achieve an improved understanding of crew coordination performance, (b) develop mechanisms for the measurement of aircrew coordination, and (c) design and test training strategies. The overall goal of the research program was to develop and institutionalize a systematic methodology for the design, delivery, and evaluation of crew resource management (CRM) training.

The purpose of this chapter is to provide an overview of our research program and of the CRM training methodology used within naval aviation. Specifically, in this chapter we: (a) provide a brief overview of our program of research; (b) present six drivers we used as an overarching framework during the design, delivery, and evaluation of CRM training; (c) describe our CRM methodology; and (d) offer a set of lessons learned for professionals involved in designing, delivering, and/or evaluating resource management training. Although the lessons learned we present were generated from naval aviation, we believe they have benefits for other environments where coordination among multiple crew members is necessary for effective performance.

A PROGRAM OF CRM TRAINING RESEARCH

The goal of this program, which was conducted over a period of 10 years, was to design, implement, and evaluate a systematic training approach tailored specifically to the needs of naval aviators to effectively coordinate their resources (Prince & Salas, 1993, 1999; Salas, Fowlkes, Stout, Milanovich, & Prince, 1999; Salas, Prince, et al., 1999). Our first step toward this goal was to assess existing CRM training research and programs in other domains. In doing this, we found diverse ideas about CRM across the various communities that use CRM training. Therefore, the first thing we did was agree on a definition of CRM. For our purposes in describing CRM training research and the CRM training methodology, we adopt the definition of *CRM* as "a set of teamwork competencies that allow the crew to cope with situational demands that would overwhelm any individual crew member" (Salas, Prince, et al., 1999, p. 163). This definition was not intended to replace definitions of CRM that had been specified in other disciplines and communities but rather to complement them while also serving as a guide for CRM training research within the Navy. One of the main thrusts behind our effort was geared toward linking CRM practice to training research literature by capitalizing on training principles and guidelines derived from a wide array of investigations (see Cannon-Bowers & Salas, in press, for a review of these training research findings). It was our belief that, when well designed and delivered, CRM training could be effective. Therefore, we conceptualize CRM *training* as:

a family of *instructional strategies* that seek to improve teamwork *in the cockpit* by applying well-tested training *tools* (e.g., performance measures, exercises, feedback mechanisms) and appropriate training *methods* (e.g., simulators, lectures, videos) targeted at specific *content* (i.e., teamwork knowledge, skills and attitudes)." (Salas, Prince, et al., 1999, p. 163; italics and boldface added)

This definition, for us, offered a framework that allowed us to focus on the *training* component of CRM. Within the definition are certain key phrases that we would like to briefly address. First, the "family of instructional strategies" refers to the many forms of training strategies that are available and that can benefit CRM training (e.g., event-based training, cross-training). Second, we encompass "teamwork in the cockpit" in a broad sense, to include both intracockpit and intercockpit relations. Third, training "tools" are the how-to's that have been designed for teams. Fourth, many forms of training "methods" or delivery exist that can be strategically tailored to reflect the most appropriate method of delivery, depending on training content. Finally, CRM training "content" can be streamlined to focus on the knowledge, skills, and abilities (KSAs) that are most relevant to CRM.

Our research effort involved a systematic approach of investigations and subsequent evaluations. To begin, we examined naval mishap and incident reports, interviewed subject matter experts (SMEs) to identify target areas (i.e., skills and behaviors) related to effective and ineffective performance and reviewed existing training programs. We then used this information to design scenarios and conduct simulator studies, which enabled us to better understand the relationship between the target behaviors and aircrew coordination performance. Next, we investigated different techniques for training the identified behaviors that were associated with *effective* aircrew performance. On the basis of those findings we conducted several prototype training programs (see Fowlkes, Lane, Salas, Franz, & Oser, 1994 and Prince & Salas, 1993, for a detailed description of the training programs). Evaluations of these programs suggested that CRM training structured around the principles we used (i.e., identifying target areas, identifying and understanding the relationship between behaviors and performance, and training techniques) had a positive effect on aircrew performance (see Salas, Prince, et al. (1999) for a breakdown of our CRM evaluation studies).

Because of the successful results found through the evaluations of our prototype training programs, we documented the significant lessons learned into a systematic methodology that could be used to facilitate the design of CRM training programs for naval aviation (see Table 14.1). The CRM training methodology was used to support the development and implementation of CRM training throughout naval aviation. To institutionalize the lessons learned and findings from our research efforts, an annual requirement

TABLE 14.1
A Methodology to Design and Deliver Crew Resource Management Training

Step	Description	Product(s)
1. Identify operational/mission requirement	Review existing training curriculum including course master material lists, instructor guides, and standard operating procedures (SOPs); interview aviation SMEs; observe crews performing missions; review relevant mishap/accident reports.	• mission-specific context/examples • General understanding of coordination demands within the task
2. Assess team training needs and coordination demand	Use same data sources as in Step 1, with emphasis on identifying deficiencies I existing team training and specifying all tasks required to perform missions that involve a teamwork element.	• Coordination-demand analyses • List of tasks requiring coordination
3. Identify teamwork competencies and KSAs	Link team training needs to a theory of team performance that allows delineation of competencies (our emphasis initially was on skills) required to perform each of the team tasks identified in Step 2.	• Set of KSAs
4. Determine team training objectives	For each teamwork KSA, develop a training objective that can be empirically evaluated to determine whether if was accomplished.	• List of targets for training • Full list of generic and task-specific training objectives
5. Determine instructional delivery method	The method for accomplishing the instruction should be specified (e.g., information, demonstration, and/or practice and feedback) in this step. Consideration should be given to costs and availability of simulators.	• Lectures • Videos • Role play exercises • PC-based methodologies • High-fidelity simulators • Accident reviews as analysis • Training curriculum

6. Design scenario exercises and create opportunities for practice	Design scenarios or exercises in which events are embedded to provide trainees an opportunity to demonstrate each of the required KSAs identified in objectives in which accomplishment requires practice and feedback.	• Valid realistic scenario(s) including all relevant peripheral support
7. Develop performance assessment/measurement tools	In conjunction with scenario design, develop measures that can reliably assess whether each of the KSAs was demonstrated at an observable behavioral level.	• Behaviorally based checklists • Subjective evaluation forms • Outcome metrics and criteria
8. Design and tailor tools for feedback	Design and/or tailor measurement tools for use in debrief in which trainees are made aware of which required team behaviors they did and did not perform successfully. This tool should also help instructors diagnose the causes of poor performance and provide guidance for specific improvement in future operations.	• Instructor training • Debriefing checklists and guides
9. Evaluate the extent of improved teamwork in the cockpit	Design experiments to assess the effectiveness of the training. Because of operational constraints, quasi-research methods may need to be applied.	• Reaction data • Learning data • Knowledge acquisition data • Transfer data • Performance data

Note. (SME = subject matter expert; KSA = knowledge, skills, and abilities. From "A Methodology for Enhancing Crew Resource Management Training" by E. Salas, C. Prince, C. A. Bowers, R. J. Stout, R. L. Oser, and J. A. Cannon-Bowers, 1999, *Human Factors,* 41, pp. 161-172. Copyright 1999 by Lawrence Erlbaum Associates. Inc., Reprinted with permission.

for naval aviators to receive aircrew coordination training based on the methodology was formally established in a naval aviation instruction (OPNAVINST 3710.7Q, 1998).

DRIVERS OF THE CRM METHODOLOGY

The CRM methodology is based on six underlying drivers that guide our training research, design, delivery, and evaluation. The drivers are crucial in that they provide us with an overarching framework around which to organize our efforts throughout the research program. The drivers also provide us with a mechanism to easily assess whether the development of the CRM methodology is progressing in an effective manner. Specifically, the six drivers behind our methodology are that CRM training should be (a) research based, while maintaining a practical focus; (b) derived from requisite competencies; (c) reflective of the tasks of the targeted team; (d) focused on crew-member-specific requirements; (e) integrated throughout the training pipeline; and (f) continuously evaluated and validated. Although these drivers are presented as separate entities, they are not independent of each other. On the basis of our research and application of CRM training that has implemented these drivers, we suggest that they have considerable application for other training environments where effective coordination is necessary for effective performance.

Driver 1: Training Design and Delivery
Should Be Research Based

Although an extensive amount of literature exists in the areas of teamwork and team training, much of this literature has gone unnoticed and unused by individuals in many areas of organizational practice (Salas, Cannon-Bowers, & Blickensderfer, 1997; Salas, Rhodenizer, & Bowers, in press). In fact, Salas et al. (in press) noted that practitioners cited three main problems with CRM training: (a) flaws in training content that are due to the lack of viable theories, (b) the use of unsystematic and atheoretical evaluation criteria to determine the training's effectiveness, and (c) a lack of empirical evidence.

In devising the CRM program, we incorporated concepts from various psychological disciplines. Specifically, principles derived from industrial-organizational psychology (e.g., assessment centers, Maher, 1983; training, Goldstein, 1993), human factors psychology (e.g., situation awareness, Endsley, 1995; decision making, Orasanu, 1993), cognitive psychology (e.g., mental models, Rouse & Morris, 1986; practice schedules, Schmidt & Bjork, 1992), and social psychology (e.g., group performance, Foushee & Helmreich, 1988;

social learning theory, Bandura, 1977) were coupled with lessons learned from team training research from nonaviation domains. The review and integration of these multidisciplinary fields provided a sound scientific basis for the design and delivery of CRM training.

Driver 2: Training Content Should Be Competency Based

Although the "right" attitude can be an important factor in aircrew coordination performance, effective performance is also behavioral and cognitive in nature. Through discussions with the fleet and observations of varied CRM training success, we determined that a multidimensional approach was needed for the CRM program to move toward performance improvements.

Recently, Cannon-Bowers, Tannenbaum, Salas, and Volpe (1995) performed a comprehensive review of the team performance literature and developed a framework that identified specific competencies for teams in a variety of settings. They determined that effective team performance requires team members to possess and use knowledge-based competencies (e.g., have an understanding of the directions and rules of how to perform the task), skill-based competencies (e.g., ability to perform the task), and attitude-based competencies (e.g., willingness to perform the task). Furthermore, Cannon-Bowers et al. purported that team competencies can be team generic or team specific and task generic or task specific. For example, a team whose membership remains the same and whose members perform the same tasks over time (e.g., aircrews, sports teams) would be considered team specific and task specific. Determining requisite team competencies is essential in identifying optimal training strategies for each type of team that is being trained.

Because aviation team training should be "targeted at specific content (i.e., teamwork knowledge, skills and attitudes)" (Salas, Prince, et al., 1999, p. 163), the requirement to focus on specific aircrew coordination competencies was a driving factor in the design, development, and implementation of CRM training content. In our case, specific competencies were determined through analyses of the teamwork skills and behaviors required in the targeted naval aviation community. An initial needs analysis conducted with active-duty helicopter pilots identified seven competency areas (i.e., decision making, assertiveness, mission analysis, communication, leadership, adaptability and flexibility, and situational awareness) as the primary components of aircrew coordination (Prince & Salas, 1991, 1993). Further research with aviators from three fixed wing communities (i.e., fighter aircraft, e.g., F-14; attack aircraft, e.g., A-6; and training aircraft, e.g., T-44) revealed that the seven competency areas identified by Prince and Salas (1993) were considered to be important across communities (see Stout, Prince, Baker, Bergondy, & Salas, 1992). In sum, through systematic needs analyses with aviation SMEs, we identified requisite competencies for optimal aviation team performance. Behaviors composing these competencies thus became our training focus.

Driver 3: Training Should Reflect Relevant Tasks of the Targeted Mission

Although the delineation of general competencies was an important starting point, the varied nature of naval aviation missions required that the general competencies be translated into mission-specific requirements. Too often, attempts are made to transfer programs implemented in one community to another community (that is not necessarily similar to the one for which the training was designed) without redesigning the training to focus on those competencies needed to meet the specific performance requirements in the "new" community. Although approaches like this may result in a useful program, they also have the potential to lead to a negative transfer of the training to the operational environment.

In an effort to focus on critical competencies required for mission-specific performance, we prescribe that training should be tailored to reflect the mission (e.g., by phase of flight, by mission type, and by differing rules of engagement) of the specific community for which it is being designed. For example, the specific competencies associated with whether the mission involves an emergency or a nonemergency, or is conducted as part of a peacekeeping operation or during war, are likely to vary.

During the design and development of CRM training for naval aviation, we identified the operational and mission requirements for the community receiving the training. We conducted specific analyses to determine the competencies (i.e., decision making, assertiveness, mission analysis, communication, leadership, adaptability and flexibility, and situational awareness) and tasks (i.e., mission, collective, individual, and occupational) necessary to effectively perform across a variety of missions and mission phases for both routine and nonroutine tasks (Bowers, Morgan, Salas, & Prince, 1993).

To ensure the training was mission specific, we conducted critical-incident interviews with aircraft platform personnel. The interviews resulted in the identification of examples of the competencies associated with tasks that are most critical to mission performance. The interviews were conducted with SMEs from the specific community that planned to implement the training. Next, we queried the aircrews to determine the "criticality" and "importance to train" of the competencies associated with the tasks deemed most critical to mission performance by means of questionnaires given to SMEs from each aircraft platform. In sum, to ensure the training content reflected tasks from the targeted community, we surveyed and conducted interviews with aviators and a portion of other aircraft platform personnel.

Driver 4: Training Should Focus on Each Team Member's Role in the Team

The specific competencies associated with different crew member positions (e.g., pilot, copilot, weapon systems operator, crew chief) are different within a given aircraft and across different aircraft. For example, copilots in military aircraft have the primary responsibility of providing backup for the pilot in a manner similar to their civilian counterparts. However, in other two-person-crewed aircraft, although weapon systems operators provide backup to the pilot, their primary responsibility is to operate the systems necessary for employment of the weapons during tactical engagements.

In an effort to identify specific competencies associated with each crew member position, we observed crews from different aviation communities performing in full-mission simulations (Prince & Salas, 1993), after which we conducted structured interviews with more than 200 aviators representing different crew positions across a wide range of naval aviation platforms. The SME interviews resulted in the identification of examples of the competencies associated with specific crew member positions in a variety of aircraft platforms. So, in addition to analyzing the competencies necessary for the missions we also focused on the specific competencies required by each crew member's position.

Throughout all phases of the research effort, representatives from each crew member position were involved in the training design, development, and implementation process. During our development of the CRM training methodology for naval aviation, operational aircrews were instrumental in ensuring the accuracy and relevance of training content. They participated through various means, including: (a) performing mission-oriented scenarios; (b) completing surveys to assess the importance of the competencies and the criticality of performing training on the competencies; (c) reviewing training materials to ensure operational relevance; and (d) identifying implementation issues, thus allowing us to develop and investigate a variety of training approaches focusing on competencies required for effective performance in crew-member-specific positions (cf. Bowers et al., 1993; Bowers, Baker, & Salas, 1994). Specific procedures for involving the operators in the CRM design, development, and implementation process—for example, critical incident interviews, survey-based needs analyses, and review of materials—are forwarded in the CRM training methodology (see Table 14.1).

Driver 5: Training Should Be Integrated/Seamless

In addition to defining the competencies as a function of specific missions and crew member positions, we also performed additional analyses to determine when, how, and where the training needed to occur within the career of the operators. The results of the analyses suggested that CRM training should be integrated throughout each stage of an aviator's entire career. In naval aviation, an aviator's career begins with basic training, then proceeds to individual and unit-level training, followed by air wing integration training and, finally, to fleet exercises. For example, pilots who operate fighter/attack aircraft first complete basic training, then receive specific fighter/attack training and move through intermediate strike training and, finally, advanced strike training.

Furthermore, the training must be incorporated into all facets of the curricula (e.g., classroom, simulations, flight). Training researchers (Salas, Cannon-Bowers, Rhodenizer, & Bowers, 1999; Salas, Prince, et al., 1999) have found that effective CRM training programs typically include four phases, consisting of information (e.g., lecture, seminars), demonstration (e.g., videos to support behavioral modeling by showing effective and ineffective behaviors), practice (e.g., role play, simulators), and feedback (e.g., from teammates, instructors). Additionally, integration should include the use of several training strategies, methods, and tools (Salas & Cannon-Bowers, 1997).

Many early CRM training programs primarily followed a lecture format (Helmreich & Foushee, 1993). Aside from using only a lecture or seminar format, effective and ineffective behaviors were sometimes demonstrated through videotapes. Few training opportunities were available for aircrews to practice and receive feedback related to the competencies required for effective performance. The limited number of opportunities for this to occur is due in part to the resources necessary for practice and feedback (i.e., simulation) and the operational resources of the operators to perform in real-world situations. Optimally, teams should be given opportunities to practice the skills and behaviors they have learned using simulations (e.g., scenarios and role plays; Baker, Prince, Shrestha, Oser, & Salas, 1993; Beard, Salas, & Prince, 1995). Within naval aviation there are many devices that can be used for practice, including procedures trainers, operational flight trainers, and tactical trainers.

In addition to practice, research has shown that providing feedback to team members improves their performance (Johnston, Smith-Jentsch, & Cannon-Bowers, 1997; Salas & Cannon-Bowers, 1997; Smith-Jentsch, Zeisig, Acton, & McPherson, 1998). Studies of CRM implementation have shown that providing feedback has a demonstrated positive effect on crew performance and positive change in their attitudes and behaviors (Fowlkes et al., 1994; Prince & Salas, 1993; Salas, Fowlkes, et al., 1999; Stout, Salas, & Kraiger, 1997).

The CRM training methodology prescribes the long-term goal to integrate CRM throughout all phases of naval aviation team members' careers. Specifically, our longitudinal approach consists of three phases: familiarization CRM, integrated CRM, and recurrency CRM. The familiarization phase provides trainees with preliminary CRM training and is typically a stand-alone course. This phase provides information, demonstrations, and limited opportunities to practice the critical skills and behaviors associated with the seven competency areas through classroom exercises. Integrated CRM training provides trainees with CRM incorporated throughout an existing training syllabus (i.e., lectures, simulator exercises, aircraft flights). This phase emphasizes practice and feedback of the critical skills and behaviors. Supporting information and demonstrations to facilitate practice and feedback of the critical skills and behaviors are also incorporated into integrated CRM. All integrated CRM practice-and-feedback sessions occurring in the simulator should be recorded to facilitate the trainee feedback process. Recurrency CRM training provides trainees with CRM training on a recurring basis after the completion of integrated CRM. The emphasis of recurrency CRM is twofold: first, to refresh specific mission and operational information related to the critical skills and behaviors required for successful performance, and second, to provide training on new coordination aspects (based on new equipment or missions). Supporting demonstrations and practice and feedback (as appropriate) are provided to continue the transfer of information during recurrency CRM. It is important to recognize that each phase of CRM training focuses on the CRM instructional approaches (i.e., information, demonstration, practice, feedback) and the critical skills and behaviors, while building on previous phases of training. Finally, it is interesting to note that the integrated approach to CRM within naval aviation parallels efforts being conducted in civilian aviation communities under the Federal Aviation Administration's (FAA's) Advanced Qualification Program (FAA, 1991).

Driver 6: Training Should Be Continuously Evaluated and Validated

As with any training, the effectiveness of the interventions needs to be tested. Kirkpatrick (1976) presented a multifaceted evaluation approach that considers several levels of evaluation, including trainee reactions, extent of learning, extent of performance change, and impact on organizational effectiveness. Except for impact on organizational effectiveness, the multifaceted approach was adopted within CRM during the initial validation of actual training programs in several "testbed" communities. An example of a multifaceted approach was described by Stout, Salas, and Kraiger (1997), who used cognitive knowledge assessments, attitude measurements, and behavioral assessments to measure team performance.

We assessed CRM training programs within naval aviation in terms of attitudes, skills, cognitions, and behaviors. Effective CRM is expected to have an impact on performance and therefore needs to be measured in regard to all of these factors. This multifaceted approach to measurement and validation permits a more detailed assessment than would be permitted through the use of only one factor. During the initial validation of our CRM training program, comparisons were made between an experimental group, which received training according to the CRM methodology and a control group, which did not receive training based on the CRM methodology. Each team's performance was assessed by (a) objective raters who were unaware of the experimental conditions and (b) a variety of performance measures (e.g., Targeted Acceptable Responses to Generated Events or Tasks [TARGETs]). Effective performance was defined by fleet SMEs on the basis of the competencies identified through our analysis and through a review of training manuals and operating procedures for the community of interest (Fowlkes et al., 1994).

The evaluation process was grounded in the development of scenarios that were based on requisite skills and linked to training objectives (Baker et al., 1993; Prince, Oser, Salas, & Woodruff, 1993). Using four aircraft communities and a total of 55 crews, a multifaceted approach was used to conduct formal evaluations of naval aviation CRM training (Salas, Prince, et al., 1999). Data were collected with various performance assessment and measurement tools (e.g., behaviorally based checklists, subjective evaluation forms, and outcome metrics and criteria) in conjunction with scenario design (Baker & Salas, 1992; Brannick, Salas, & Prince, 1997; Fowlkes et al., 1994). Crews that received training based on the CRM methodology demonstrated 8% to 20% more behaviors associated with effective performance compared to crews that did not receive such training (see Salas, Fowlkes, et al., 1999; Smith-Jentsch, Salas, & Baker, 1996, for empirical evidence; Stout, Salas, & Fowlkes, 1997; Stout, Salas, & Kraiger, 1997).

In sum, these six drivers guided the development of our CRM training methodology. In the next section we present each component of the methodology.

THE CRM TRAINING METHODOLOGY

The CRM training methodology is a series of steps, designed specifically with the goal of structuring the design, delivery, and evaluation of CRM for naval aviation training. The CRM training methodology consists of nine components (Salas, Prince, et al., 1999): (a) identify operational/mission requirements, (b) assess team training needs and coordination demand, (c) identify teamwork competencies and KSAs, (d) determine team training objectives, (e) determine instructional delivery method, (f) design scenario exercises and create opportunities for practice, (g) develop performance assessment/measurement tools, (h) design and tailor tools for feedback, and (i) evaluate the extent of improved teamwork in the cockpit. An overview of the CRM training methodology is presented in Table 14.1.

At first glance, the methodology contains many components similar to that of other systematic frameworks for instructional and training design. Although it builds on these previous frameworks, the CRM training methodology is unique in three important ways. First, a major component of the methodology is an emphasis on practice and feedback in simulated operational environments. The other frameworks provide general approaches that can be applied to any aspect of training, including those in which a set of instructions (e.g., lessons, lectures, computer-based training) composes the curriculum. Second, the methods and tools have been researched and tailored to meet the specific needs found in team training environments. In comparison, the other frameworks have been primarily applied to individual training environments. Third, and most important, the CRM methodology has resulted in quantifiable performance improvements in environments where individuals and teams performing complex tasks were the training focus (e.g., Dwyer, Oser, & Fowlkes, 1995; Fowlkes et al., 1994; Johnston et al., 1997).

Finally, it is our belief that the success of programs developed using the CRM training methodology was due in large part to tailoring the training to include principles from each of the six drivers presented earlier.

LESSONS LEARNED FROM CRM TRAINING EXPERIENCES

The following lessons learned are based on our experiences and observations made throughout the years in researching and implementing the drivers and methodology presented in the previous discussion. These lessons learned have been generalized so that they can be applied to other team training environments. While it is possible to reflect on each lesson learned in isolation, our experiences from efforts to provide CRM training for naval aviation suggest that the best results are achieved when all lessons learned are considered in an integrated manner.

Lesson I: Exploit the Science of Training

One of the most difficult issues between scientists and practitioners is translating and applying research results into practice. Of course, it is easy to pick up a magazine or a manual offering the latest and greatest plan to design and deliver effective training programs. But how do you know if it will work? A great deal of applied research has concentrated on training and training effectiveness issues, resulting in research-based, empirically supported steps and guidelines for designing and delivering training that works. Professionals who exploit the science of training have been benefiting from, and will continue to benefit from, considering these findings when designing their training programs.

Lesson 2: A Training Program Is Not a Panacea

Training has been shown to improve performance and reduce human error, but training, even when developed and used most appropriately, cannot solve every problem. Many factors play a part in the effectiveness of a training program. Variables such as pretraining motivation, management support of training, and posttraining environment have been shown to play a part in posttraining transfer (Quiñones, 1995; Smith-Jentsch, Jentsch, Payne, & Salas, 1996). Additionally, organizational influences such as climate and culture may contribute to the effectiveness of resource management training programs (see chap. 15). As a result, even the most carefully designed training program may not achieve the desired results if factors external to the training are present. Professionals who understand that training is not a panacea will have an enhanced awareness of issues outside the realm of their training programs that may effect overall training effectiveness.

Lesson 3: Technology Alone Does Not Guarantee Effective Learning

The use of various types of technology can greatly enhance the quality and transfer potential of a training program. However, training effectiveness cannot be leveraged solely on the technology that is used to train. Salas, Cannon-Bowers, et al. (1999) addressed common myths about training in organizations, including the (lack of) utility of a training approach that focuses on technology while disregarding the elements that support a solidly designed program. Specifically, "the underlying principles used in the training development are more important than the technology" (p. 140). Basically, although technologies such as simulation have been shown to enhance a well-designed program, one cannot assume that effective learning is guaranteed just because the program is outfitted with high-tech "bells and whistles."

Lesson 4: Focus on Training to Needs, Not Needing to Train

From our research and experiences within naval aviation, we learned that there was support for a need to train, but simply knowing that training is needed is not enough to design an effective program. Often, organizations require their employees to attend certain types of training without considering the specific areas where training is actually needed. The optimal training program will be designed with a focus on the specific KSAs that need to be trained.

Lesson 5: One Size Does Not Fit All

One training program may not be all things to all people. Consider the differences along a continuum between training novices (e.g., new employees), who are learning terms and concepts for the first time, and experts (e.g., long-time employees), who possess and continually use their existing base of KSAs. Obviously, along the continuum different approaches to training (i.e., methods, tools, and strategies) will be more effective in one circumstance than another. Professionals who understand that one training program does not fit all training needs will be able to refine their resource management training programs on the basis of specific needs and desired outcomes.

Lesson 6: Help the Experts Help Themselves

Our practical experience and research findings support the belief that in designing any type of training program the learner's level of skill must be considered. Program capabilities should be designed to match the learner's capabilities. As such, a learner-centered training approach is designed to engage the learner while drawing on his or her experiences. Because of this, it is our belief that training programs can benefit from standardizing the training approach to fit with the learner's level of expertise. Professionals involved with training design should consider the benefits of helping the experts help themselves.

Lesson 7: Measure for Meaning, Not For Marks

Performance measurement is an essential aspect of developing and maintaining resource management programs. Earlier in this chapter we discussed the contributions that practice and feedback can make to a training program. The benefits of implementing practice-and-feedback strategies into the training process emphasize the criticality of measuring the process as well as the outcome. When learning is assessed throughout the training program (process) in addition to at the conclusion (outcome), it is possible to target specific areas in need of remediation and modify the training approach to hone in on the areas most in need of extra attention. Professionals who measure their approach to training for meaning throughout the entire program and not just for "marks" at the end can benefit from a program that is more accurately tailored to address their unique training needs.

Lesson 8: Train Like You Perform

One approach to training is to practice the task using simulations and exercises. When considering approaches to training resource management skills, simulations can provide trainees with mock situations in which to practice with and receive feedback on the KSAs for enhancing their performance. Depending on various factors (e.g., available funding and time), the level of fidelity can range from high to low. Regardless of the fidelity of the simulation itself, in developing scenarios for the simulations it is essential that the scenarios outline a series of events or "targets" that have been carefully identified from SMEs and critical incidents in the transfer environment. This approach to scenario development allows the training scenarios to elicit the requisite KSAs of the transfer environment, regardless of the fidelity of the simulation. Professionals who combine carefully designed scenarios with appropriate levels of simulation fidelity can increase the likelihood that their training will reflect the KSAs that are performed in the transfer environment.

Lesson 9: Make Sure You Are Getting What You Expected

As with any training program, professionals involved with resource management training programs should ensure that the program is providing what was expected. To do this, it is necessary to continuously evaluate and validate the training program. As we discussed earlier in this chapter, a multifaceted approach to resource management training within naval aviation has helped us to monitor and modify the training to fit what was expected. The only way to do this is to evaluate the training program on multiple levels (e.g., Kirkpatrick's [1976] approach of reactions, learning, behaviors, and results).

SUMMARY AND CONCLUSIONS

It is our belief that the drivers, methodology, and lessons presented in this chapter will aid practitioners in developing programs that train individuals to perform better as crew members or team members, no matter what type of aircraft they are flying or with what type of team they are working. Although many resource management training methodologies exist, we believe ours has shown signs of success for several main reasons: (a) we have based each element of the training program on reliable science; (b) we had a vision and objectives; (c) we had the commitment from sponsors, users, researchers, and aviators to develop a useful, valid, and practical methodology; (d) we developed useful products along with the research; (e) we have provided specifications (how-to's) for building an effective resource management training program; and (f) we have established a partnership with the individuals who are actually using the programs.

Although this chapter focuses on training for teams that perform in naval aviation environments, many analogies exist between these environments and the characteristics often present in other team settings. Many teams that perform complex tasks must make tactical decisions, deal with ambiguity, coordinate different assets, and adapt to situations in which standardized procedures are not available. Examples are abundant and include emergency response teams, nuclear power plant teams, medical teams, law enforcement teams, and command-and-control teams. Because of these common factors, research results, and our experiences in developing the CRM training methodology, the drivers and lessons are believed to have considerable application to other environments where effective team coordination is necessary for optimal performance.

ACKNOWLEDGMENTS

The views expressed here are those of the authors and do not reflect the official positions of the organizations with which the authors are affiliated. Throughout the years, several agencies and individuals have contributed to this program of research and development: NAVAIR PMA-205; Office of Naval Research; David Baker; Rebecca Beard; Maureen Bergondy; Mike Brannick; Janis Cannon-Bowers; Jim Driskell; Jennifer Fowlkes; Florian Jentsch; Mike Lilienthal; Dana Milanovich; Ben Morgan, Jr.; Elizabeth Muñiz; Jerry Owens; Ashley Prince; Carolyn Prince; Lisa Shrestha; Kimberly Smith-Jentsch; Renée Stout; Bob Swezey; and Scott Tannenbaum. We sincerely appreciate their hard work and dedicated efforts in enabling this program to offer viable solutions to the naval aviation community. Also, we would like to thank the (approximately) 1,500 aviators who volunteered their time throughout the years to assist us with this program.

REFERENCES

Baker, D. P., Prince, C., Shrestha, L., Oser, R. L., & Salas, E. (1993). Aviation computer games for crew resource management training. *International Journal of Aviation Psychology, 3*, 143–156.

Baker, D. P., & Salas, E. (1992). Principles for measuring teamwork skills. *Human Factors, 34*, 469–475.

Bandura, A. (1977). *Social learning theory*. Englewood Cliffs, NJ: Prentice Hall.

Beard, R. L., Salas, E., & Prince, C. (1995). Enhancing transfer of training: Using role-play to foster teamwork in the cockpit. *International Journal of Aviation Psychology, 5*, 131–143.

Bowers, C. A., Baker, D. P., & Salas, E. (1994). Measuring the importance of teamwork: The reliability and validity of job/task analysis indices for team-training design. *Military Psychology, 6*, 205–214.

Bowers, C. A., Morgan, B. B., Jr., Salas, E., & Prince, C. (1993). Assessment of coordination demand for aircrew coordination training. *Military Psychology, 5*, 95–112.

Brannick, M. T., Salas, E., & Prince, C. (1997). *Team performance assessment and measurement: Theory, methods, and applications*. Mahwah, NJ: Lawrence Erlbaum Associates.

Cannon-Bowers, J. A., & Salas, E. (1998). *Making decisions under stress: Implications for individual and team training.* Washington, DC: American Psychological Association.

Cannon-Bowers, J. A., Tannenbaum, S. I., Salas, E., & Volpe, C. E. (1995). Defining team competencies and establishing team training requirements. In R. Guzzo & E. Salas (Eds.), *Team effectiveness and decision making in organizations* (pp. 333–380). San Francisco: Jossey-Bass.

Dwyer, D. J., Oser, R. L., & Fowlkes, J. E. (1995). Symposium on distributed simulation for military training of teams/groups: A case study of distributed training and training performance. In *Proceedings of the 39th annual Human Factors and Ergonomics Society meeting* (pp. 1316–1320).

Endsley, M. R. (1995). Measurement of situation awareness in dynamic systems. *Human Factors, 37,* 65–84.

Federal Aviation Administration. (1991). *Advanced qualification program* (Advisory Circular 120–54). Washington, DC: U.S. Department of Transportation.

Foushee, H. C., & Helmreich, R. L. (1988). Group interaction and flight crew performance. In E. L. Wiener & D. C. Nagel (Eds.), *Human factors in aviation* (pp. 189–227). San Diego, CA: Academic Press.

Fowlkes, J. E., Lane, N. E., Salas, E., Franz, T., & Oser, R. L. (1994). Improving the measurement of team performance: The TARGETs methodology. *Military Psychology, 6,* 47–61.

Goldstein, I. L. (1993). *Training in organizations* (3rd ed.). Belmont, CA: Wadsworth.

Helmreich, R. L., & Foushee, H. C. (1993). Why crew resource management? Empirical and theoretical bases of human factors training in aviation. In E. L. Wiener, B. G. Kanki, & R. L. Helmreich (Eds.), *Cockpit resource management* (pp. 3–45). San Diego, CA: Academic Press.

Johnston, J. H., Smith-Jentsch, K. A., & Cannon-Bowers, J. A. (1997). Performance measurement tools for enhancing team decision making. In M. T. Brannick, E. Salas, & C. Prince (Eds.), *Team performance assessment and measurement: Theory, methods, and applications* (pp. 311–330). Mahwah, NJ: Lawrence Erlbaum Associates.

Kirkpatrick, D. L. (1976). Evaluation of training. In R. L. Craig (Ed.), *Training and development handbook* (2nd ed., pp. 18.1–18.27). New York: McGraw-Hill.

Maher, P. T. (1983). An analysis of common assessment center dimensions. *Journal of Assessment Center Technology, 6,* 21–29.

OPNAVINST 3710.7Q (1998). Washington, DC: Chief of Naval Operations Instruction.

Orasanu, J. M. (1993). Decision-making in the cockpit. In E. L. Wiener, B. G. Kanki, & R. L. Helmreich (Eds.), *Cockpit resource management* (pp. 137–168). San Diego, CA: Academic Press.

Prince, C., Oser, R. L., Salas, E., & Woodruff, W. (1993). Increasing hits and reducing misses in CRM/LOS scenarios: Guidelines for simulator exercise development. *International Journal of Aviation Psychology, 3,* 69–82.

Prince, C., & Salas, E. (1991). The utility of low fidelity simulation for training aircrew coordination skills. *Proceedings of the International Training Equipment Conference,* 87–91.

Prince, C., & Salas, E. (1993). Training and research for teamwork in the military aircrew. In E. L. Wiener, B. G. Kanki, & R. L. Helmreich (Eds.), *Cockpit resource management* (pp. 337–366). San Diego, CA: Academic Press.

Prince, C., & Salas, E. (1999). Team process and their training in aviation. In D. Garland, J. Wise, & D. Hopkin (Eds.), *Handbook of aviation human factors* (pp. 193–213). Mahwah, NJ: Lawrence Erlbaum Associates.

Quiñones, M. (1995). Pretraining context effects: Training assignment as feedback. *Journal of Applied Psychology, 80,* 226–238.

Rouse, W. B., & Morris, N. M. (1986). On looking into the black box: Prospects and limits in the search for mental models. *Psychological Bulletin, 100,* 349–363.

Salas, E., & Cannon-Bowers, J. A. (1997). Methods, tools, and strategies for team training. In M. A. Quiñones & A. Ehrenstein (Eds.), *Training for a rapidly changing workplace: Applications of psychological research* (pp. 249–279). Washington, DC: American Psychological Association.

Salas, E., & Cannon-Bowers, J. A. (in press). The anatomy of team training. In L. Tobias & D. Fletcher (Eds.), *Handbook on research in training.* New York: Macmillan.

Salas, E., Cannon-Bowers, J. A., & Blickensderfer, E. L. (1997). Enhancing reciprocity between training theory and practice: Principles, guidelines, and specifications. In J. K. Ford & Associates (Eds.), *Improving training effectiveness in work organizations* (pp. 291–322). Mahwah, NJ: Lawrence Erlbaum Association.

Salas, E., Cannon-Bowers, J. A., Rhodenizer, L. G., & Bowers, C. A. (1999). Training in organizations: Myths, misconceptions, and mistaken assumptions. In G. R. Ferris (Ed.), *Research in personnel and human resources management* (Vol. 17, pp. 123–161). Greenwich, CT: JAI Press.

Salas, E., Fowlkes, J. E., Stout, R. J., Milanovich, D., & Prince, C. (1999). Does CRM training improve teamwork skills to work in the cockpit? Two evaluation studies. *Human Factors, 41*, 326–343.

Salas, E., Prince, C., Bowers, C.A., Stout, R., Oser, R. L., & Cannon-Bowers, J. A. (1999). A methodology for enhancing crew resource management training. *Human Factors, 41*, 161–172.

Salas, E., Rhodenizer, L., & Bowers, C. A. (in press). The design and delivery of CRM training: Using all of the resources available. *Human Factors.*

Schmidt, R. & Bjork, R. (1992). New conceptualizations of practice: Common principles in three paradigms suggest new concepts for training. *Psychological Science, 3*, 207–217.

Smith-Jentsch, K. A., Jentsch, F. G., Payne, S. C., & Salas, E. (1996). Can pretraining experiences explain individual differences in learning? *Journal of Applied Psychology, 81*, 110–116.

Smith-Jentsch, K. A., Salas, E., & Baker, D. (1996). Training team performance-related assertiveness. *Personnel Psychology, 49*, 909–936.

Smith-Jentsch, K., Zeisig, R. L., Acton, B., & McPherson, J. (1998). Team dimensional training. In J. A. Cannon-Bowers & E. Salas (Eds.), *Making decisions under stress: Implications for individual and team training* (pp. 271–297). Washington, DC: American Psychological Association.

Stout, R., Prince, C., Baker, D. P., Bergondy, M. L., & Salas, E. (1992). Aircrew coordination: What does it take? In *Proceedings of the Thirteenth Biennial Psychology in the Department of Defense symposium* (pp. 133–137).

Stout, R. J., Salas, E., & Fowlkes, J. E. (1997). Enhancing teamwork in complex environments through team training. *Group Dynamics, 1*, 169–182.

Stout, R. J., Salas, E., & Kraiger, K. (1997). The role of trainee knowledge structures in aviation team environments. *International Journal of Aviation Psychology, 7*, 235–250.

A GLOBAL PERSPECTIVE OF RESOURCE MANAGEMENT TRAINING

15

Culture, Error, and Crew Resource Management

Robert L. Helmreich
John A. Wilhelm
James R. Klinect
Ashleigh C. Merritt
The University of Texas at Austin

The latest CRM programs explicitly focus on error and its management. CRM training, in its current state, can best be described as one of the critical interventions that can be used by organizations in the interests of safety. More specifically, pilot CRM skills provide countermeasures against risk and error in the form of threat and error avoidance, detection, and management. In the period just prior to the birth of these new programs, CRM training had been successfully applied to the U.S. and Western cockpit environments, although its acceptance was not universal. As we observed these cockpit programs applied mindlessly to non-Western pilot groups and nonpilot groups such as flight attendants, maintenance personnel, dispatch, and even to nuclear power plant and refinery operations, we began to see the effectiveness of the programs slipping. We tried two approaches. We initiated a new research program into the dimensions of national culture relevant to the aviation environment and CRM training in particular. By knowing more about national cultures we could begin to design CRM programs that were culturally sensitive and that would have greater impact on line operations. The pilot culture, and that of individual organizations, also began to be understood as relevant to the success and failure of CRM programs. Simultaneously, we began to revisit the basic concepts of CRM in the hope of better explicating its goals and objectives. Perhaps there were universal objectives that could be derived that applied to pilots of all nations, and even to nonpilot groups. The marriage of these cultural research programs (What aspects of CRM should be tailored to specific organizations and cultures?) and a "back to basics" attempt to refine the goals and objectives of CRM (What are the universal goals?) produced the new generation of CRM programs that we describe as *error management CRM* (see Helmreich & Foushee, 1993; Helmreich & Merritt, 1998; Helmreich & Wilhelm, 1991; Helmreich, Merritt, & Wilhelm, 1999; Merritt & Helmreich, 1997; and Salas, Fowlkes, Stout, Milanovich, & Prince, 1999, for discussions of the evolution of CRM and outcomes of CRM training).

To better understand how CRM skills fit into the pilot's job description, we began sketching a broad conceptual model. Ultimately, the job of the pilot is to operate the aircraft in a safe manner in order to transport passengers and goods from place to place. The successful management of risk or threat is a primary task. The model we finally produced has four levels: external threats, internal threats (labeled *crew-based errors*), crew actions, and outcomes. See Fig. 15.1 for our model of threat and error management. At the first level three types of external threat may confront crews: expected risks, such as high terrain surrounding an airport; unexpected risks, in

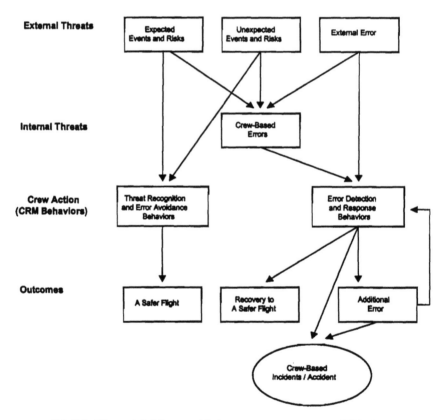

FIG. 15.1.   The model of threat and flight crew error management. CRM = Crew Resource Management.

the form of system malfunction or changing weather; and errors by external parties, for example, incorrect dispatch releases or air traffic instructions. When either an expected risk or unexpected risk is recognized, crews can use CRM behaviors for error avoidance by evaluating the threat's implications and using decision-making skills to determine a course of action. Threat recognition and error avoidance are associated with situation awareness and represent a proactive response that can be observed when groups share and evaluate the situation and include contextual factors in planning. For example, a crew may recognize the risk associated with bad weather at their destination (situation awareness) and practice error avoidance by increasing the fuel load and reconsidering their choice of an alternate airport to reduce risk and conduct a safer flight. The potential error would be to have insufficient fuel to reach a safe alternate.

Human error is inevitable, so when an error occurs (whether committed by an external agent or by the crew), it is the crew's task to detect and respond to it. The behaviors of effective error detection and management are best illustrated by cross-checking and verifying actions, evaluating the quality of decisions made, and so on. When errors are not detected or corrected, the level of risk for a flight is increased.

This model does not represent a significant departure from the original training programs that were called *cockpit resource management* in the early 1980s. First-generation CRM was developed in response to National Aeronautics and Space Administration (NASA) findings that "pilot error" was involved in the majority of air crashes and was seen as a method to reduce such error (Cooper, White, & Lauber, 1980). However, the linkage between early curricula and pilot error was unclear and, with the passage of time, the goals of CRM appear to have become lost on many participants in the training (Helmreich, Merritt, & Wilhelm, 1999). The purpose of our model is to reestablish the basic goals of CRM.[1] We also recognize that, to implement CRM optimally, the cultural context of flight operations needs to be considered.

CULTURE AND SAFETY

In aviation, the three cultures—professional, organizational, and national—can have both a positive and a negative impact on the probability of safe flight. *Safe flight* is the positive outcome of timely risk recognition and effec-

---

[1]The definition of CRM in the opening chapter of a 1993 book on CRM did not mention error but described it as the process of "optimizing not only the person-machine interface and the acquisition of timely, appropriate information, but also interpersonal activities including leadership, effective team formation and maintenance, problem-solving, decision making, and maintaining situation awareness" (Helmreich & Foushee, 1993, p. 3).

tive error management, which are universally desired outcomes. The responsibility of organizations is to minimize the negative components of each type of culture while emphasizing the positive. Both CRM and technical training form part of an error management philosophy and program.

Professional Culture and Its Manifestations

Although we recognized the existence of and some of the manifestations of the professional culture of pilots early in our investigations of flight crew behavior and attitudes, we did not immediately understand its potency as an influence on safety. In retrospect, the roots of a strong professional culture are clear—early aviation was an extremely dangerous undertaking, for those in combat, carrying the mail, or stunt flying for awed audiences. To commit to such a hare-brained endeavor required a strong sense of personal invulnerability and efficacy. The respect and envy engendered among generations of adolescents also fostered pride in being one of "the few," to borrow Winston Churchill's description of Spitfire pilots during the Battle of Britain. This image of personal disregard for danger and invulnerability reached its zenith with the early astronauts (all chosen from the ranks of test pilots) and was immortalized by Tom Wolfe in *The Right Stuff* (1979).

When we began systematically assessing pilots' attitudes about their jobs and personal capabilities, we found that the pilot culture showed great consistency among more than 15,000 pilots in more than 20 countries (Helmreich & Merritt, 1998). What distinguished pilots on the positive side was an overwhelming liking for their job. Pilots are proud of what they do, and they retain their love of the work. Fig. 15.2 shows the responses of pilots from 19 countries to the stem "I like my job." On a 5-point scale where 1 indicates *disagree strongly* and 5 indicates *agree strongly*, no group had a mean below 4.5, and several had means over 4.9.[2]

On the negative side, there was widespread endorsement of items that reflect an unrealistic self-perception of invulnerability to stressors such as fatigue. Pilots also report that their decision making remains unimpaired by in-flight emergencies and that a true professional can leave behind personal problems on entering the cockpit. These are indeed negative manifestations of the "right stuff." Unfortunately, pilots who are imbued with a sense of invulnerability are less likely to feel the need for countermeasures against error or to value the support of other crew members. We have found equally unrealistic attitudes about personal efficacy among physicians and mariners (Helmreich & Merritt, 1998). The behavioral implications of such attitudes were illustrated in a CRM seminar observed by a member of our research

---

[2]Liking for the profession is independent of attitudes about one's organization. Some of those most enthusiastic about their profession expressed passionate dislike for the organization and indicated that morale was abysmal in their airline.

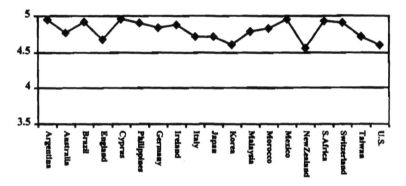

FIG. 15.2. Mean scores of pilots from 19 countries on the item, "I like my job." Higher scores indicate a stronger liking for one's job. Adapted from *Culture at Work in Aviation and Medicine*, by R. L. Helmreich and A. C. Merritt, 1998, Aldershot, England: Ashgate. Copyright 1998 by Ashgate. Adapted with permission.

team. In this session, a pilot remarked that "Checklists are for the lame and weak." Figure 15.3 shows graphically some of the positive and negative influences of pilots' professional culture on safety. As the figure illustrates, positive components can lead to the motivation to master all aspects of the job, to being an approachable team member, and to pride in the profession. On the negative side, perceived invulnerability may lead to a disregard for safety measures, operational procedures, and teamwork.

Organizational Culture and Safety

Investigations of causal factors in accidents and incidents in technology-rich domains are increasingly becoming focused on the critical role of organizational culture. John K. Lauber, the first PhD psychologist and human factors expert to serve on the National Transportation Safety Board (NTSB), spearheaded an effort to examine and identify the role of organizational culture in aviation accidents where blame would previously have centered on errors by crew members or maintenance personnel (NTSB, 1991). In England, the work of James Reason (1990, 1997) has centered on the role of organizations in industrial disasters, including nuclear power generation and petroleum refining.

A *safety culture* is the outcome that organizations reach through a strong commitment to acquiring necessary data and taking proactive steps to reduce the probability of errors and the severity of those that occur (Merritt & Helmreich, 1997). A safety culture includes a strong commitment to training and to reinforcing safe practices and establishing open lines of communication between operational personnel and management regarding threats

FIG. 15.3. Positive and negative influences of pilots' professional culture on the safety of flights. Adapted from *Culture at Work in Aviation and Medicine*, by R. L. Helmreich and A. C. Merritt, 1998, Aldershot, England: Ashgate. Copyright 1998 by Ashgate. Adapted with permission.

to safety. In our data collection we ask a number of questions about perceptions of management's commitment to safety. Table 15.1 shows the percentage of pilots from two organizations who agreed with two safety-related items.

Although the majority of pilots in each organization indicated that they know the proper channels for communicating safety concerns, the percentage is substantially lower in Airline B. More telling are the differences in the percentage who said they believed their safety suggestions would be acted on. This ranges from 68% in Airline A to 19% in Airline B, but even in Airline A there is obvious skepticism about the organization's commitment to safety.

Organizational practices clearly determine the pride that individuals have in working for an organization. These attitudes undoubtedly exert an influence, although indirectly, on safety and compliance. In one airline, 97% of the pilots agreed with the statement "I am proud to work for this organization,"

TABLE 15.1
Percentage of Pilots Who Agreed With Two Safety Items on the Flight Management Attitudes
Questionnaire

Item	Airline A	Airline B
I know the correct safety channels to direct queries.	85	57
My safety suggestions would be acted on.	68	19

whereas at another fewer than 20% agreed. Similar variability was found in attitudes regarding trust in senior management. The organizational culture is important, because when it is strong and positive, pilots and other groups may more readily accept new concepts such as CRM and its associated training.

National Culture in Aviation

The view has been widespread in aviation that the cockpit is a culture-free zone, one in which pilots of all nationalities accomplish their common task of flying safely from one point to another. Data, however, have begun to accumulate suggesting that there are substantial differences in the way pilots conduct their work as a function of national culture and that the areas of difference have implications for safety (Helmreich & Merritt, 1998; Johnston, 1993; Merritt, 1996; Merritt & Helmreich, 1996a; Merritt & Helmreich, 1996b; Sherman, Helmreich, & Merritt, 1997).

Geert Hofstede's (1980, 1991) four-dimensional model of culture has proved to be a useful starting place to examine the effects of national culture on flightdeck behavior. We took his survey of work attitudes as a benchmark and augmented his questions with a new set of items that were more directly relevant to the aviation environment (Helmreich & Merritt, 1998). Three of Hofstede's four dimensions replicated and proved to be conceptually relevant to team interactions in the cockpit.

The first dimension, *Power Distance* (PD), reflects the acceptance by subordinates of unequal power relationships and is defined by statements indicating that juniors should not question the decisions or actions of their superiors and the nature of leadership (i.e., consultative vs. autocratic). Figure 15.4 shows mean scores on our measure of PD, the Command Scale, of pilots from 22 nations. High scores on the scale indicate high PD and acceptance of a more autocratic type of leadership. In high-PD cultures safety may suffer from the fact that followers are unwilling to make inputs regarding leaders' actions or decisions. Countries such as Morocco, the Philippines, Taiwan, and Brazil have the highest scores, indicating the highest acceptance of unequally distributed power. At the other end of the power continuum are countries such as Ireland, Denmark, and Norway, with the United States also scoring at the low end of the distribution.

The second dimension, *Individualism–Collectivism*, defines differences between individualistic cultures, in which people define situations in terms of costs and benefits for themselves, and more collectivist ones, in which the focus is on harmony within one's primary work or family group. The concept of teamwork and communication may be more easily achieved by collectivists than by those with a more individualistic orientation. The United States and Australia score highest in individualism, whereas many Latin American and Asian cultures rank as highly collectivist.

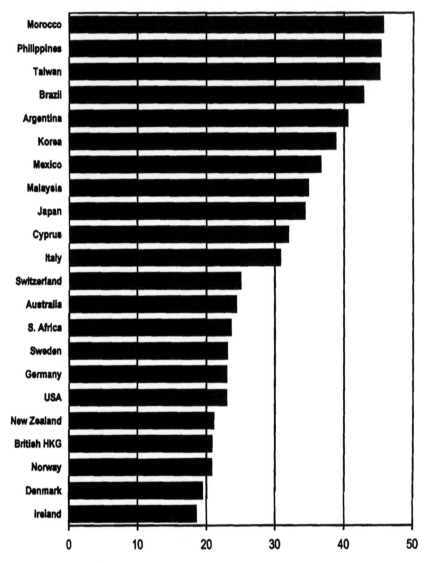

FIG. 15.4.   Mean scores of pilots from 22 countries on the Flight Management
Attitudes Questionnaire measure of Power Distance (scale range: 0–100). S. =
South; HKG = Hong Kong. Adapted from *Culture at Work in Aviation and Medicine*, by R. L. Helmreich and A. C. Merritt, 1998, Aldershot, England: Ashgate.
Copyright 1998 by Ashgate. Adapted with permission.

The third dimension, called Uncertainty Avoidance (UA) by Hofstede, replicated only when it was redefined to focus on the beliefs that written procedures are needed for all situations and that an organization's rules should never be broken, even when doing so might be in the organization's best interest (Helmreich & Merritt, 1998). This dimension, which we call *Rules and Order*, can have both positive and negative implications. Individuals who are high on this dimension may be least likely to deviate from procedures and regulations but may be less creative in coping with novel situations. Individuals low on this dimension may be more prone to violations of procedures but may be better equipped to deal with conditions not covered by procedures. On the redefined measure, Taiwan, Korea, and the Philippines scored highest, whereas Anglo cultures, such as the United Kingdom, Ireland, and the United States, scored very low. Figure 15.5 shows the means on one scale item: "Written procedures are required for all in-flight situations."

One of the unexpected findings from our cross-cultural research was the magnitude of differences in attitudes about automation—both preference for automation and opinions regarding its use (Sherman, Helmreich, & Merritt, 1997). In particular, pilots from high-PD cultures are both more positive about automation and more likely to use it under all circumstances. We (e.g., Helmreich & Merritt, 1998) have suggested that computers may be anthropomorphized in some cultures as a high-status, electronic crew members who are not to be questioned, a strategy that is clearly inappropriate in many situations. Figure 15.6 shows ordered means on a composite measure of preference for and reliance on automation.

There are not "good" and "bad" national cultures with regard to the prevalence of human error and the universal goal of safety. Each culture has elements with both positive and negative implications for effective group function as it affects these universal goals. However, there are organizational cultures that actively discourage safety initiatives and eschew efforts to build a safety culture. Ron Westrum (1992) referred to such cultures as *pathological* in their rejection of information that might avert catastrophe. In such organizations the primary defenses are the positive aspects of the professional and national cultures and the diligence of regulatory agencies. Ultimately, though, it is the responsibility of organizations to promote a safety culture and to maximize the positive and minimize the negative aspects of professional and national cultures.

THREAT AND ERROR IN FLIGHT OPERATIONS

Errors have been extensively studied in the laboratory, in training programs, and in *postmortem* analyses of crew behavior in accidents and incidents. Similarly, systematic evaluations of the nature of external threats to safety are

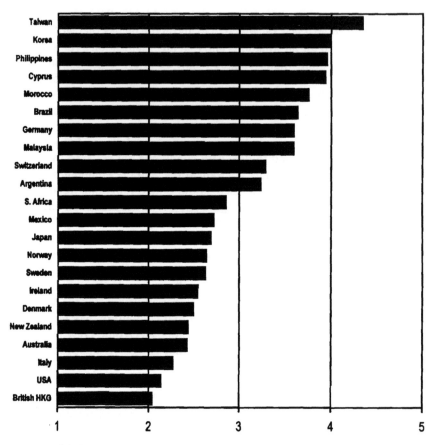

FIG. 15.5.   Mean scores of pilots from 22 countries on the Flight Management
Attitudes Questionnaire item "Written procedures are required for all in-flight
situations" (scale range 1 = *strongly disagree* to 5 = *strongly agree*). S. = South;
HKG = Hong Kong. Adapted from *Culture at Work in Aviation and Medicine*, by
R. L. Helmreich and A. C. Merritt, 1998, Aldershot, England: Ashgate. Copy-
right 1998 by Ashgate. Adapted with permission.

most frequently conducted when they are associated with adverse events.
There is a dearth of systematic empirical data on the kinds, frequency, management, and resolution of threats and errors in *normal* flight operations. If
safety efforts are to be optimally effective, such information is essential. In
an effort to fill this gap, our research group has started a new program to
examine threat and error in line audits of normal operations (described in
the *Line Audits* section). An addendum to the Line/Line Operational Simulations (LOS) Checklist, a form for the collection of systematic data during line
flights (Line/LOS Error Checklist LLEC; Helmreich, Klinect, Wilhelm, & Jones,

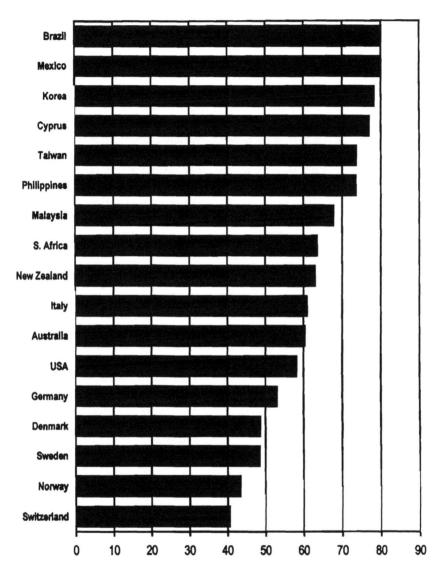

FIG. 15.6.   Mean scores of pilots from 18 countries on the Flight Management Attitudes Questionnaire Automation Preference and Reliance Scale (scale range 0–100). Adapted from *Culture at Work in Aviation and Medicine*, by R. L. Helmreich and A. C. Merritt, 1998, Aldershot, England: Ashgate. Copyright 1998 by Ashgate. Adapted with permission.

1999) was developed to record threats and crew-based errors and crew behaviors during normal flights.

Between July 1997 and November 1998 we conducted formal studies to investigate CRM, threat, and error management at three airlines using the LLEC.[3] The first was a commuter airline, where we observed 123 flight segments. The second and third were major airlines, with a focus on international, long-haul operations. In these audits we observed approximately 100 segments in each airline. For a detailed look at results of these audits, see the article by Klinect, Wilhelm, and Helmreich (1999). Of the flights observed, 72% experienced one or external threats (such as adverse weather, high terrain, mechanical malfunctions, language problems with air traffic control, etc.), with an average of 1.91 per flight and a range from 0 to 11. Although many of the situations experienced were not serious in themselves, they did increase the level of risk and the probability of error. When a large number of external threats are associated with a particular flight, demands on the crew are greatly increased. Earlier audits have revealed that conditions of high complexity with off-normal conditions may either stimulate crews to superior performance or lead to performance breakdowns (Hines, 1998).

Now consider the right side of the threat-and-error model presented in Fig. 15.1—the side that deals with crew-based error and error management. We operationally define this type of error as *'crew action or inaction that leads to deviation from crew or organizational intentions or expectations.''* Violations of formal requirements such as regulations, standard operating procedures (SOPs), and policies are included in this definition. We are indebted both to James Reason (1990, 1997) and Patrick Hudson (1998), whose work has greatly influenced our efforts. Although we recognize the distinction made by Reason and Hudson between errors and violations, we have labeled violations *intentional noncompliance errors,* because we realize that the intent in violations is usually to shortcut what is seen as an unnecessary procedure or regulation or to use a more effective strategy. In developing a model of crew-based error we found that the usual taxonomies and classifications of flight crew error management did not fit our data well. This led us to develop a revised taxonomy of cockpit crew error that we feel may be of value for both research and operational evaluation. This model is shown in Fig. 15.7.

We use a five-way classification of error. As discussed earlier, violations such as completing checklists from memory or failing to observe a sterile

---

[3]Excluded are cognitive errors that do not result in observable behaviors or verbalizations. It should also be noted that those observed are experts (as opposed to novices used in laboratory research or those in training) and that the crews' behaviors are highly consequential. Observers, especially on long flights, were not present at all times on the flightdeck (normally taking a rest period during cruise at altitude). As a result, the recorded incidence of error is a conservative estimate of the actual frequency.

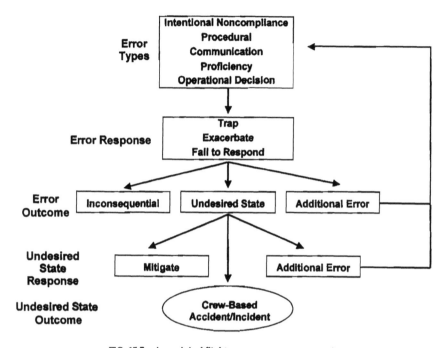

FIG. 15.7.   A model of flightcrew error management.

cockpit are called *intentional noncompliance errors* and consist of cases in which crews choose to shortcut or ignore procedures. *Procedural errors* include slips, lapses, and mistakes in the execution of regulations or procedures in which the crew intended to follow procedures but made an error in execution. Specific procedural errors observed include making incorrect entries in the flight management computer and unintentionally skipping items on checklists. *Communications errors* occur when information is incorrectly transmitted or interpreted. These include errors not only within the flight crew but also in interactions with air traffic control, such as incorrect readbacks. The fourth classification, *proficiency errors*, is reflected in events where one or more crew members lack the knowledge to perform a needed action such as a flight management computer procedure or lack necessary stick and rudder skill to properly fly the aircraft. The final category consists of *operational decision errors*, which are discretionary decisions not covered by SOPs where crews make a decision that unnecessarily increases the level of risk on the flight. These often reflect deviations from policy in cases where there are no formal procedures. Examples include (a) all crew members focusing their attention on reprogramming the flight management computer on final approach, which is discouraged by the company's automation philosophy, and (b) the crew accepting an air traffic control command that leads to an unstable approach. Table 15.2 shows the percentage of each error type observed.

Error responses that can be made by the crew are limited. Three responses to crew error are identified: (a) *trap*—the error is detected and managed before it becomes consequential; (b) *exacerbate*—the error is detected, but the crew's action or inaction leads to a negative outcome; (c) *fail to respond*—the crew fails to react to the error because it is either undetected or ignored.

After an error occurs and the crew responds, there is an outcome that can be classified into one of four categories. An *undesired aircraft state* is a condition in which the aircraft is unnecessarily placed in a condition that increases risk to safety. It includes incorrect navigation, undesirable fuel state, unstable approach, long landing, and so on. An outcome is *inconsequential* when an error is discovered and trapped without leading to an undesired state. Undetected or ignored errors can also be inconsequential when they have no adverse effects on the safe completion of the flight (luck?). *Additional error* refers to an outcome in which the initial error leads to (or is closely associated with) a subsequent one, either through no response or an exacerbating response on the part of the crew. For example, failure to run a landing checklist may lead to a failure to lower landing gear. After entering an undesired aircraft state, the condition can be managed by a crew response that corrects (mitigates) the error or in a manner that exacerbates the severity by leading to another error or to an accident or incident. For some undesired aircraft states the crew may not have the option to respond—the state is the end of the sequence. An example of this would be a long landing. If one error causes another error at any point, we can start again at the top of the model; the situation represents the classic "error chain." Figure 15.8 gives examples of errors classified using this methodology. As the figure shows, intentional noncompliance errors were the most frequently observed (54%), followed by procedural (29%), communications (6%), operational decision making (6%), and proficiency errors (5%). As we discuss later, the consequences of each error type and their distribution across organizations differ widely.

TABLE 15.2
Percentage of Each Error Type Observed in Line Audits

Type of Error	% of All Errors
International noncompliance	54
Procedural	29
Communications	6
Proficiency	5
Operational decision making	6

Trapped Error

Error Type - Procedural Error
During pre-departure, FO punched a wrong waypoint into the Flight Management Computer.
Error Response –Trap
The crew caught the error during the crosscheck.
Error Outcome – Inconsequential

Ignored error

Error Type – Intentional Noncompliance
F/O performs the After Takeoff Checklist from memory.
Error Response – Fail to respond (ignored)
Captain notices the SOP violation but says nothing.
Error Outcome – Additional error
F/O failed to retract the landing gear.

Exacerbated Error

Error Type – Communication
F/O told the Captain to turn down the wrong runway.
Error Response – Exacerbate
Captain turned down the runway.
Error Outcome – Undesired State
The aircraft is on the wrong runway
Undesired State Response - Mitigate
Undesired State Outcome - Recovery
After reviewing the taxi chart, the crew taxied to correct runway.

Undetected Error

Error Type – Procedural
Asked to level off at 22,000 feet, the Captain double clicked the altitude hold button on the mode control panel [engaged it, then disengaged it] and it was never engaged.
Error Response – Fail to respond (undetected)
The crew did not notice the error.
Error Outcome – Undesired state
Crew flew through the assigned altitude.
Undesired State Response – Mitigate
Undesired State Outcome – Recovery
The altitude deviation was noticed by the captain, who returned the aircraft to the proper altitude and mode.

FIG. 15.8. Examples of errors classified. FO and F/O = first officer; SOP = standard operating procedure.

The distribution of errors, however, is not symmetrical across flight segments. At least one error was committed on 64% of the flights we observed. An average of 1.84 errors were recorded per flight, with a range of 0 to 14. The distribution of errors by phase of flight is summarized in Table 15.3. The highest percentage of errors, 39%, occurred during the approach-and-landing phase of flight. Boeing's compilation of worldwide jet accidents between

TABLE 15.3
Distribution of Observed Crew Error by Phase of Flight

Phase of Flight of Error	% of Errors
Preflight	23%
Take off/climb	24%
Cruise	12%
Approach and landing	39%
Taxi/park	2%

1959 and 1997 revealed that 55% of all accidents occurred during this phase of flight (Boeing Commercial Airplane Group, 1998). The British civil aviation global accident database shows that 70% of accidents occur in the approach-and-landing phase, but it also includes non-jet and air taxi operations (Civil Aviation Authority, 1998). The data validate the importance of taking proactive steps to reduce risk and error in this phase of flight (Khatwa & Helmreich, 1999).

Of course, not all errors become consequential. Fortunately, the error tolerance of the aviation system is such that most flights, even in the face of threat and flight crew error, end uneventfully. We operationally defined as consequential errors those that resulted in either an additional error or in an undesired aircraft state. Additional crew errors and undesired aircraft states can result from either a failure to respond to an error (i.e., one that is undetected or ignored by the crew) or from crew actions that exacerbate the error. The least consequential error type was intentional noncompliance. For this type of error, only 2% of observed errors became consequential. Pilots show good judgment in choosing to violate regulations that have a low probability of becoming consequential. At the other extreme, 69% of proficiency errors and 43% of operational decision-making errors resulted in consequential outcomes. Although these occurred relatively infrequently, they were often consequential. Intermediate were communications errors, which were consequential 13% of the time, and procedural errors, which were consequential 23% of the time. Because procedural errors occur relatively frequently, they account for a high proportion of consequential outcomes.

In addition, errors that occurred at some phases of flight were more potent in causing consequential outcomes than those that occurred in others. Descent/approach/land errors were the most potent—39% become consequential. Thus, not only are more errors made in this critical phase of flight, but they also become more consequential. Takeoff errors and cruise errors became consequential 12% of the time, and preflight errors become consequential 7% of the time. We have not observed enough errors in the taxi/park phase to adequately judge their potency. One of the main findings of our study is the striking differences among organizations on our measures of threat and error management. The demonstrated between-organization (and between-fleet) differences have several important implications. The first is that organizations cannot assume that their operation will correspond to normative data from the industry. The high degree of variability observed corresponds to differences in the operating environment and, most important, demonstrates the power of organizational cultures and subcultures (Reason, 1997).

IMPLEMENTING THREAT AND ERROR MANAGEMENT

Early CRM advocates fell into the trap of thinking and asserting that CRM would be a universal panacea for the problem of human error. This did not happen. In today's more restricted, but realistic model, CRM is seen as a tool that can be used to build a safety culture in the framework of the three cultures that influence flight operations.

Policy, Trust, and Data

A credible organizational policy that recognizes the inevitability of human error and elucidates a credible commitment to error management is essential for effective CRM. This policy must be built on trust and a nonpunitive stance toward error. Rather than seeking to blame and punish those who err, management needs to understand the roots of error in the organization and to develop an array of defenses against future recurrences. We are in no way advocating that organizations tolerate the intentional violation of their rules or those of the regulatory agency. No organization can expect to survive if it allows its employees to disregard procedures and safety standards.

   To specify needed actions and to determine if safety efforts are effective, organizations must have current and accurate data on the state of their operations and the nature and number of threats and errors in their operation. To obtain complete and accurate data requires a high level of trust on the part of employees. They must be willing to share their mistakes without fear of reprisal. Their trust must also include the belief that management will act on safety issues when they are uncovered. If this trust is established, organizations can obtain meaningful data and use them both to guide the development of appropriate training and as a yardstick for assessing trends in performance and error.

Sources of Data on Organizational
Performance and Error

Because the accident rate in commercial aviation is extremely low, surrogate measures must be used as safety and organizational effectiveness indicators (Helmreich, Chidester, Foushee, Gregorich, & Wilhelm, 1990). One indicator is pilot performance during formal evaluations by company evaluators or the regulator (e.g., the Federal Aviation Administration [FAA] in the United States). Although these data demonstrate that the evaluated pilots have the ability to perform their jobs, they do not reveal how the pilots behave when not under surveillance. Pilots, having above-average intelligence and valuing their jobs, can adhere strictly to rules when being checked and are also in a state of higher vigilance during evaluation. Although the checks may not be diagnostic of system performance, they do have great value for modeling and reinforcing appropriate behaviors.

Another organizational indicator is performance in training, but this is also an imperfect predictor of behavior during line operations because it also measures the *ability* of the individual or crew to perform appropriately while under surveillance.[4] Because of these limitations, organizations need to develop alternative sources of data that minimize the jeopardy–best-behavior problem. We will describe three other sources of data that organizations can use to gain understanding of the efficacy of their safety and training efforts and to plan the most effective use of their resources. Our view of the data necessary to manage error effectively parallels that of Capt. Daniel Maurino of the United Nations' International Civil Aviation Organization. On the basis of his global experience with air transport, Maurino (1998a, 1998b) concluded that the most valuable data on the health of operations come from the monitoring of normal operations.

Line Audits. We have collaborated in the conduct of line audits in a number of airlines (three were the source of the error data discussed earlier). It is our belief, and that of the participating airlines, that such data provide a reasonably accurate and comprehensive picture of line operations. The key to success of an audit is the credible assurance to crews that all observations are without jeopardy and that no identifiable information on any crew will be revealed to management or regulators. In practice, we have trained a group of expert observers from the airline (pilots from training, flight standards, the union, etc.) in the use of our Line/LOS Error Checklist (Helmreich, Klinect et al., 1999). Using this form, systematic evaluations of crew CRM

---

[4]This is more of a problem in low Uncertainty Avoidance cultures such as the United States where individuals do not feel compelled to adhere to procedures under all conditions. In high UA countries, performance in training is likely to be a much better predictor of line performance.

skills are made at various phases of flight, along with threat and crew error, and their management. The team of observers samples flights in all fleets and types of operations, usually for a period of a month. That a realistic picture of the operation is being captured is shown by the fact that observers frequently see violations of SOPs and regulations. For example, as part of a line audit we observed instances of failure to complete (or even use) checklists. This was particularly prevalent in one fleet of one airline. Neither line checks nor FAA inspections had suggested that this might be a problem. The line audit database gives clear guidance to management as to what to emphasize in training and indicates where problems of leadership or poor safety norms may be present. Analyses of the aggregated, deidentified data from line audits give the industry insights into ubiquitous problems such as the use of flightdeck automation, the variability of performance in the system, and standardization of procedures and practices (Helmreich, Hines, & Wilhelm, 1996; Helmreich & Merritt, 1998; Hines, 1998).

Confidential Survey. Organizations can augment line audit data with confidential surveys, often with an instrument such as the Flight Management Attitudes Questionnaire (FMAQ; Merritt, Helmreich, Wilhelm, & Sherman, 1996). Surveys provide insights into perceptions of the safety culture and illuminate aspects of teamwork among flight crews and other organizational elements, including maintenance, ramp, and cabin crews. At the most detailed level survey data also indicate the level of acceptance of fundamental concepts of CRM among line crews. They also show where differences may have developed between operational units of organizations, such as fleets and bases. Data from surveys can be used effectively to guide curriculum development for recurrent training by helping the organization target the most important operational issues.

Incident Reporting Systems. Incidents provide invaluable information about points of potential vulnerability in the aviation system. Confidential incident-reporting systems, such as NASA's Aviation Safety Reporting System (ASRS) and the British Airways Safety Information System (BASIS), are very useful for the overall system. In the United States, the Aviation Safety Action Programs (ASAP; FAA; 1997) concept was designed to give organizations more complete data on incidents in their own operations. ASAP encourages participation by providing crew members with protection from regulatory reprisal for many types of incidents and rapid feedback about organizational efforts to prevent their recurrence. Each reported incident is reviewed by a team (including representatives of management, the pilots' union, and the FAA) that develops a plan of action along with feedback to the reporter. American Airlines has the longest experience with ASAP and is receiving reports at a rate of more than 3,500 per year. As long as crews feel safe in submitting information to programs such as ASRS, BASIS, and ASAP, the data can give organizations an invaluable early warning system about potential threats to safety. Our research group, in cooperation with several U.S. airlines, has initiated a project to develop a new ASAP form to probe more deeply into human factors issues in incidents (Jones & Tesmer, 1999). The object of this effort is to generate data that can be combined with those from other sources—such as audits, surveys, and training records—to provide organizations with a more comprehensive view of their operations and better guidelines for operations and training.

We also recognize the value of data collected during normal operations from flight data recorders under programs such as the FAA's Flight Operations Quality Assurance (FOQA). Such data provide critical information on the nature and location of instances in which normal flight parameters are exceeded. A limitation of flight recorder data is that they provide no insight into why events occurred and the human factors issues associated with them. Line audits, confidential surveys, and incident reporting systems can augment FOQA programs and lead to a better understanding of causal factors.

Using Data Proactively for Safety. The data collected in support of safety can be directly used in safety and error reduction initiatives. By examining the categories of error observed in their own observations, organizations obtain a valid report card on the effectiveness of their operation that different elements of the organization can use to plan necessary action. For example, a high frequency of operational decision errors may suggest a need for additional SOPs. Conversely, a large number of noncompliance errors may indicate inappropriate or too many and too complex SOPs (see also Reason, 1997, for a discussion of SOPs and compliance).

Focusing CRM on Threat and Error

CRM courses have matured from the presentation of general concepts of team interaction and personal styles to become much more technical and operationally relevant training programs. Encouraging progress has been made toward the seamless integration of CRM and technical training that was identified as a major goal at the second NASA CRM conference in 1986 (Orlady & Foushee, 1987). One of the outcomes of this movement toward defining CRM in terms of specific behaviors has been a trend toward proceduralization of CRM, requiring interpersonal behaviors and communications as part of technical maneuvers. The positive side of this is clear guidance for crews as to expected behaviors and, concurrently, the ability to assess and reinforce their practice. There are several negative aspects of proceduralization. One is the possible loss of understanding of CRM's broader, safety goals when it becomes a set of required actions appended to technical maneuvers (Helmreich, Merritt, & Wilhelm, 1999). The second, clearly identified by Reason (1997), is that the proliferation of procedures may serve to *reduce* compliance. As more and more well-intentioned procedures find their way into operations, they may lose impact and significance, almost inviting violations. A third negative aspect of proceduralization is that it may cause CRM programs to lose sight of important, phase-independent skills such as leadership and team building.

Placing CRM in the framework of threat recognition, error avoidance, and error management should help maintain awareness of the organization's commitment to safety. The formal review of known risks and off-normal conditions can be made part of a crew's preparation. This type of review also represents a readily observable behavior that can be assessed and reinforced by training and checking personnel. One of the major venues for decision making should be the formulation and sharing of error avoidance strategies in response to recognized threats. Similarly, detection and management behaviors are usually observable and can be evaluated and reinforced. As Tullo and Salmon (1998) noted, monitoring and assessing these behaviors present a new challenge for instructors and evaluators, especially those dealing with behavior in normal operations.

CRM training should address the limitations of human performance, a problem evidenced by the high level of denial of personal vulnerability that is characteristic of the professional culture of pilots and other demanding professions. This denial works to the detriment of threat recognition and acceptance of the inevitability of error. There is empirical evidence that these attitudes can be modified by training (see Helmreich & Merritt [1998] for an example of attitude change about the effects of fatigue on performance). Awareness of human limitations should result in greater reliance on the redundancy and safeguards provided by *team* instead of individual actions. This training can best be accomplished by providing understandable information about the psychological and physiological effects of stress, with examples drawn from aviation experience. The narrowing of attentional focus under stress provides a compelling example of the deleterious effects of stress. Positive examples of using CRM in crises—for example, the performance of the crew of United Airlines flight 232 after losing aircraft control following the disintegration of an engine—can build acceptance of team concepts (Predmore, 1991). It is also important to define the nature and types of cognitive errors to which all humans are prey (e.g., Reason, 1990). Making these slips and omissions salient to pilots through everyday, operational examples can also foster endorsement of the use of CRM countermeasures against error.

National Culture and CRM

Although threat recognition and error management are universally valued, this does not imply that the same CRM training will work as well in Turkey as in Texas. The rationale provided to flight crews for error management and the description of relevant behavioral countermeasures need to be in a context that is congruent with the culture. For example, assertiveness on the part of junior crew members can be accepted as an effective strategy and practiced comfortably in individualistic, low-PD cultures such as the United States. In contrast, simply advocating the use of assertion by juniors in many high-PD cultures is likely to be seen as a bizarre and unworkable proposal. On the other hand, assertive behavior could be acceptable if it is seen as a means of protecting the organization (or in-group) and as a means of saving the face of the captain by keeping him from making a consequential error. However, there is still much to be learned about fitting training strategies to cultures. We see testing the error management approach as a challenge for both researchers and practitioners and an area where cross-cultural collaboration will be essential.

IMPLEMENTING CRM IN OTHER DOMAINS

The Operating Room

One of the more advanced applications of CRM concepts has been in medicine—specifically, the function of teams in operating rooms (OR) and emergency rooms (see chap. 13 for a detailed discussion). Most of the original impetus came from anesthesiologists such as David Gaba at Stanford University and Hans-Gerhard Schaefer at the University of Basel/Kantonsspital who saw parallels between the OR environment and the cockpit (Gaba & DeAnda, 1988; Helmreich & Schaefer, 1994; Helmreich & Davies, 1996). In reality, the OR is a more complex environment than the cockpit with multiple groups composed of anesthesiologists and surgeons (both attendings and residents), nurses, orderlies, and, of course, a patient. In the OR the lines of authority between the surgical and anesthetic teams are unclear, and this in itself can be a source of conflict (Helmreich & Schaefer, 1994).

The development of resource management for the OR has taken a rather different course from that in aviation. Whereas aviation programs had their foundation in formal awareness training, normally in an interactive seminar context with subsequent training and reinforcement in full mission simulation line-oriented flight training, the medical programs were built around simulation with only cursory discussion of more global issues (Davies & Helmreich, 1996). A major limitation of most programs has been a focus only on the anesthesia team rather than the full OR complement, usually with an actor paid to role-play the surgeon (see Wilkins, Davies, & Mather, 1997, for a discussion of simulator training in anesthesia). Such programs involve part-task rather than full-mission simulation, which is problematic because our observations suggest that most of the observed difficulties in the OR come at the interface between teams (Helmreich & Davies, 1996; Sexton et al., 1997a).

A more comprehensive approach to training was developed by the late Hans-Gerhard Schaefer and his colleagues at the University of Basel/Kantonsspital. The group there focused on building a complete OR simulator that allows members of the surgical team as well as the anesthesia team to conduct meaningful work (laparoscopic surgery) and captures the richness of interteam interactions (Helmreich & Schaefer, 1997). As yet, systematic data have not been developed to validate the impact of such training programs on medical participants, although self-reports indicate that the experience is perceived as valuable by those who receive it (Sexton et al., 1997b).

Shipboard

Another logical venue for the application of CRM is in maritime operations. The NTSB (e.g., NTSB, 1993) has been urging maritime operators to adopt *bridge resource management* as a parallel to CRM on the basis of accidents showing maritime human factors problems to be similar to those in aviation. Robert L. Helmreich has collaborated with the Danish Maritime Institute and the Risoe Institute in Denmark to develop a new instrument conceptually similar to the FMAQ. Results of preliminary surveys of mariners in several countries show similar human factors issues and a professional culture as prone to the denial of personal vulnerability as those of aviation and medicine (Helmreich & Merritt, 1998).

In both the medical and maritime environments the tendency has been to lightly adapt cockpit training programs to build awareness of human factors issues. This approach, which disregards the realities of the environments and the cultures involved, is reminiscent of early CRM programs, which also lacked specificity and relevance. One exception to this has been training developed by the Danish Maritime Institute, which is similar to later generation CRM programs and augments seminar training with the marine equivalent of full-mission simulation using a high-fidelity ship simulator (e.g., Andersen, Soerensen, Weber, & Soerensen, 1996).

Overall, it is our belief that error management concepts are highly applicable to other domains in which teamwork and technology are required. We feel equally strongly that programs must be embedded in organizational threat and error management efforts.

SUMMARY AND GUIDELINES FOR ORGANIZATIONS IMPLEMENTING MORE OPERATIONALLY FOCUSED PROGRAMS

CRM is not a fully realized concept. Cultural effects are not fully understood, and the nature of errors in the normal, operational environment must be further explored. There is also need for the involvement of the research community to develop new approaches and build databases showing the nature and frequency of errors and the multiple strategies involved in coping with threat and responding to error.

The actions necessary to develop and apply resource management programs go far beyond the design and delivery of training programs. If an organization is not receptive to the training initiative and contextual effects are ignored, programs are unlikely to achieve the desired outcomes and may inappropriately raise questions about the training itself. The individuals charged with developing and delivering training need to establish with senior management a common view of goals and to obtain commitment that the program will be supported with appropriate resources. The following guidelines are essential and can provide a checklist for program development:

- *Guideline 1: Build trust.* Senior management, in cooperation with employee groups, must establish a relationship of trust that will encourage and reward individuals and teams that share safety-related information.
- *Guideline 2: Adopt a nonpunitive policy toward error.* Management's policy toward error must be to elicit information without punishing those who make errors while trying to accomplish their jobs in accordance with regulations and SOPs.

Effective programs clearly involve management at the highest levels and extend beyond the purview of program developers and trainers. The task is much more daunting than simply developing a training program for operational personnel. There is little to be gained and much to be lost by initiating training that falls short of expectations. On the other hand, with the appropriate level of commitment, programs can be initiated that should yield measurable improvements in safety and efficiency—and the side benefit of better organizational communication and morale.

- *Guideline 3: Provide training in error avoidance, detection, and management strategies for crews.* With the supportive infrastructure mentioned in Guidelines 1 and 2, formal training can give crews the tools and countermeasures to error that they need to optimize flight operations.

- *Guideline 4: Provide special training in evaluating and reinforcing error avoidance, detection, and management for instructors and evaluators.* Key personnel who are responsible for training and the evaluation of performance need special training in the concepts and assessment of threat and error management. It is essential that error management be evaluated and reinforced not only in training but also in line operations. The major change here is in formally recognizing that error, in itself, is part of system operations and that effective error management can represent effective crew performance.
- *Guideline 5: Demonstrate a willingness to reduce error in the system.* The organization must establish mechanisms to deal with safety-related information and to make changes necessary to reduce or mitigate error.
- *Guideline 6: Collect data that show the nature and types of threat and error.* Organizations must commit resources necessary to obtain and analyze data showing its operational status. These data sources can include line audits, surveys, incident reports, and training evaluations.

ACKNOWLEDGMENTS

Research supporting this article was funded by Federal Aviation Administration Grants 92-G-017 and 99-G-004. We are deeply indebted to Capts. Bruce Tesmer and Sharon Jones, who made major contributions to the development of the conceptual models presented but are not responsible for any errors therein. Thomas Chidester has also contributed significantly to our thinking about the meaning of training for safety in aviation, and we thank him for his insights. The work reported here was completed while Ashleigh Merritt was a doctoral candidate and postdoctoral fellow at the Aerospace Crew Research Project at the University of Texas at Austin. She is now a project manager at Dedale, Roissy CDG Cedex, France. Working for John Wreathall & Co., Ashleigh is now applying her skills as a systems psychologist in the nuclear domain.

REFERENCES

Andersen, H. B., Soerensen, P. K., Weber, S., & Soerensen, C. (1996). *A study of the performance of captains and crews in a full mission simulator.* Roskilde, Denmark: Risoe National Laboratory.
Boeing Commercial Airplane Group. (1998). *Statistical summary of commercial jet airplane accidents: Worldwide operations 1959–1997.* Seattle, WA: Author.
Civil Aviation Authority. (1998). *Global fatal accident review: 1980–1996.* London: Author.
Cooper, G. E., White, M. D., & Lauber, J. K. (Eds). (1980). *Resource management on the flightdeck: Proceedings of a NASA/Industry workshop* (NASA CP-2120). Moffett Field, CA: National Aeronautics and Space Administration, Ames Research Center.

Davies, J. M., & Helmreich, R. L. (1996). Simulation: It's a start. *Canadian Journal of Anaesthesia, 43,* 425–429.

Federal Aviation Administration. (1997). *Advisory circular 120-66: Aviation safety action programs.* Washington, DC: Author.

Gaba, D. M., & DeAnda, A. (1988). A comprehensive anesthesia simulating environment re-creating the operating room for research and training. *Anesthesiology, 69,* 387–394.

Helmreich, R. L. (1997, May). Managing human error in aviation. *Scientific American,* pp. 62–67.

Helmreich, R. L., Chidester, T. R., Foushee, H. C., Gregorich, S. E., & Wilhelm, J. A. (1990). How effective is cockpit resource management training? Issues in evaluating the impact of programs to enhance crew coordination. *Flight Safety Digest, 9,* 1–17.

Helmreich, R. L., & Davies, J. M. (1996). Human factors in the operating room: Interpersonal determinants of safety, efficiency and morale. In A. A. Aitkenhead (Ed.), *Clinical anaesthesiology: Safety and risk management in anaesthesia* (pp. 277–296). London: Balliere Tindall.

Helmreich, R. L., & Foushee, H. C. (1993). Why crew resource management? Empirical and theoretical bases of human factors training in aviation. In E. Wiener, B. Kanki, & R. Helmreich (Eds.), *Cockpit resource management* (pp. 3–45). San Diego, CA: Academic Press.

Helmreich, R. L., Hines, W. E., & Wilhelm, J. A. (1996). *Issues in crew resource management and automation use: Data from line audits* (University of Texas Aerospace Crew Research Project Tech. Rep. 96-2). Austin: University of Texas.

Helmreich, R. L., Klinect, J. R., Wilhelm, J. A., & Jones, S. G. (1999). *The Line/LOS Error Checklist, Version 6.0: A checklist for human factors skills assessment, a log for off-normal events, and a worksheet for cockpit crew error management.* (University of Texas Team Research Project, Tech. Rep. 99-01). Austin: University of Texas.

Helmreich, R. L., & Merritt, A. C. (1998). *Culture at work in aviation and medicine: National, organizational, and professional influences.* Aldershot, England: Ashgate.

Helmreich, R. L., Merritt, A. C., & Wilhelm, J. A. (1999). The evolution of crew resource management. *International Journal of Aviation Psychology, 9*(1), 19–32.

Helmreich, R. L., & Schaefer, H.-G. (1994). Team performance in the operating room. In M. S. Bogner (Ed.), *Human error in medicine* (pp. 225–253). Hillside, NJ: Lawrence Erlbaum Associates.

Helmreich, R. L., & Schaefer, H.-G. (1997). Turning silk purses into sows' ears: Human factors in medicine. In L. Henson, A. Lee, & A. Basford (Eds.), *Simulators in anesthesiology education* (pp. 1–8). New York: Plenum.

Helmreich, R. L., & Wilhelm, J. A. (1991). Outcomes of crew resource management training. *International Journal of Aviation Psychology, 1,* 287–300.

Hines, W. E. (1998). *Teams and technology: Flight crew performance in standard and automated aircraft.* Unpublished doctoral dissertation, The University of Texas at Austin.

Hofstede, G. (1980). *Culture's consequences: International differences in work-related values.* Beverly Hills, CA: Sage.

Hofstede, G. (1991). *Cultures and organizations: Software of the mind.* Maidenhead, England: McGraw-Hill.

Hudson, P. T. W. (1998, June). *Bending the rules: Violation in the workplace.* Invited keynote address delivered to Society of Petroleum Engineers, 1998 International Conference on Health, Safety and Environment in Oil and Gas Exploration and Production, Caracas, Venezuela.

Johnston, N. (1993). CRM: Cross-cultural perspectives. In E. Wiener, B. Kanki, & R. Helmreich (Eds.), *Cockpit resource management* (pp. 367–398). San Diego, CA: Academic Press.

Jones, S. G., Tesmer, B. (1999). A new tool for investigating and tracking human factors issues in incidents. In R. S. Jensen & L. Rakovan (Eds.), *Proceedings of the tenth international symposium on aviation psychology* (pp. 696–701). Columbus: Ohio State University.

Khatwa, R., & Helmreich, R. L. (1999). Analysis of critical factors during approach and landing in accidents and normal flight. *Flight Safety Digest, 18*(1–2), 1–212.

Klinect, J. R., Wilhelm, J. A., Helmreich, R. L. (1999). Threat and error management: Data from line operations safety audits. In R. S. Jensen & L. Rakovan (Eds.), *Proceedings of the Tenth International Symposium on Aviation Psychology* (pp. 683–688). Columbus: Ohio State University.

Maurino, D. (1998a). Human factors training would be enhanced by using data obtained from monitoring normal operations. *ICAO Journal, 53,* 17–23.

Maurino, D. (1998b). Forward. In R. L. Helmreich & A. C. Merritt (Eds.), *Culture at work: National, organizational, and professional influences* (pp. xiii–xxiv). Aldershot, England: Ashgate.

Merritt, A. C. (1996). *National culture and work attitudes in commercial aviation: A cross-cultural investigation.* Unpublished doctoral dissertation, The University of Texas at Austin.

Merritt, A. C., & Helmreich, R. L. (1996a). Creating and sustaining a safety culture: Some practical strategies. In B. Hayward & A. Lowe (Eds.), *Applied aviation psychology: Achievement, change and challenge* (pp. 20–26). Sydney, Australia: Avebury Aviation.

Merritt, A. C., & Helmreich, R. L. (1996b). Human factors on the flightdeck: The influences of national culture. *Journal of Cross-Cultural Psychology, 27,* 5–24.

Merritt, A. C., & Helmreich, R. L. (1997). CRM: I hate it, what is it? (Error, stress, culture). In *Proceedings of the Orient Airlines Association air safety seminar* (pp. 123–134). Metro Manila: The Philippines.

Merritt, A. C., Helmreich, R. L., Wilhelm, J. A., & Sherman, P. J. (1996). *Flight Management Attitudes Questionnaire 2.0 (International) and 2.1 (USA/Anglo;* University of Texas Aerospace Crew Research Project Tech. Rep. 96–4). Austin: University of Texas.

National Transportation Safety Board. (1991). *Aircraft accident report: Britt Airways, Inc., In-flight structural breakup, Eagle Lake, Texas, September 11, 1991* (Rep. No. NTSB AAR-92/04). Washington, DC: Author.

National Transportation Safety Board. (1993). *Marine accident report: Grounding of the United Kingdom Passenger Vessel RMS Queen Elizabeth 2 Near Cuttyhunk Island, Vineyard Sound, Massachusetts, August 7, 1992* (Rep. No. NTSB MAR-93/01). Washington, DC: Author.

Orlady, H. W., & Foushee, H. C. (1987). *Cockpit resource management training.* (Pub. No. NASA CP-2455). Moffett Field, CA: National Aeronautics and Space Administration, Ames Research Center.

Predmore, S. C. (1991). Microcoding of communications in accident investigation: Crew coordination in United 811 and United 232. In R. S. Jensen (Ed.), *Proceedings of the sixth international symposium on aviation psychology* (pp. 350–355). Columbus: Ohio State University.

Reason, J. (1990). *Human error.* New York: Cambridge University Press.

Reason, J. (1997). *Managing the risks of organisational accidents.* Aldershot, England: Ashgate.

Salas, E., Fowlkes, J. E., Stout, R. J., Milanovich, D. M., & Prince, C. (1999). Does CRM training improve teamwork skills in the cockpit? Two evaluation studies. *Human Factors, 41,* 326–343.

Sexton, B., Marsch, S., Helmreich, R. L., Betzendoerfer, D., Kocher, T., & Scheidegger, D. (1997a). Jumpseating in the operating room. In L. Henson, A. Lee, & A. Basford (Eds.), *Simulators in anesthesiology education* (pp. 107–108). New York: Plenum.

Sexton, B., Marsch, S., Helmreich, R. L., Betzendoerfer, D., Kocher, T., & Scheidegger, D. (1997b). Participant evaluations of team oriented medical simulation. In L. Henson, A. Lee, & A. Basford (Eds.), *Simulators in anesthesiology education* (pp. 109–110). New York: Plenum.

Sherman, P. J., Helmreich, R. L., & Merritt, A. C. (1997). National culture and flightdeck automation: Results of a multination survey. *International Journal of Aviation Psychology, 7,* 311–329.

Tullo, F., & Salmon, T. (1998). The role of the check airman in error management. In R. S. Jensen & L. Rakovan (Eds.), *Proceedings of the Ninth International Symposium on Aviation Psychology* (pp. 511–513). Columbus: Ohio State University.

Westrum, R. (1992). Cultures with requisite imagination. In J. Wise, D. Hopkin, & P. Stager (Eds.), *Verification and validation of complex systems: Human factors issues* (pp. 401–416). Berlin: Springer-Verlag.

Wilkins, D. G., Davies, J. M., & Mather, S. J. (1997). Simulator training in anaesthesia. *Current Opinion in Anaesthesia, 10,* 481–484.

Wolfe, T. (1979). *The right stuff.* New York: Farrar, Straus & Giroux.

FINAL OBSERVATIONS

16

Research and Practices of Resource Management in Organizations: Some Observations

Eduardo Salas
Clint A. Bowers
University of Central Florida

Eleana Edens
Federal Aviation Administration

Resource management training in organizations has existed for more than 20 years. It started as an intervention in the airline industry as a way of mitigating human error in the cockpit (see Wiener, Kanki, & Helmreich, 1993). Its aim was (and continues to be) to increase safety in the skies. The investment by the airlines, the National Aeronautics and Space Administration (NASA), the Federal Aviation Administration (FAA), and the military, and the effort put forth by the aviation industry, scientists, training developers, and many others, is unquantifiable. Suffice to say that the investment is in the millions of dollars, and the hours spent designing, developing, implementing, and evaluating resource management training are countless. So, what have we learned after all this investment? What is the value of resource management training to organizations? What is the contribution of resource management to organizational effectiveness? What has been learned from the airline experience that transfers to other organizations? As we can see from the preceding chapters, the science and practice of resource management training has evolved and matured and is beginning to answer these questions.

Much progress has been made. Resource management training has evolved from a simple seminar offered in a classroom to an integrated instructional system that includes practice in simulators and feedback on required teamwork competencies (Helmerich, 1999; Salas, Fowlkes, Stout, Milanovich, & Prince, 1999). Resource management training has matured to a set of well-organized and theoretically based principles, tools, guidelines, and methods that create sound instructional strategies that organizations can apply and use (Salas, Rhodenizer, & Bowers, 1999). The purpose of this chapter is to glean from the chapters in this book several observations about what we know about the state-of-the-science and practice of resource management training.

SOME OBSERVATIONS

Taken together, the chapters in this volume suggest that the field of crew resource management (CRM) has a long history of systematic progress. The field has indeed blossomed into a coherent science offering practical solutions to a complex problem: human error. The lessons learned and guidelines offered in these chapters suggest that this field is ready to make significant contributions to a wider range of domains, industries, and organizations. The future is promising. This is not to say that there are not still deficiencies in the science that need to be addressed, or that there is no room for improvement in how we design, develop, deliver, and evaluate resource management training—more progress is needed, and it will come as we continue to systematically tackle these deficiencies with the cooperation of sponsors, users, and fellow scientists.

In reviewing and integrating the chapters in this volume, we can make a number of observations. The trends, interconnections, and similarities in how the authors discuss and describe their experiences in conducting research or applying resource management training in organizations allow us to make some general observations of CRM training. We next delineate a few general observations (perhaps conclusions) from this volume.

Observation 1: Resource Management Is Not and Should Not Be Limited to Aviation

Resource management now is applied in many organizations as a way to improve teamwork, reduce human error, and increase efficiency. Resource management has been implemented in the medical, maintenance, and nuclear power industries; in military and oil offshore operations; and in other industrial settings. CRM is now a well-accepted training strategy to improve key human performance competencies (e.g., assertiveness, communication, decision making) of teams working in complex and dynamic environments.

Observation 2: The Science of Training, Group Performance, Human Factors, and Industrial-Organizational Psychology Can Help Design and Deliver Resource Management Training

Resource management users and developers can now rely on science. Funding provided by NASA, the FAA, and the military has fostered theories, proven principles, useful guidelines, and practical methods. These are derived from a variety of disciplines: human factors, aviation, cognition, and industrial-organizational psychology. Instructional developers can now tap this wealth of information to guide their individual design and delivery efforts.

Observation 3: All Resource Management Training Is Not Created Equal

One-size programs simply do not fit all situations. The task characteristics and the nature of the interdependence determine the appropriate training. This underscores the importance of conducting training-needs analyses (Goldstein, 1993) and coordination analysis (Bowers, Morgan, & Salas, 1991; Salas & Prince, 1993; Levine & Balker, 1991). These tools yield valuable information about how, what, why, and when to design and deliver resource management training. Regardless of the job for which training is being developed, it is critical to identify the key elements for training and assessment. In our research we have found that training needs are different among even similar aircraft (Bowers, Baker, & Salas, 1994). When trying to translate CRM principles in other environments, it is likely that conducting needs analyses is even more important.

Observation 4: Tools Have Been Created to Design and Deliver Resource Management Training

There are now useful tools for instructional developers to design and deliver resource management training. They range from how to analyze the tasks to how to evaluate resource management interventions. Although these tools will be refined as we learn more, they offer users and instructional designers viable methods and approaches that should yield valuable information on how, when, why, and what to train.

Observation 5: Resource Management Training Can Be Effective in Reducing Human Error and Increasing Mission Performance and Teamwork

Data suggest that a well-designed resource management training program can enhance human performance (see Helmreich & Foushee, 1993; Salas et al., 1999; Stout, Salas, & Fowlkes, 1997). Although more data are needed, and a direct, cause-and-effect link between an intervention and organizational results (e.g., safety data, profit) accidents is still elusive, the data suggest that a well-designed resource management training can produce desired goals. More evaluation is definitely needed and should be encouraged by all parties involved.

Observation 6: Resource Management Training Is a Global Psychological Intervention

Resource management is now a universal psychological intervention aimed at managing human error and enhancing teamwork (e.g., Johnston, 1993). It has been applied in Europe, Latin America, Asia, and Africa, in a variety of domains. We think that this trend will continue.

Observation 7: Success Isn't Guaranteed—Evaluation Is Critical

Resource management training must be evaluated each time it is delivered. Evaluation should consider trainee reaction, learning, and skill development as well as the impact the training has on the organization. Although this last objective is easier said than done, we need to keep trying to show how resource management training affects the "bottom line" in organizations: reducing accidents, improving service, increasing productivity, reducing errors, and the like.

Observation 8: Resource Management Training Cannot Be Isolated From the Organization It Supports

Resource management training is part of an organizational system (see chap. 15). The organization influences and shapes how the intervention will be designed and delivered. What the organization does and the signals it sends about the value of resource management training will dictate the fate of the intervention (Guzzo & Salas, 1995; Kozlowski & Salas, 1997). Therefore, if the organizational system around the training is not considered, the intervention may fail; so, resource management must be an organizational goal (see chap. 12). Like all training, resource management can have varying degrees of success. It is prone to failure if it is perceived as something employees have to do. The likelihood of success can be increased by having the top levels of management endorse its principles. Emphasis should be given to maximizing the contribution of all team members.

THE FUTURE . . .

A few observations about the future of resource management training are needed. The future looks promising, but we need to learn more. As already stated, progress has been made—however, more progress is needed. We need to know more about the boundaries of resource management; that is, we need data that help us see what transfers from setting to setting and what does not. It would be useful to have a set of guidelines that help instructional developers, users, and organizations decide what needs to be done to ensure the success of a resource management training intervention.

We also need to begin to standardize (if possible) the language we use to communicate what resource management comprises. We have many labels for the same behaviors or skills, which create some confusion to practitioners and prevent the accumulation of knowledge. We hope that with some dialogue, the resource management community (users, sponsors, scientists) can reach at least some agreements.

We must go deeper into our understanding of practice and feedback in resource management training. We need to know the kinds of practice that are sufficient to produce desired learning objectives. Therefore, we need to conduct more empirical work on the nature of practice and feedback.

Technology can only help the design and delivery of resource management training. We need to explore (and exploit) recent advances in intelligent tutoring systems, automated performance measurement, distance learning, and computer-based simulation. These technologies can facilitate the delivery of resource management training. As long as we keep the focus on learning these technologies will provide a valuable aid to training developers and users.

Finally, we need more evaluations. We hope that by now all parties in the resource management community understand the value of evaluation. We simply need more, and better, evaluations.

CONCLUDING REMARKS

Resource management training is here to stay. It is a concept that continues to evolve, and its full potential has not been realized. We hope this volume pushes resource management training to the next level—one where it is widely applied to solve organizational problems across domains, industries, organizations, and cultures.

REFERENCES

Bowers, C. A., Baker, D. P., & Salas, E. (1994). Measuring the importance of teamwork: The reliability and validity of job/task analysis indices for team-training design. *Military Psychology, 6*, 205–214.

Bowers, C. A., Morgan, B. B., Jr., & Salas, E. (1991). The assessment of aircrew coordination demand for helicopter flight requirements. In R. S. Jensen (Ed.), *Proceedings of the Sixth International Symposium on Aviation Psychology* (pp. 308–313). Columbus: Ohio State University.

Goldstein, I. L. (1993). *Training in organizations* (3rd ed.). Belmont, CA: Brooks/Cole.

Guzzo, R. A., & Salas, E. (1995). *Team effectiveness and decision making in organizations*. San Francisco: Jossey-Bass.

Helmreich, R. L., & Foushee, H. C. (1993). Why crew resource management? Empirical and theoretical basis of human factors training in aviation. In E. L. Wiener, B. G. Kanki, & R. L. Helmreich (Eds.), *Cockpit resource management* (pp. 3–45). San Diego, CA: Academic Press.

Helmreich, R. L., Merritt, A. C., & Wilhelm, J. A. (1999). The evolution of crew resource management training in commercial aviation. *International Journal of Aviation Psychology, 9*(1), 19–32.

Helmreich, R. L., Weiner, E. L., & Kanki, B. G. (1993). The future of crew resource management in the cockpit and elsewhere. In E. L. Wiener, B. G. Kanki, & R. L. Helmreich (Eds.), *Cockpit resource management* (pp. 479–499). San Diego, CA: Academic Press.

Johnston, N. (1993). CRM: Cross-cultural perspectives. In E. L. Wiener, B. G. Kanki, & R. L. Helmreich (Eds.), *Cockpit resource management* (pp. 367–393). San Diego, CA: Academic Press.

Kozlowski, S. W. J., & Salas, E. (1997). An organizational systems approach for the implementation and transfer of training. In J. K. Ford & Associates (Eds.), *Improving training effectiveness in work organizations* (pp. 247–289). Mahwah, NJ: Lawrence Erlbaum Associates.

Levine, E. L., & Balker, C. V. (1991, April). Team task analysis: A procedural guide and test of the methodology. In E. Salas (Chair), *Methods and tools for understanding teamwork: Research with practical implications*. Symposium conducted at the Sixth Annual Conference of the Society for Industrial and Organizational Psychology, St. Louis, MO.

Salas, E., Fowlkes, J. E., Stout, R. J., Milanovich, D. M., & Prince, C. (1999). Does CRM training improve teamwork skills in the cockpit?: Two evaluation studies. *Human Factors, 41*, 326–343.

Salas, E., Prince, C., Bowers, C. A., Stout, R., Oser, R. L., & Cannon-Bowers, J. A. (1999). A methodology to enhance crew resource management training. *Human Factors, 41*, 161–172.

Stout, R. J., Salas, E., & Fowlkes, J. E. (1997). Enhancing teamwork in complex environments through team training. *Group Dynamics, 1*, 169–182.

Wiener, E. L., Kanki, B. G., & Helmreich, R. L. (1993). *Cockpit resource management*. San Diego, CA: Academic Press.

Author Index

A

Acton, B., 48, 54, 158, 164, 292, 301
Aiken, L. H., 269, 281
Ainsworth, I. K., 19, 30
Air Transport Association, 197, 215
Air Transport Association, AQP
Subcommittee, 131, 132
Algina, J., 174, 187
Allen, D., 163, 163
Alliger, G. M., 170, 173, 186, 187
Allnutt, M. F., 64, 71
Alonso, R. R., 81, 93
American Medical Association
Guides Red Flags, 96, 126
Ammerman, H. L., 35, 51, 52
Andersen, G., 271, 281
Andersen, H. B., 327, 329
Anderson, H. N., 81, 92
Anderson, J. R., 172, 187
Anonymous, 269, 279
Argyle, M., 78, 92
Arvey, R. D., 12, 15, 29
Austin, J. T., 174, 187
Avermaete, van, J. A. G., 229, 231

B

Baddeley, A., 221, 231
Baker, D. P., 34, 36, 37t, 43, 46, 47,
52, 54, 77, 78, 81, 85, 88, 94, 131,
133, 134, 136t, 137, 141, 142, 143,
159, 163, 289, 291, 292, 294, 299,
301, 339
Baker, E., 77, 94
Baldwin, T. T., 88, 92

Balker, C. V., 337, 340
Ballough, J., 238, 243, 260, 262
Balzer, W. K., 137, 144
Bandura, A., 149, 150, 155, 163, 289,
299
Banner, M. J., 273, 280
Baraket, T., 149, 150, 164
Barnett, A. I., 278, 281
Barnett, J., 89, 93
Bauman, M., 43, 52
Beard, R. L., 47, 52, 292, 299
Beaubien, J. M., 185t, 203, 215
Becker, E. S., 35, 51, 52
Behson, S. J., 50, 54, 91, 94
Bennett, W., 170, 186
Benson, R., 36, 53, 77, 92, 131, 144
Bergen, L. J., 35, 51
Bergondy, M. L., 289, 361
Bernardin, H. J., 137, 139, 142, 144
Berninger, D., 239, 264
Berwick, D. M., 276, 279
Betzendoerfer, D., 327, 331
Birnbach, R. A., 131, 132, 143
Bjork, R. A., 66, 71, 288, 301
Blair, M. D., 35, 53
Blickensderfer, E. L., 27, 30, 50, 52
Blume, B. J., 82, 94
Bobko, P., 35, 53
Boeing Commercial Airplane
Group, 320, 329
Boehm-Davis, D. A., 19, 30, 90, 93,
133, 134, 144, 176, 177, 178, 182,
185t, 187, 198t, 203, 215
Borman, W. C., 132, 144, 170, 187,
188
Botvin, G. J., 77, 94

Subject Index